DEVELOPMENTS IN
EARTH SURFACE PROCESSES 7

CATCHMENT DYNAMICS AND RIVER PROCESSES

Mediterranean and Other Climate Regions

DEVELOPMENTS IN EARTH SURFACE PROCESSES

Volumes 1 and 3 are out of print

2. WEATHERING, SOILS & PALEOSOLS
   I.P. MARTINI and W. CHESWORTH (Editors)
4. ENVIRONMENTAL GEOMORPHOLOGY
   M. PANIZZA
5. GEOMORPHOLOGICAL HAZARDS OF EUROPE
   C. EMBLETON and C. EMBLETON-HAMANN (Editors)
6. ROCK COATINGS
   R.I. DORN
7. CATCHMENT DYNAMICS AND RIVER PROCESSES
   C. GARCIA and R.J. BATALLA (Editors)

Developments in Earth Surface Processes, 7

CATCHMENT DYNAMICS AND RIVER PROCESSES

Mediterranean and Other Climate Regions

Edited by

Celso Garcia
Department of Earth Sciences, University of the Balearic Islands
Palma de Mallorca, Spain

Ramon J. Batalla
Department of Environment and Soil Sciences, University of Lleida
Lleida, Spain

ELSEVIER

Amsterdam – Boston – Heidelberg – London – New York – Oxford
Paris – San Diego – San Francisco – Singapore – Sydney – Tokyo

ELSEVIER B.V.
Radarweg 29
P.O. Box 211, 1000 AE, Amsterdam
The Netherlands

ELSEVIER Inc.
525 B Street, Suite 1900
San Diego, CA 92101-4495
USA

ELSEVIER Ltd
The Boulevard, Langford Lane
Kidlington, Oxford OX5 1GB
UK

ELSEVIER Ltd
84 Theobalds Road
London WC1X 8RR
UK

First edition 2005

ISBN-13: 978-0-52084-5
ISBN-10: 0-444-52084-8

∞ The paper used in this publication meets the requirements of ANSI/NISO Z39.48-1992 (Permanence of Paper).
Printed in The Netherlands

Foreword

Catchment Dynamics and River Processes is a book that looks primarily at the effects of water in fluvial systems and on slopes mainly in Mediterranean climate zones, but also in a few other climate regions. The 12 chapters presented discuss various aspects of general Mediterranean hydrology, erosion, transport, and deposition, including gullying, mass movement stability of saturated terraces, soil properties following forest fires, and landscape disturbance and sequestration of organic carbon. In addition, the dynamics of gravel-bed rivers, avulsions on alluvial fans, the effects of dams and water diversions, and obstruction-forced pools all receive attention as well.

In general the papers in this Festschrift in honor of Professor Maria Sala constitute a gathering of ideas and methodologies that provide ways to think about and to measure common landscape processes. This new book also constitutes our latest addition to the series on **Developments in Earth Surface Processes**, and as such, the work of the 27 scientists on these pages represents themes of possible use to hydrologists, engineers, planners and others working in various drainage basins.

Some readers may think that any book of edited papers is just another collection of miscellany, but in fact, this method of presentation in geomorphology represents the best way to present the diversity in our field. Furthermore, edited papers on a theme, produced by experts in the discipline, gives us a highly manageable tool to address any problem. Such a technique allows us to investigate from many different perspectives, and to find solutions to problems in a variety of fashions. This volume then, gives the reader a view of state-of-the-art research in catchment dynamics, particularly for regions with winter precipitation and summer drought. River processes seem to dominate much of our planet, because catchment dynamics and fluvial actions are so important to the works of humanity, even though mass-movement processes are spatially more extensive. The result in this book is a concentration on surface processes of interest to a significant proportion of the world's people.

John F. Shroder
Editor-in-Chief
Developments in Earth Surface Processes

DEDICATION

Maria Sala, Professor of Physical Geography at the University of Barcelona, has devoted her academic career to the study of the functioning of terrestrial landscapes based on the interrelation between their main elements: water, soil, sediments and vegetation. The degradation of natural systems, especially in Mediterranean drainage basins, has driven her scientific interest since the mid-1970s, the time when she started research into the geomorphologic processes in the nowadays well-known Tordera drainage basin. Her pioneering work in the field of experimental process geomorphology in Catalonia and Spain and the global vision of her scientific approach has been acknowledged worldwide, which constitutes a perennial contribution to a better understanding of natural processes and the interrelationship between the environment and humans.

Preface

Professor Maria Sala is retiring from her academic duties in September 2005. The idea for this volume was developed as a humble tribute to Maria Sala by her former graduate students. With this primary idea in mind, several colleagues and friends, who have collaborated with her or her Ph.D. students, joined us in the task and helped in defining a topic and, what is more important, generously contributed to this volume. Their support and encouragement is gratefully acknowledged.

Maria Sala introduced experimental and field-based studies of soil and fluvial processes into Spain during the late 1970s and early 1980s. Research on this broad topic has grown remarkably since then. This book shows some of these advances and documents the latest research on it, although with a particularity: it gives special treatment to research on Mediterranean climate regions, an ever-present issue in Maria Sala's research career. The theme of this book is the latest research on slope and river processes with a special emphasis on rivers and catchments with a Mediterranean climate. Chapters cover a gamut of topics describing research and applications of studies, mainly in Spain, but also in Israel, the USA, Canada, the UK and New Zealand.

The book is a collection of papers giving state-of-the-art research on natural and anthropogenic processes in soil erosion, catchment hydrology, suspended sediment and bedload dynamics in gravel-bed rivers, and present-day diagnosis and future key-paths for river management. The first two papers by Gallart et al. examine through long-term experimental research the hydrological response, erosion processes and sediment dynamics, within active badlands and land-use changes, in the Vallcebre catchments of the eastern Pyrenees. They show the importance of saturated areas in runoff generation and the importance of relatively small areas of badlands on the sediment supply and sediment yield in the catchments. Calvo-Cases et al. analyse the results obtained after 7 years of measuring runoff and sediment movement in a series of limestone areas under Mediterranean semi-arid and sub-humid conditions, and show some clear differences in the process behaviour and the role of the factors controlling hillslope dynamics. Harvey and Gutiérrez-Elorza analyse gully systems and their coupling characteristics in the Barranco de la Clamor, Ebro Basin. They show how the system appears to be vulnerable to regime switching following changes in the runoff and erosion rates of the gullied zones and changes in its coupling characteristics. Zgaier and Inbar use two methods to determine the influence of soil saturation on the stability of abandoned agricultural hillslope terraces in Western Galilee. Úbeda et al. analyse the long-term effects on soil properties in three areas after forest fires of different intensity and demonstrate that after 7 years, significant differences still persist in all the soil properties in the experimental areas. Gomez and Trustum examine landscape disturbance, forest clearances and carbon sequestration in alluvium bordering steepland rivers in New Zealand. Batalla et al. provide a valuable revision and synthesis of a decade of studies undertaken during the 1990s on sediment transport in the Mediterranean Tordera River. The authors focus their work on assessing the temporal variability of sediment transport and on the relative importance of each of the main lithologies in the area on the total sediment yield of the catchment. Church and Hassan propose that, due to catchment-scale control of sediment supply, a characteristic condition of upland cobble–gravel channels is that they usually move only limited amounts of bed material and are relatively stable. Harris Creek in British Columbia represents the normal fluvial regime for such settings. Woodsmith and Hassan address the classic question of pool maintenance in fluvial geomorphology, and present a compelling analysis of the effect of flow obstruction on pool-forming processes at varying discharges. Kondolf and Batalla examine the hydrological effects of dams and water diversions on rivers in California as a basis for understanding why rivers undergo morphological and ecological changes, such as encroachment of vegetation into the formerly active channel, in response to reduced flood scour and sediment supply. Werrity et al. identify and map zones with contrasting sensitivities in a way that shows boundaries as undefined, dynamic and evolving, and propose management strategies for the Spey–Feshie confluence in Scotland.

Celso Garcia and Ramon J. Batalla
Palma de Mallorca and Lleida, October 2004

Acknowledgements

Several individuals have contributed to the production of this book. We would like to thank those professionals who contributed by providing peer reviews for each chapter: Basil Gomez, Trevor Hoey, Moshe Inbar, Mike Kirkby, Graham Leeks, Tom Lisle, María Martínez-Mena, David Montgomery, John A. Moody, Roy Morgan, Phil Owens, Jean Poesen, M. Concepción Ramos, Olav Slaymaker, Norman Smith, Nicola Surian, Martin C. Thoms, Stefan Uhlenbrook, Gert Verstraeten, Mark Voltz, Des Walling and Alan Werrity.

John F. Shroder, Editor in Chief of the Elsevier series on Developments in Earth Surface Processes, provided guidance and help in the early stages of the book and edited our final submissions. Ms. Tonny Smit, Administrative Editor at Elsevier, was the patient person always on the other side of the email to solve doubts and problems.

Contents

List of Contributors

Eva Arnau-Rosalen Department of Geography, University of Valencia, 46010 Valencia, Spain

J. Carles Balasch Department of Environment and Soil Sciences, University of Lleida, 25198 Lleida, Spain

Ramon J. Batalla Department of Environment and Soil Sciences, University of Lleida, 25198 Lleida, Spain

Sara Bernia Department of Physical Geography, University of Barcelona, 08028 Barcelona, Spain

Andrew Black Department of Geography, University of Dundee, Dundee, DD1 4HN, UK

Carolina Boix-Fayos Department of Soil and Water Conservation, CEBAS-CSIC, Apartado 164, 30100 Murcia, Spain

Adolfo Calvo-Cases Department of Geography, University of Valencia, 46010 Valencia, Spain

Xavier Castelltort Institute of Earth Sciences 'Jaume Almera' (CSIC). Lluis Solé i Sabarís s/n, 08028 Barcelona, Spain

Michael Church Department of Geography, The University of British Columbia, Vancouver, British Columbia, Canada V6T 1Z2

Francesc Gallart Institute of Earth Sciences 'Jaume Almera' (CSIC). Lluis Solé i Sabarís s/n, 08028 Barcelona, Spain

Celso Garcia Department of Earth Sciences, University of the Balearic Islands, 07122 Palma de Mallorca, Spain

Basil Gomez Geomorphology Laboratory, Indiana State University, Terre Haute, Indiana, 47809 USA

Mateo Gutiérrez-Elorza Seccion de Geomorfología, Departamento de Geología, Facultad de Ciencias, Universidad de Zaragoza, 51009 Zaragoza, Spain

Adrian M. Harvey Department of Geography, The University of Liverpool, Roxby Building, Liverpool, L69 7ZT, England

Marwan A. Hassan Department of Geography, The University of British Columbia, Vancouver, British Columbia, Canada V6T 1Z2

Trevor B. Hoey Department of Geography and Geomatics, Centre for Geosciences, University of Glasgow, G12 8QQ, UK

Moshe Inbar Dept. of Geography, The University of Haifa, Haifa 31999, Israel

G. Mathias Kondolf Department of Landscape Architecture and Environmental Planning, University of California, Berkeley, CA 94720-2000, USA

Jérôme Latron Institute of Earth Sciences 'Jaume Almera' (CSIC), Lluis Solé i Sabarís s/n, 08028 Barcelona, Spain

Pilar Llorens Institute of Earth Sciences 'Jaume Almera' (CSIC), Lluis Solé i Sabarís s/n, 08028 Barcelona, Spain

David Regüés Pyrenean Institute of Ecology (CSIC), Campus Aula Dei, Avda. Montañana 1005. Apdo 202, 50080 Zaragoza, Spain

Albert Rovira Department of Environment and Soil Sciences, University of Lleida, 25198 Lleida, Spain

Elisabeth Simelton Earth Sciences Centre, Physical Geography, University of Göteborg, PO Box 460, Göteborg, 405 30 Sweden

Montse Soler Institute of Earth Sciences 'Jaume Almera' (CSIC), Lluis Solé i Sabarís s/n, 08028 Barcelona, Spain

Noel A. Trustrum Institute of Geological and Nuclear Sciences, PO Box 30 368, Lower Hutt, New Zealand

Xavier Úbeda Departament of Physical Geography, University of Barcelona, 08028 Barcelona, Spain

Alan Werritty Department of Geography, University of Dundee, Dundee, DD1 4HN, UK

Richard D. Woodsmith USDA, Forest Service, Pacific Northwest Research Station, 1133 N. Western Ave., Wenatchee, WA 98801, USA

Ali Zgaier Dept. of Geography, The University of Haifa, Haifa 31999, Israel

List of Figures

List of Tables

Catchment Dynamics and River Processes: Mediterranean and Other Climate Regions
Celso Garcia, Ramon J. Batalla, Editors

Chapter 1

Catchment dynamics in a Mediterranean mountain environment. The Vallcebre research basins (southeastern Pyrenees) I: hydrology

Francesc Gallart*, Jérôme Latron[1] and Pilar Llorens

Institute of Earth Sciences "Jaume Almera" (CSIC), Lluis Solé i Sabarís s/n, 08028 Barcelona, Spain

Abstract

Hydrological response and processes were studied in a set of Mediterranean middle mountain catchments subject to land-use changes over a period of 14 years. The Vallcebre catchments are located in the Pyrenean ranges, built up by sedimentary rocks and covered with loamy soils. The vegetation cover is pastures and pine forests, mostly occupying former agricultural terraces. Some relatively small heavily eroded landscapes (badlands) occur in the catchments, playing relevant hydrological and geomorphic roles. Annual precipitation is about 924 mm and potential evapotranspiration is about 700 mm, but one main season with water deficit occurs in summer, and eventually another in winter. Forest rainfall interception represents about 24% of precipitation. Tree transpiration is more sensitive to soil water stress than grass transpiration. During most of the year, subsurface flows along hillslopes drive the spatial organisation of soil moisture and the occurrence of saturated areas, although this spatial organisation is also controlled by the patterns of vegetation cover and the terraced microtopography. During the dry periods, subsurface flow ceases, saturated areas disappear and the spatial variation of soil moisture decreases. Spatial differences in soil moisture are also small in very wet conditions. Stream flow from these catchments is dominated by storm flow, and the runoff generating mechanisms show a clear seasonal pattern, controlled mainly by the soil moisture and the extent of saturated areas. During the dry periods, in spite of the high rainfall intensities, runoff is produced only on impervious areas and badlands. At the end of the dry periods, some large rainfall events generate significant runoff because of the perched saturation of shallow soil horizons. Thereafter, runoff generation is dominated by the role of saturated areas with a significant contribution from baseflow. Tests performed with several hydrological models demonstrated their capacity to simulate runoff main events during wet periods, but evidenced the need for an increased model complexity to simulate summer small events, the first events after the dry season, and the overall water balance.

Keywords: Mediterranean region, soil moisture, runoff generation, evapotranspiration, rainfall interception, land use changes

---

*Corresponding author. *E-mail address:* fgallart@ija.csic.es (F. Gallart).
[1]Present address: Pyrenean Institute of Ecology (CSIC), Avda. Montañana 1005, 50059 Zaragoza, Spain.

1. Introduction

Sub-humid Mediterranean mountains share the hydrological processes from both wet and dry environments and are the source of water necessary for human life and activity in the drier downstream areas. The study of the hydrological functioning of these areas, besides its fundamental interest, may help to anticipate the hydrological consequences of both climate and land cover change, as well as to design land-use strategies that might counteract these changes.

The Vallcebre catchments of the southeastern Pyrenees were instrumented in 1989 to address these questions, as well as the erosion and sediment transport issues examined in the second part (Gallart et al., 2005). This is a scientifically driven research whose results are expected to be useful in the long term, following the methodology established for the International Hydrological Decade 1965–1975 (Toebes and Ouryvaev, 1970) renewed in some aspects for the IAHS Decade on Predictions in Ungauged Basins (PUB): 2003–2012 (Sivapalan, 2003). Along with the monitoring of weather, precipitation and stream flow, several state variables (soil moisture and tensiometry, water table) and processes (rainfall interception, tree transpiration, grass evapotranspiration) are measured permanently or during several-year long periods.

This chapter reviews the main results obtained after over 14 years of study in the area, most of them addressed in Ph.D. theses and published in several scientific journals. The characteristics of climate and precipitation, the hydrological role of vegetation, the spatial and temporal patterns of soil moisture, and the runoff generation mechanisms are the main aspects addressed, along with a summary of the main results derived from the application of several hydrological models.

1.1. Site characteristics

The Vallcebre catchments are located in the headwaters of the Llobregat River, on the southern margin of the Pyrenees (Catalonia, northeastern Spain) at 42°12N, 1°49E (Fig. 1.1), with altitudes between 1100 and 1700 m. The importance of convective precipitations, the marked water deficit in summer, and the deep impact of human activity on the landscape are the main characters of this area that are representative of Mediterranean environments.

The bedrock is composed of red clayey smectite-rich mudstones with massive limestone beds of continental facies attributed to the Palaeocene (Aepler, 1968). The soft mudstones are prone to landsliding and erosion by water, leading to the occurrence of intensely dissected landscapes with poor vegetation cover (badlands), that occupy less than 3% of the catchment area but play a dominant role in sediment production (see Gallart et al., 2005). Soil thickness varies greatly, depending on lithology, geomorphology and the changes induced by terracing. Limestone areas are overlain by discontinuous soils up to 40 cm thick; soils on hillslopes over clayey rocks are up to 80 cm thick and agricultural terraces can have soils thicker than 3 m. Topsoils are silty loam but show high infiltration capacities due to their well developed structure, which results from high contents in organic matter in the shallow horizons (mean 12%), and also presumably from the relatively recent agricultural works (Rubio, 2003). Soil cracking during summer is

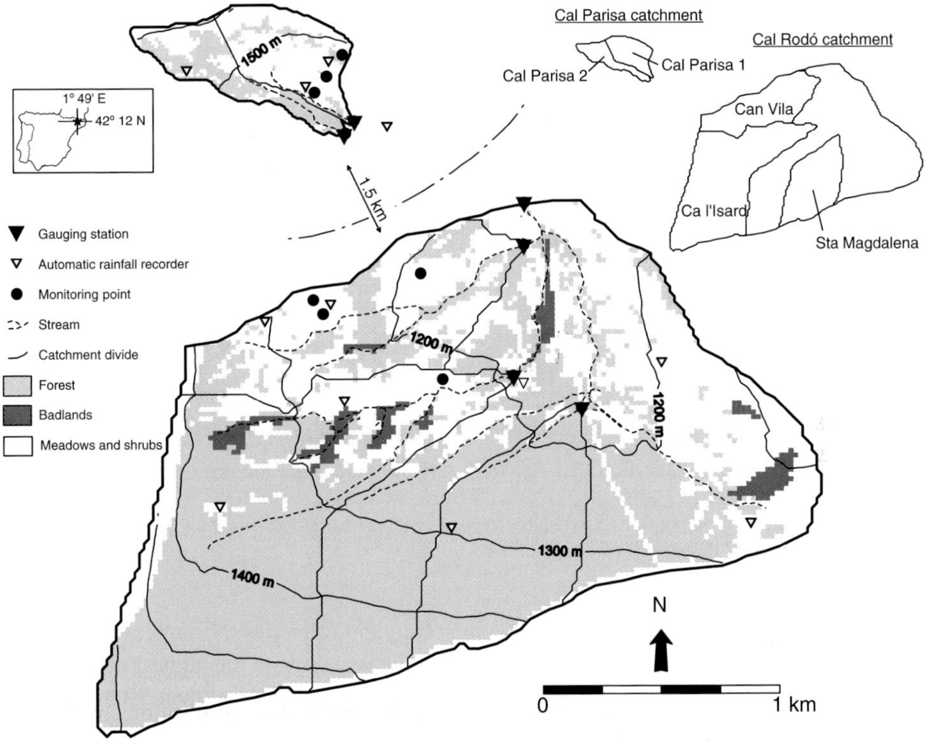

Figure 1.1. Map of the Vallcebre catchments.

favoured by smectite clay content (near 10%) and enhances infiltration capacity, whereas soil crusting was only observed on bare badland surfaces (Regüés et al., 1995). Nevertheless, hydraulic conductivity declines rapidly in depth, inducing the formation of shallow semi-permanent aquifers, despite the high topographic gradients. The arrangement of agricultural terraces also results in high spatial variability of hydraulic conductivities, which tend to be much higher near the outer edges (3×10^{-4} m s^{-1}) than for the internal parts (2×10^{-7} m s^{-1}), where terrace construction exposed the parent material (Haro et al., 1992). Some of these terraces disrupt the continuity of shallow hillslope aquifers during wet conditions, leading to the formation of small saturated areas in their internal parts. A network of ditches is therefore usually associated with the terraces, which act to drain these saturated areas and convey overland flow (Llorens, 1991; Gallart et al., 1994).

Mean annual temperature at 1440 m a.s.l. is 7.3°C, and the average values for July and January are 15 and 1°C, respectively. The differences between minima and maxima are 13 and 9°C in July and January, respectively. The growing season (mean temperature higher than 7°C) lasts for about 7 months, from early May to early October. At an annual time scale, precipitation (924 mm) exceeds potential evapotranspiration (700 mm), but June and July, as well as February and March, are months when potential evapotranspiration typically exceeds precipitation (Fig. 1.2). Winter is the season with less

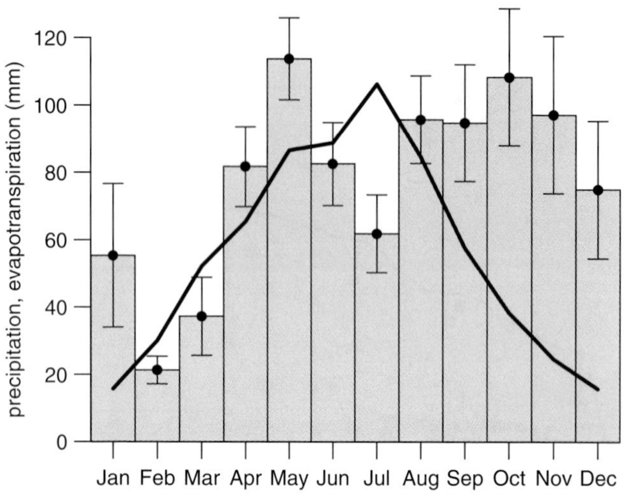

Figure 1.2. Monthly precipitation (bars) and reference evapotranspiration (thick line) at Vallcebre. Error bars represent standard deviations of precipitation.

precipitation, spring is the season with the most frequent rainy days, summer is the season with the highest rainfall intensities, and autumn is the season with the highest monthly and daily precipitation totals. Snowfalls are occasional and represent less than 5% by volume (Gallart et al., 2002). The average number of successive days without precipitation ($p < 0.4$ mm) is about 10 in winter and about 3 for most of the remaining year, although it raises to 6 in late July (Latron, 2003). During the period between September 1994 and December 1997, the highest precipitation in 24 h was 131 mm, and the highest rainfall intensity in 20 min was 77 mm h^{-1}, although 80% of the events had a mean rainfall intensity lower than 6 mm h^{-1} (Latron, 2003).

Climax vegetation is woodland of *Quercus pubescens* with bushes of *Buxus sempervirens* on the rocky outcrops, and *Pinus sylvestris* on the colder north-facing aspects. Most of the gentle hillslopes were deforested in the past and terraced for agricultural use (mainly cereals and potatoes) and subsequently underwent a progressive abandonment during the second half of the 20th Century. Former agricultural terraces are now covered by meadows of mesophile grasses (*Aphyllantion*) with hydrophile patches (*Molinia coerulea*), and are used for both grass harvesting or extensive cattle and sheep grazing. Spontaneous afforestation by *Pinus sylvestris* has occurred following abandonment. Marginal areas, which were the first to be abandoned, nowadays have a rather dense forest cover, whereas those areas abandoned later exhibit forest patches and scattered trees (Poyatos et al., 2003).

An analysis of vegetation changes in the last 40 years (Poyatos et al., 2003) showed that in 1957, forests of diverse density covered 39% of the catchment and grassland and crops 28%, whereas in 1996 the relative areas of these covers changed into 64 and 18%, respectively. On terraced areas (39% of the catchment surface), grasslands and crops decreased from 74 to 47% in these 40 years, due to the abandonment of farming and stockbreeding.

1.2. Instrumental design

Instrumentation of the research catchments started in 1989 (Llorens and Gallart, 1992; Balasch et al., 1992), and significantly improved since 1994 (Gallart et al., 1997). The pluviometric network consists of 12 tipping-bucket rain gauges connected to data-loggers that record 0.2 mm precipitation increments at a temporal resolution of 1 s. These rain gauges, located 1 m above the ground surface, are calibrated for a large range of rainfall intensities by means of a dynamic calibration method. Two standard automatic weather stations were installed in the respective catchment clusters, and are used to estimate the reference evapotranspiration (evapotranspiration from a standard grass plot, after Smith, 1992). One of these stations was equipped with two aspirated psicrometers (Pt100-based) at two heights above the ground (0.45 and 1.45 m) in order to employ the Bowen Ratio Energy Balance (BREB) method to estimate the actual evapotranspiration from the grass-covered plot (Poyatos and Llorens, 2003).

All the stream gauging stations have control structures where water level and temperature measurements are logged at intervals between 2 and 60 min; self-cleaning concrete box weirs are installed at gauging stations Cal Rodó and Ca l'Isard in order to avoid clogging by the abundant suspended and bedload sediment transported in these streams, V-notch steel weirs are installed at both Can Vila and Santa Magdalena stations and steel H-flumes are used at both Cal Parisa 1 and Cal Parisa 2 stations.

Soil–water content has been measured in the Vallcebre catchments since 1993 using the Time-Domain Reflectometry (TDR) method at nine profiles distributed in the main geo-ecological units (seven in the Cal Parisa catchment and two in Can Vila subcatchment). These profiles consist of sets of four vertical 20 cm-long probes permanently installed in the ground at 0–20, 20–40, 40–60 and 60–80 cm depth. These are read every week with a Tektronix 1502-C cable tester (Rabadà and Gallart, 1993).

A network of soil tensiometers was installed in late 1996. Depths to water table are measured weekly at four old wells and four continuously recording piezometers were instrumented in the catchments in late 1995, to study the dynamics of the water table during rain events.

The Ramon Poch experimental forest plot in the Cal Parisa basin has been monitored since 1993 to evaluate the water balance of a representative afforestation patch. This 198 m^2 plot has a monospecific cover of *Pinus sylvestris* with no understorey. Stand density is about 2400 stems ha^{-1}. The plot was instrumented for the continuous monitoring of rainfall interception (Llorens et al., 1997), tree transpiration (Oliveras and Llorens, 2001) and soil water potential, as well as for periodical measurement of soil moisture. Throughfall is measured by three clusters of three 1 m^2 troughs connected to tipping bucket devices, and stem flow is measured in three clusters of trees provided with rings, which are also connected to tipping bucket devices. All measurements are recorded by the same data logger, which store interception and sap flow data at 5 and 15 min intervals, respectively.

2. Results and discussion

2.1. Rainfall interception

Mean throughfall in the experimental forest plot during the period July 1993–August 2000 represents about 74% of bulk rainfall, whereas stemflow accounts for only 2%, leading

Table 1.1. Classification of the rainfall interception events depending on their duration, rainfall intensity and atmospheric conditions.

Class	Interception rate (%)	Duration (h)	Rainfall intensity (mm h^{-1})	VPD (mb)
A	15	26 (long)	1.6 (low)	0.3 (wet)
B	13	5 (short)	7.6 (high)	1.1 (dry)
C	49	12 (medium)	1.2 (low)	2.0 (very dry)

VPD means vapour pressure deficit. Data from Llorens et al. (1997).

to a rainfall interception rate of 24% of bulk rainfall (Llorens et al., 2003). Observation of the role of meteorological conditions on rainfall interception at the event scale, analysed by Llorens et al. (1997), lead to the identification of three types of events, considering their duration, the differences in atmospheric conditions and rainfall intensity (Table 1.1). Long events with low rainfall intensities and wet atmospheric conditions (class A) are the most common events measured. Short events with high rainfall intensities and dry atmospheric conditions (class B) and medium events with low rainfall intensities and very dry atmospheric conditions (class C) are less common but more characteristic of the Mediterranean climate. The application of a common rainfall interception model provided acceptable results (Llorens, 1997). No clear seasonal control on rainfall interception rates has been observed at the event scale, as a consequence of the compensation between the characteristics of the events, principally its magnitude and the atmospheric conditions (Gallart et al., 2002).

2.2. *Forest and pasture transpiration*

Tree transpiration during the studied vegetative periods (May–September from 1995 to 2000) showed important differences between wet and dry summers (Llorens et al., 2003). During a wet summer (1995), mean soil water content was about 0.28 $cm^3\ cm^{-3}$. Relative transpiration of pine trees was about 45% (ratio between actual and reference evapotranspiration), with a marked temporal correlation with reference evapotranspiration, because there was enough water available for transpiration. During a dry summer, as in 1998, mean soil water content was only 0.18 $cm^3\ cm^{-3}$. The great dependence of tree transpiration on soil water content clearly broke the relationship between transpiration and evaporative demand, with values of the relative evapotranspiration lower than 25%. In these conditions, trees showed a continuous decrease in transpiration rates, with respect to atmospheric demand, due to soil water depletion.

The study of actual evapotranspiration from a terrace with a mesophilous pasture (Poyatos and Llorens, 2003) showed that during the mild and warm season (mean air temperature higher than 7°C), the relative evapotranspiration rate is close to unity when volumetric soil moisture was higher than 0.30 $cm^3\ cm^{-3}$, whereas it decreases down to 35% when soil moisture is about 0.20 $cm^3\ cm^{-3}$. During the colder season, in spite of the high soil water availability, the relative evapotranspiration is low (about 30%) as a consequence of grass senescence, and increases progressively with decreasing albedo in springtime.

2.3. Soil water

Gallart et al. (1994) showed that frequently saturated areas occur on downslope locations as a consequence of the outcrop of the shallow water table. The spatial pattern of these saturated areas is partly coherent with the topographic index values (Beven and Kirkby, 1979), derived from a 20 m-mesh DEM, although some saturated areas did occur elsewhere, commonly in the internal parts of terraces.

The analysis of records of soil water content (Gallart et al., 1997) showed that soils under forest cover are typically drier than under grass, due to rainfall interception, and that underground water transfer that feeds frequently saturated areas is interrupted during dry periods. Llorens et al. (2003), using a longer time series, observed that spatial soil moisture variability is higher during intermediate wetness conditions and decreases during both wet and dry conditions.

Three representative soil moisture profiles were selected for a more detailed analysis of soil moisture regime through the year (Gallart et al., 2002). The first profile is located in a frequently saturated area downslope, covered by hydrophile grass. The two others are located in a mid-slope position, covered by pine trees and mesophilic grass, respectively.

The first profile shows a marked intra-annual variability (between 0.35 and 0.59 $cm^3\ cm^{-3}$). In winter, the soil profile is always saturated and remains so until the end of spring. In June, the soil water content decreases rapidly due to the increasing evapotranspiration demand and the interruption of the subsurface water transfer. The lowest soil water content is reached by the end of July. Later, following the rainfall inputs of August and the successive months, soil water content tends to increase until the end of November. Saturation during the first part of the year and its breakdown in late June is predictable, as water content during this period presents low interannual variability, but the depth and duration of the dry period are much less predictable, as these depend on large rainfall events restoring the underground water transfer.

The other two profiles show a less pronounced intra-annual variability (between 0.23 and 0.38 $cm^3\ cm^{-3}$). Soil water content decreases after the second fortnight of January and there is no increase in soil water content until early May. The summer drought includes June and July with the lowest soil water content during the second fortnight of July. Finally, there is a progressive wetting-up of the soils from August to the end of the year. The inter-annual variability for these two profiles was small and regular throughout the year.

The seasonal trend of the profiles described illustrates a catchment hydrodynamics pattern characterised by saturated areas surrounded by wet soils in early winter, which suffer a progressive drying until middle spring, although some saturated areas are still present. After an increase in wetness during late spring, the summer drought dries out the saturated areas. Finally, autumn is characterised by the wetting of the whole catchment, this being more pronounced in the saturated areas and more irregular and delayed in the forest-covered profiles.

2.4. Runoff generation

From the very first results (Llorens, 1991; Llorens and Gallart, 1992) it was apparent that the response of these catchments is much more driven by antecedent conditions than by rainfall intensities. More detailed results (Latron et al., 1997, 2000, 2003; Gallart et al., 2002) showed that during the year, the dominant runoff generation mechanisms change gradually, as a result of both varying catchment antecedent wetness conditions and changing rainfall events characteristics (intensity and volume).

Using information on rainfall, stream flow, soil tensiometry and depths to water table, three main kinds of runoff events have been identified in the Vallcebre catchments (Latron, 2003):

(1) Summer runoff events, which tend to occur as a result of short-duration, high-intensity convective storms over dry soils and deep water table. Runoff coefficients for this type of event are very low (typically less than 1%), as runoff is restricted largely to the poorly permeable rocky and badland areas of the catchment, resulting in a flashy hydrograph and low peak flow rates. Soil moisture responds to these events, but usually the water table does not.

(2) Wetting-up transition events, which usually occur in autumn and eventually in spring, as a result of prolonged, lower intensity frontal rainfall over dry catchments with a deep water table, which rises up days after the rainfall event. Runoff coefficients are intermediate (3–15%) and recession limbs are relatively short. Overland flow is produced over areas saturated "from above" where temporary perched aquifers occur due to limited permeability of the deep soil horizons.

(3) Wet events, which occur in late autumn or early winter when large rainfall events occur on wet soils with a shallow water table, which quickly responds to precipitation. Runoff coefficients are high (10–60%) and recession limbs contribute significantly to the flow volume. Overland flow occurs mainly on areas saturated "from below".

An early analysis of the role of the terraces on runoff generation (Gallart et al., 1994) suggested that the saturated areas created by the terraces, along with the role of drainage ditches, would enhance the storm runoff by contribution from saturated areas of the catchment for a given water reserve, compared with the response for a non-terraced catchment, but this hypothesis has never been tested. On the other hand, the delays between rainfall and runoff peaks in the more terraced sub-basins Cal Parisa (0.18 km^2) and Can Vila (0.56 km^2), are, respectively, 1.5 and 2.5 h (Llorens, 1991; Latron, 2003). These delays were very long when compared with the common response of catchments where saturation overland flow is the main runoff-generating mechanism (Anderson and Burt, 1990). The role of the scattered situation of the saturated areas (Latron and Gallart, 2002) linked to the terraces, and the pathways of water from these terraces to the main drainage net may explain this delayed response and deserve more attention.

2.5. Baseflow contribution

As stated before, the water table was very dynamic during "wet" events, which accounted for most of the total runoff. This fact was further explored to perform a first estimation of groundwater contribution to runoff, assuming that if we can obtain a robust function between the depth to the water table and stream discharge during periods without overland flow or active evapotranspiration, this relationship might be extrapolated during floods to obtain the contribution from groundwater. We expect that this approach will afford a minimum contribution from groundwater, because it does not take into account the presumable role of perched aquifers or the deformations of the water table that may increase its contribution to stream flow during events (Sklash and Farvolden, 1979).

During recession periods, exponential regressions were obtained between the depth to the water table (measured at one point) and stream runoff at the two main outlets, using instantaneous values measured at noon (Latron et al., 2000; Gallart et al., 2002):

$$\text{For Cal Rodó:}\quad Q = 2.99\,\mathrm{e}^{-0.020z},\quad n = 455,\quad r^2 = 0.71,\quad p < 0.01$$

$$\text{For Can Vila:}\quad Q = 7.94\,\mathrm{e}^{-0.022z},\quad n = 448,\quad r^2 = 0.73,\quad p < 0.01$$

where Q is runoff (mm day^{-1}), and z is the depth to the water table (cm).

Using these relationships, daily base flow discharges have been obtained from the depth to the water table. These base flow estimates are subject to some uncertainty due to the scatter of the above relationships, but represent a minimum assumption approach. Thus the difference between the base flow estimates and the discharge measured at the outlet may be interpreted as storm (overland) flow. The validity of this approach is discussed below.

Following this procedure, overland flow represents 63% of total flow for the Cal Rodó catchment and 50% for the Can Vila sub-catchment (over a period of 26 months). This difference is consistent with the characteristics of the catchments, as the Can Vila sub-catchment has thicker soils and less impervious area than the Cal Rodó catchment.

Figure 1.3 shows a sequence of both observed stream flow and estimated base flow during 1996 for the Can Vila sub-catchment. This graph illustrates the seasonal arrangement of the classification of runoff events summarised before:

- During the dry period (early July–mid October), the low water table yields a base flow contribution almost nil. Summer showers produce very limited runoff and are unable to raise the water table.
- At the end of the dry period (October), a large rainfall event caused a moderate flow response with an intermediate runoff coefficient (by 10%), as well as a clearly delayed rise in the water table. The base flow estimated from the water table level during this event was limited during the flood peak but increased significantly a few days later, when it was higher than the measured stream flow.
- During wet periods (spring and late autumn), the water table is high and its variations are well correlated with rainfall events. Estimated base flow contribution is important, showing peaks during the events and maintaining flow even during late winter, a period with little precipitation.

Even if the flow separation method used here is not fully adequate at the event scale, as shown by some inconsistencies between measured stream flow and estimated base flow for some events, it allows a primary estimation of the base flow contribution through the year.

The comparison of this method with other methods based on environmental tracers and on modelling approaches is under way and suggests that it provides lower groundwater contributions than the other methods (Latron et al., 2004).

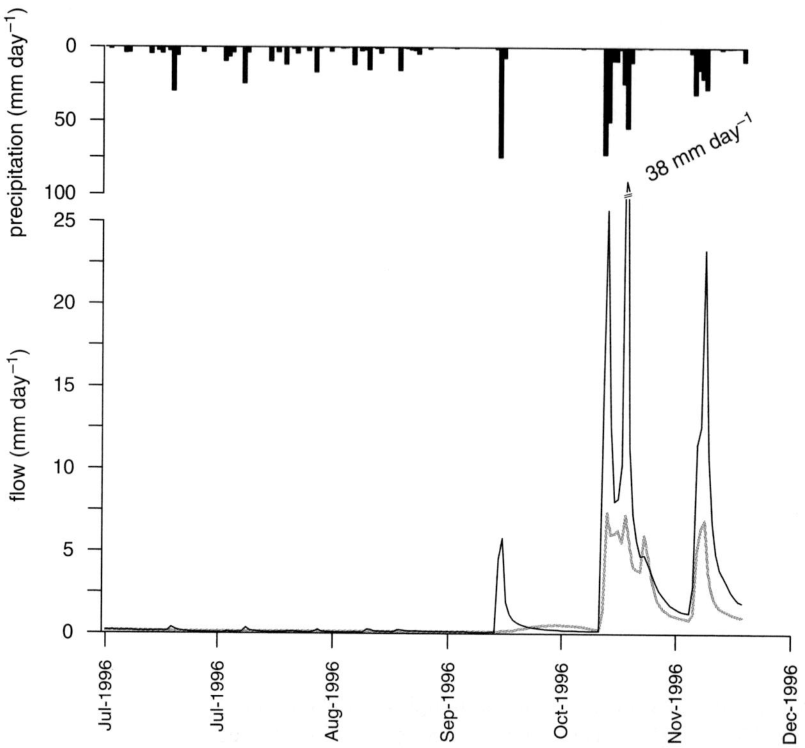

Figure 1.3. Daily precipitation (black bars), daily flow (black line) and daily baseflow (grey line) at the Can Vila catchment. Baseflow was estimated from water table levels (see text).

One of the major problems faced with such an approach, however, is linked to the representativeness of the water table measured at one single point. In the Cal Rodó catchment, the analysis of data from the piezometric network (Latron, 2003) has shown that the correlation between water-table depths measured at different points throughout the catchment is relatively good during wet and dry periods, but decreases during transition periods.

The relationship between water table and baseflow from one side and the extension of saturated areas from the other has been analysed during different wetness conditions (Latron and Gallart, 2002; Latron, 2003). The results demonstrated that the extent of saturated areas in the Can Vila catchment is nearly proportional to stream baseflow, and has an exponential relationship with the water table level measured at the above-mentioned point.

2.6. Hydrological modelling

Five different hydrological models were tested with data from Vallcebre catchments in a comparative work (Table 1.2), using separate calibration and validation data records (Buchtele et al., 1998; Gallart et al., 2000). SHETRAN (Ewen et al., 2000), a physically

Table 1.2. Nash and Sutcliffe (1970) efficiency criterion for the daily flow simulations obtained with a sample of models.

	Cal Rodó		Can Vila	
	Calibration	Validation	Calibration	Validation
SHETRAN	0.90	0.72	0.96	0.81
TOPCAT	0.95	0.45	0.93	0.63
TOPKAPI	–	–	0.87	0.65
BROOK	0.85	0.62	0.86	0.79
SACRAMENTO	0.60	0.42	0.56	0.51

Data from Gallart et al. (2000).

based, distributed model, provided the best results for a wide range of antecedent conditions. TOPCAT (Quinn et al., 1998), a simplification of the TOPMODEL approach, provided acceptable results during wet periods, but underestimated flow for other conditions. TOPKAPI (Ciarapica and Todini, 1998, 2002), a new physically based model with both lumped and distributed versions, overestimated flow especially during dry conditions, presumably owing to an oversimplistic parameterisation for these early runs of the model. BROOK (Federer, 1993), a physically based one-dimensional model underestimated most of the events although it showed an acceptable robustness. Finally, SACRAMENTO (Burnash, 1995), a conceptually lumped, soil-moisture accounting model provided the worst results, although this failure may be attributed to the complex parameterisation of this model or to the inappropriateness of the scale at which the model was applied.

The distributed modelling system SHETRAN was further tested with a deeper analysis of the internal functioning of the model (Anderton et al., 2002a,b). SHETRAN provided generally reasonable results, in terms of efficiency as well as in terms of runoff coefficients of the simulated flow (Latron et al., 2003), for the three kinds of events (dry conditions, transition period and wet conditions) in the calibration period. It was only for the wet conditions events that the simulated runoff coefficient was significantly lower than the observed one. For the events corresponding to the evaluation period, results were satisfactory for wetting-up, transition, and wet-conditions events, with efficiency values of 0.89 and 0.86, respectively. For the wet-conditions event, the runoff coefficient of the simulated flow showed the same underestimation as for the calibration period, mainly due to the difficulty of the model in properly reproducing the hydrograph recession.

The SHETRAN model succeeded also in reproducing many of the aspects of hydrological functioning in the Cal Rodó catchment reasonably well. The model captured the magnitude of runoff event peaks fairly accurately; changes in catchment-scale soil water reserve were reasonably well reproduced, and the dynamics of the water table in valley floor areas were well represented. However, some aspects of simulated process operation have been shown to be less satisfactory. In particular, the underestimation of hydrograph recessions was the most obvious shortcoming in model performance, and was responsible for the marked decline in efficiency values when moving from the calibration to evaluation periods. Further evaluation indicated that the underestimation of flow recessions appears to have been linked to overestimation of evapotranspiration. It appears that the water that should have drained from the catchment in the days following rainfall was in fact retained in the soil profile in the simulation. However, despite this, changes in water reserve for both Cal Rodó and Can Vila were generally well simulated, at least at the weekly interval of the available data. Thus, it appears that, instead of contributing to runoff, the excess soil moisture generated by underestimation of recession flows was consumed by an overestimation of evapotranspiration from the soil profile.

There are a number of possible explanations for this. The first is that, although the current model structure and grid resolution are appropriate, the combination of parameter values used in the soil moisture and root-zone, evapotranspiration models is inappropriate. However, there is also the possibility that the scale of model application is insufficient to capture important micro-scale heterogeneity in process operation in the catchments. It may be that neglecting the micro-scale spatial heterogeneity in soil moisture, and runoff pathways arising from the presence of terracing as well as drainage ditches (Gallart et al., 1994), is one of the reasons for the incorrect simulation of recession flows, and the consequent over-availability of soil moisture for evapotranspiration. Finally, it may be that other processes governing runoff generation and soil moisture dynamics in the catchment, such as flow through macro-pores, are not incorporated explicitly within SHETRAN.

The first of these issues has been explored further in a sensitivity analysis of the SHETRAN soil water model for Can Vila (Anderton et al., 2002a). This demonstrated that, by changing soil hydraulic parameters within reasonable limits, a number of simulations could be produced that reproduced different aspects of the catchment response (stream hydrographs, water reserve and phreatic level). However, no one part of the parameter space yielded simulations that gave optimum results in terms of all responses. Presumably, the evapotranspiration model was not sufficiently well parameterised, and therefore the exploration of soil hydraulic parameters was not sufficient to obtain a model able to simulate the different aspects of the hydrological functioning.

Finally, a daily water balance model (SIMBAL) has been developed to examine the expectable implications of climate and land cover change on water yield. This model simulates rainfall interception and transpiration from different vegetation covers, as well as stream baseflow and overland flow, using the simplest equations possible (Llorens, 1993). The model was parameterised with the information obtained at the Can Vila catchment and at the experimental plots, along with data gathered from the literature. Runoff events as well as soil water reserve and water balance were well simulated. The application of this model to the Can Vila catchment, for an increase of forest cover from the present 30% to a prospective 90%, predicts a reduction between 14 and 18% of the annual runoff, depending on the characteristics of the forest cover (Llorens et al., 2002).

3. Conclusion

During most of the year, the Vallcebre catchments behave as temperate wet ones; soil water content does not restrict plants transpiration, the evaporation of precipitation intercepted by canopies is driven by the duration of rainfall events, the subsurface flow along hillslopes feeds both saturated areas in downslope locations and stream baseflow, and the major part of stormflow corresponds to saturation excess overland flow over saturated areas during rainfall events.

Nevertheless, the increase of evapotranspirative demand during summer produces the depletion of the soil water that is not sufficiently replenished by precipitation; the limited soil water restricts plants transpiration and ceases the recharge of subsurface flow along hillslopes, producing the drying out of saturated areas and baseflow decline. Also during summer, the intensity of precipitations increase, rainfall interception events become more complex, and badlands are the only contributing areas to storm flow.

A large degree of spatial complexity must be added to this temporal arrangement: patches of forest and pasture influence local water balance and soil moisture, the agricultural terraces disturb the shallow aquifers, increase the spatial variability of soil properties and condition the formation of saturated areas, yet the artificial drainage ditches modify overland flow.

The above picture has been composed using field observations and different kinds of measurements, and several models were able to reproduce, at different extents, the observed behaviour. Nevertheless, none of the tested models attempted to cope with the complexity of some characteristics such as the terraced topography and the patchiness of vegetation. In fact, the models used, through the utilisation of "effective" parameters, overrode these sources of complexity. Although most likely any model will be able to cope with all the complexity of any catchment (Beven, 2001), the progress on the knowledge of the actual hydrological functioning of catchments is necessary for the progress of both applied and scientific hydrology (Dunne, 1998; Beven, 2001; Sivapalan et al., 2003).

Acknowledgements

This research was supported by the Agreement between the CSIC and the Spanish Ministry of Environment (RESEL network of the LUCDEME project), as well as the PROHISEM project (REN2001-2268-C01-01/HID), funded by the Spanish Government, and the TempQsim project (EVK1-CT-2002-00112) funded by the European Commission.

The authors are indebted to Maria Sala for her noteworthy support at the beginning of the works. Data and support provided by O. Avila, X. Huguet, R. Poch, R. Poyatos, D. Rabadà, D. Regüés, C. Rubio, C. Salvany and M. Soler, among others, are acknowledged. The comments made by Uhlenbrook S. improved the quality of the chapter.

References

Aepler, R., 1968. Das Garumnium der Mulde von Vallcebre und ihre Tektonik (Spanien, Provinz Barcelona), Diplomarbeit. Naturwissenschaftlichen Fakultät der freien Universität Berlin, Berlin, p. 101.

Anderson, M.G., Burt, T.P., 1990. Subsurface runoff. In: Anderson, M.G., Burt, T.P. (Eds), Process Studies in Hillslope Hydrology. Wiley, Chichester, pp. 365–400.

Anderton, S., Latron, J., Gallart, F., 2002a. Sensitivity analysis and multi-response, multi-criteria validation of a physically-based distributed model. Hydrol. Processes 16 (2), 333–353.

Anderton, S., Latron, J., White, S., Llorens, P., Salvany, C., Gallart, F., O'Connell, P.E., 2002b. Internal evaluation of a physically-based distributed model using data from a Mediterranean mountain catchment. Hydrol. Earth Syst. Sci. 6 (1), 67–83.

Balasch, J.C., Castelltort, F.X., Llorens, P., Gallart, F., 1992. Hydrological and sediment dynamics network design in a Mediterranean mountainous area subject to gully erosion. In: Bogen, J., Walling, D.E., Day, T. (Eds), Erosion and Sediment Transport Monitoring Programmes in River Basins, IAHS Publication 210. IAHS, Wallingford, UK, pp. 433–442.

Beven, K., 2001. How far can we go in distributed hydrological modelling? Hydrol. Earth Syst. Sci. 5 (1), 1–12.

Beven, K.J., Kirkby, M., 1979. A physically-based variable contributing area model of basin hydrology. Hydrol. Sci. Bull. 24 (1), 43–69.

Buchtele, J., Buchtelová, M., Gallart, F., Latron, J., Llorens, P., Salvany, C., Herrmann, A., 1998. Rainfall-runoff processes modelling using the Sacramento and Brook models in the Cal Rodó catchment (Pyrenees, Spain). In: Elias, V., Littlewood, I.G. (Eds), Catchment Hydrological and Biogeochemical Processes in the Changing Environment, IHP-V Technical Documents in Hydrology n°37. UNESCO, Paris, pp. 15–22.

Burnash, J.C.R., 1995. The NWS river forecast system catchment modeling. In: Singh, V.P. (Ed.), Computer Models of Watershed Hydrology. Water Resource Publications, Colorado, pp. 311–366.

Ciarapica, L., Todini, E., 1998. TOPKAPI – Un modello afflussi-deflussi applicabile dalla scala di versante alla scala di bacino, Atti del XXVI Convegno di Idraulica e Costruzioni Idrauliche, Catania, 9–12 Settembre 1998, Volume II, pp. 49–60.

Ciarapica, L., Todini, E., 2002. TOPKAPI: a model for the representation of the rainfall -runoff process at different scales. Hydrol. Processes 16 (2), 207–229.

Dunne, T., 1998. Wolman Lecture: hydrologic science...in landscapes...on a planet...in the future. In Hydrologic Sciences: Taking Stock and Looking Ahead. National Academy Press, Washington, DC, pp. 138.

Ewen, J., Parkin, G., O'Connell, P.E., 2000. SHETRAN: a coupled surface/subsurface modeling system for 3D water flow and sediment and solute transport in river basins. J. Hydrol. Eng. 5, 250–258.

Federer, C.A., 1993. BROOK90 – A Simulation Model for Evapotranspiration, Soil Water and Streamflow. USDA Forest Service, Durham, NH.

Gallart, F., Llorens, P., Latron, J., 1994. Studying the role of old agricultural terraces on runoff generation in a Mediterranean small mountainous basin. J. Hydrol. 159, 291–303.

Gallart, F., Latron, J., Llorens, P., Rabadà, D., 1997. Hydrological functioning of Mediterranean mountain basins in Vallcebre, Catalonia: some challenges for hydrological modelling. Hydrol. Processes 11, 1263–1272.

Gallart, F., Latron, J., Llorens, P., Salvany, C., Anderton, S., Quinn, P., O'Connell, P.E., White, S., Ciarapica, L., Todini, E., Buchtele, J., Herrmann, A., 2000. Intercomparison of hydrological models in a small research catchment in the Pyrenees. In: Hoeben, R., Van Herpe, Y., de Troch, F.P (Eds), CD-ROM Proceedings, Ghent, Belgium.

Gallart, F., Llorens, P., Latron, J., Regüés, D., 2002. Hydrological processes and their seasonal controls in a small Mediterranean mountain catchment in the Pyrenees. Hydrol. Earth Syst. Sci. 6 (3), 527–537.

Gallart, F., Balasch, C., Regüés, D., Soler, M., Castelltort, X., 2005. Catchment dynamics in a Mediterranean mountain environment: the Vallcebre research basins (south eastern Pyrenees) II: temporal and spatial dynamics of erosion and stream sediment transport. In: Garcia, C., Batalla, R.J. (Eds), Catchment Dynamics and River Processes: Mediterranean and Other Climate Regions. Elsevier, Amsterdam, pp. 17–29.

Haro, S., Fernández, J.F., Josa, R., Gallart, F., 1992. Papel hidrológico y geomorfológico de las propiedades del suelo en una zona pirenaica de campos abandonados (Cal Parisa, Vallcebre). In: López-Bermúdez, F, Conesa, C., Romero, M.A. (Eds), Estudios de Geomorfología en España, Vol. 1. SEG, Murcia, pp. 243–250.

Latron, J., 2003. Estudio del funcionamiento hidrológico de una cuenca mediterránea de montaña (Vallcebre, Pirineos Catalanes). Ph.D. Thesis, Facultat de Geologia, Universitat de Barcelona, Spain, p. 269.

Latron, J., Gallart, F., 2002. Seasonal dynamics of runoff variable contributing areas in a Mediterranean mountain catchment (Vallcebre, Catalan Pyrenees). In: Holko, L., Miklánek, P., Parajka, J., Kostka, Z. (Eds), Interdisciplinary Approaches in Small Catchment Hydrology: Monitoring and Research. Slovak Commitee for Hydrology-NC (IHP, UNESCO), Institute of Hydrology, Slovak Academy of Sciences, Bratislava, Slovakia, pp. 173–179.

Latron, J., Llorens, P., Gallart, F., 1997. Studying spatial and temporal patterns of runoff generation processes in a mountain Mediterranean basin (Vallcebre, Catalonia). In: Viville, D., Littlewood, I.

(Eds), Ecohydrological Processes in Small Basins, IHP-V Technical Documents in Hydrology, 14. UNESCO, Paris, pp. 1–5.

Latron, J., Gallart, F., Salvany, C., 2000. Analysing the role of phreatic level dynamics on the streamflow response in a Mediterranean mountainous experimental catchment (Vallcebre, Catalonia). In: Elias, V., Littlewood, I.G. (Eds), Catchment Hydrological and Biochemical Processes in the Changing Environment, IHP-V Technical Documents in Hydrology, 37. UNESCO, Paris, pp. 107–111.

Latron, J., Anderton, S., White, S., Llorens, P., Gallart, F., 2003. Seasonal characteristics of the hydrological response in a Mediterranean mountain research catchment (Vallcebre, Catalan Pyrenees): field investigations and modelling. In: Servat, E., Najem, W., Leduc, C., Shakeel, A. (Eds), Hydrology of Mediterranean and Semiarid Regions, IAHS Publication 278. IAHS, Wallingford, UK, pp. 106–110.

Latron, J., Llorens, P., Butturini, A., Gallart, F., 2004. Assessment of runoff generation dynamics in a Mediterranean catchment using hydrometric, geochemical and modelling approaches. Progress in Surface and Subsurface Water Studies at the Plot and Small Basin Scale. 10th., ERB Assembly, Turin October 2004.

Llorens, P., 1991. Resposta hidrològica i dinàmica de sediments en una petita conca pertorbada de muntanya Mediterrània. Ph.D. Thesis, Facultat de Geografia i Història, Universitat de Barcelona, Spain, p. 276.

Llorens, P., 1993. Hydrological implications of afforestation of abandoned lands: water balance simulation of a small Mediterranean mountainous basin. Acta Geol. Hispanica 28 (2–3), 131–138.

Llorens, P., 1997. Rainfall interception by a Pinus sylvestris forest patch overgrown in a Mediterranean mountainous abandoned area II. Assessment of the applicability of Gash's analytical model. J. Hydrol. 199, 346–359.

Llorens, P., Gallart, F., 1992. Small basin response in a Mediterranean mountainous abandoned farming area: research design and preliminary results. Catena 19, 309–320.

Llorens, P., Poch, R., Latron, J., Gallart, F., 1997. Rainfall interception by a Pinus sylvestris forest patch overgrown in a Mediterranean mountainous abandoned area. I- Monitoring design and results down to the event scale. J. Hydrol. 199, 331–345.

Llorens, P., Latron, J., Oliveras, I., 2002. Modelización del efecto del cambio global en la hidrología superficial. Ejemplo de aplicación a una cuenca mediterránea de montaña. In: García, F., Berné, J.L. (Eds), 3ª Asamblea Hispano-Portuguesa de Geodesia y Geofísica. Tomo III. Universidad Politécnica de Valencia, Valencia, pp. 1679–1681.

Llorens, P., Oliveras, I., Poyatos, R., 2003. Temporal variability of water fluxes in a Pinus sylvestris forest patch in Mediterranean mountain conditions. (Vallcebre research catchments, Catalan Pyrenees). In: Servat, E., Najem, W., Leduc, C., Shakeel, A. (Eds), Hydrology of Mediterranean and Semiarid Regions, IAHS Publication 278. IAHS, Wallingford, UK, pp. 101–105.

Nash, I.E., Sutcliffe, I.V., 1970. River flow forecasting through conceptual models. J. Hydrol. 10, 282–290.

Oliveras, I., Llorens, P., 2001. Medium-term sap flux monitoring in a Scots pine stand: analysis of the operability of the heat dissipation method for hydrological purposes. Tree Physiol. 21, 473–480.

Poyatos, R., Llorens, P., 2003. Influencia de la humedad del suelo y del período de actividad vegetal en la evapotranspiración de un pastizal montano mesófilo. Actas del VII Congreso Nacional de la Asociación Española de Ecología Terrestre, Barcelona, Spain, pp. 182–195.

Poyatos, R., Latron, J., Llorens, P., 2003. Land-use and land-cover change after agricultural abandonment. The case of a Mediterranean mountain area (Catalan Prepyrenees). Mt. Res Dev. 23 (4), 52–58.

Quinn, P., Gallart, F., Latron, J., Russell, K., 1998. Nesting localized patch models and data within catchment models and data. In: Kovar, K., Tappeiner, U., Peters, N.E., Craig, R.G. (Eds), Hydrology, Water Resources and Ecology in Headwaters, IAHS Publication 248. IAHS, Wallingford, pp. 275–281.

Rabadà, D., Gallart, F., 1993. Monitoring soil water content variability in the Cal Parisa basin (Alt Llobregat) with TDR. Experimental design and first results. Acta Geol. Hispanica 28 (2–3), 85–93.

Regüés, D., Pardini, G., Gallart, F., 1995. Regolith behaviour and physical weathering of clayey mudrock in a gullied area, as dependent on seasonal weather conditions. Catena 25, 199–212.

Rubio, C., 2003. Estudi i modelització de les característiques hidrodinàmiques de perfils representatius d'una conca de recerca (Vallcebre, alt Llobregat), Memòria de Recerca. Universitat Autònoma de Barcelona, Spain, p. 82.

Sivapalan, M., 2003. Prediction in ungauged basins: a grand challenge for theoretical hydrology. Hydrol. Processes 17, 3163–3170.

Sivapalan, M., Takeuchi, K., Franks, S.W., Gupta, V.K., Karambiri, H., Lakshmi, V., Liang, X., Mcdonnell, J.J., Mendiondo, E.M., O'Connell, P.E., Oki, T., Pomeroy, J.W., Schertzer, D., Uhlenbrook, S., Zehe, E., 2003. IAHS decade on predictions in ungauged basins (PUB), 2003–2012: shaping an exciting future for the hydrological sciences. Hydrol. Sci. J. 48 (6), 857.
Sklash, M.G., Farvolden, R.N., 1979. The role of groundwater in storm runoff. J. Hydrol. 43, 45–65.
Smith, M., 1992. Report on the Expert Consultation of Revision of FAO Methodologies for Crop Water Requirements. Food and Agriculture Organisation of the United Nations, Rome, p. 60.
Toebes, C., Ouryvaev, V., (Eds), 1970. Les bassins répresentatifs et expérimentaux. UNESCO, Paris, p. 380.

Catchment Dynamics and River Processes: Mediterranean and Other Climate Regions
Celso Garcia, Ramon J. Batalla, Editors

Chapter 2

Catchment dynamics in a Mediterranean mountain environment: the Vallcebre research basins (southeastern Pyrenees) II: temporal and spatial dynamics of erosion and stream sediment transport

Francesc Gallart[1,*], J. Carles Balasch[2], David Regüés[3], Montse Soler[1] and Xavier Castelltort[1]

[1]*Institute of Earth Sciences "Jaume Almera" (CSIC), Lluis Solé Sabarís s/n, 08028 Barcelona, Spain*
[2]*Department of Environment and Soil Sciences, University of Lleida, 25198 Lleida, Spain*
[3]*Pyrenean Institute of Ecology (CSIC), Campus Aula Dei, Avda. Montañana 1005, Apdo 202, 50080 Zaragoza, Spain*

Abstract

Erosion processes and sediment dynamics were studied in a set of Mediterranean mountain catchments with very active badlands during a period of over 10 years. The relatively small badlands produced most of the sediments, whereas both the dense vegetation and the old soil conservation structures impeded significant erosion from the main area of the catchments. The dynamics of the badland surfaces are driven by intense frost weathering in winter, and the subsequent compaction and erosion of the regolith throughout the year. Infiltration rates measured by means of rainfall experiments were found to depend on regolith moisture and density and reached the lowest values in autumn, whereas regolith detachability by splash was higher in autumn and spring. Nevertheless, erosion rates at the plot scale (1430 m^2) were especially high between May and September due to the role of high intensity rainstorms. The main runoff events at the small catchment scale (1.3 km^2) occurred between November and January due to the large rainfall events of moderate intensity and the wet antecedent conditions; secondary events occurred in late spring and scattered flash floods occurred during summer. The main sediment transport events occurred in November and December, although less important events occurred in late spring and late summer. As little as 3% of the events produced 73% of sediment yield during 5 years. There was, therefore, a temporal shift of the sediment conveyance with increasing spatial scale: the regolith was produced during winter and, subsequently, eroded during summer, when intense rainstorms produced runoff only on the badland surfaces; nevertheless, most of this sediment was deposited in the channels because the catchment was dry and did not contribute to stream discharge. Subsequently, large flood events produced outside of the badlands transported the sediment deposited on the channels during the previous summer and eventually that from preceding years.

Keywords: Mediterranean region, badlands erosion, sediment yield, temporal dynamics, spatial dynamics, sediment storage

---

*Corresponding author. *E-mail address:* fgallart@ija.csic.es (F. Gallart).

1. Introduction

Badlands are commonly considered as landscapes characteristic of dry areas. Nevertheless, they also occur in wetter areas where high-topographic gradients, bedrock weakness and high-intensity rainstorms, which are rather frequent in Mediterranean environments, coexist. *Humid badlands* (Gallart et al., 2001) develop in areas, usually mountainous in character, with annual precipitation exceeding 700 mm and with frequent rainstorms during summer. There is enough water to allow dense vegetation cover, including continuous permanent pastures, but the growth of vegetation on badland surfaces is checked more by the high erosion rates than by dryness (Crosaz and Dinger, 1999; Richard and Mathys, 1999; Regüés et al., 2000b).

The area of Vallcebre was selected in 1989 for monitoring the dynamics of badlands and their contribution to catchment sediment yield as it was found to be the principal area contributing to the sediment discharge to the reservoir of La Baells. Field observations, as well as preliminary measurements, suggested loads of heavy fine sediment from the badlands, where erosion processes were subject to strong seasonal controls, particularly the role of freezing on weathering and of intense summer rainstorms on erosion. (Clotet and Gallart, 1986; Balasch et al., 1992). Nevertheless, there was little knowledge of the dynamics at the catchment scale and the possible role of temporary stores of sediments.

The temporary storage of sediments is a known process that may play a relevant role in the sediment budget (e.g. Walling, 1983), but in the case of fine sediments, the more common place of deposition is on the hillslopes (e.g. Steegen et al., 2000) or in floodplains (e.g. Walling et al., 1998). Storage of fine sediments within the channels may be relevant in large rivers (Meade et al., 1985) but in mountain streams they are normally limited to the interstices of larger elements, as they are usually transported long distances in few jumps (Bonniwell et al., 1999).

Along with the ongoing monitoring of precipitation, weather, stream discharge and suspended sediment concentrations, several experiments and temporal observations have been performed in this area. The purpose of this chapter is to analyse the geomorphic processes and sediment dynamics at different spatial and temporal scales, summarising the results obtained with studies carried out in this area.

1.1. Site characteristics

The Vallcebre area is located in the headwaters of the Llobregat basin, in the south-eastern Pyrenees, at altitudes between 1100 and 1650 m a.s.l. Mean annual precipitation is about 924 mm, and mean annual temperature at 1440 m a.s.l. is 7.3°C (Gallart et al., 2002). Freezing occurs about 100 days per year (Regüés et al., 1995). Bedrock in this area consists of a nonmarine formation attributed to the late Cretaceous period (Feist and Colombo, 1983) and dominated by smectite-rich clays that facilitate mass movements and badland erosion. The most common plant communities bordering badlands are pastures and *Pinus sylvestris* forests, both of them ranging from xerophilous to mesophilous types in relation to slope aspect and altitude. South facing areas were deforested and terraced for agricultural purposes in the past, and at present they hold pastures chiefly formed by short shrubs and xerophilous grasses. In contrast, more mesophilous pastures and denser pine stands develop on north-facing slopes.

1.2. Instrumental design

Figure 2.1 shows the location of the instruments in the catchment. Precipitation was recorded with the help of tipping-bucket gauges (0.2 mm of resolution) connected to small data-loggers that record the time of every bucket movement with a resolution of 1 s. Regolith temperature was recorded during a period of 3 years with the help of thermistors connected to data-loggers (Regüés et al., 1995). Early observations on erosion rates were made with the help of plastic bags and erosion pins (Clotet and Gallart, 1986; Clotet et al., 1988).

Rainfall simulation experiments were performed repeatedly during 3 years with a light nozzle-based spray simulator suitable for rough terrain (Calvo et al., 1988) that uses a small (0.225 m^2) target plot. As the size of this plot was very small, runoff was not considered to contribute to sediment detachment, and the sediment collected during the experiments was attributed to detachment by splash (Regüés and Gallart, 2004). Plot erosion measurements were performed during a 3-year period with a 9-slot divisor device, which collected the sediments from a natural small catchment of 1430 m^2 (Castelltort and Balasch, 1993; Castelltort, 1995).

Three gauging stations were instrumented for monitoring sediment discharge:

- Can Vila (0.56 km^2) is a sub-catchment with some eroded landscapes (0.9% of the catchment area) that do not show the characteristic deeply incised features

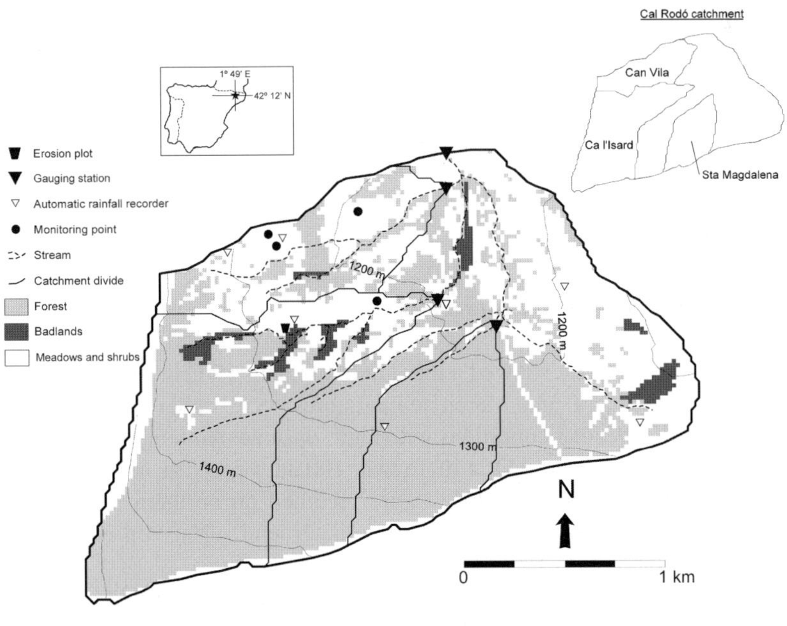

Figure 2.1. Location of the Vallcebre catchments.

characteristic of badlands. The gauging station was provided with a V-notch control, water level and temperature sensors, an OBS-3 D&A infra-red backscattering turbidity sensor and an automatic ISCO 2700 water sampler. A DT Data Electronics data-logger was used to record the readings as well as to trigger the water sampler during events.

- Ca L'Isard (1.32 km^2) is the sub-catchment with the more extended and active badlands (4.5% of the area). The gauging station was provided with a concrete rectangular flume, with similar equipment as described for the Can Vila station, as well as a Mobrey MSM 40 ultrasonic beam attenuation suspended sediment sensor, that allows the continuous recording of sediment concentrations up to 240 g l^{-1}.
- Cal Rodó (4.17 km^2) is the main catchment that embraces the former two ones, with badland areas representing a small part of the area (2.8%). The gauging station is provided with a concrete multi-level rectangular weir, and instruments similar to those described for the Ca l'Isard station.

The automatic water samplers were first installed to obtain samples of sediment concentrations when these were beyond the range of the turbidimeters. For this purpose, the samplers were triggered when the turbidity exceeded a selected threshold, and they were programmed to take samples at fixed time intervals. When the ultra-sonic sensors were installed, these were triggered by both stage and turbidity thresholds, to save electrical power during inactive periods, and the samplers were programmed to take samples during events at two different time intervals for the rising and recession limbs of the hydrographs. The water samples are used to re-calibrate the continuous recording sensors, to fill gaps due to malfunctioning, and to obtain grainsize analysis of suspended sediments and measurements of dissolved loads. The intake of the sampler tubes were located in areas of high turbulence at 10–15 cm above the stream beds; temporal variability of suspended sediments is very high, and is assumed to be a much higher source of error than the vertical gradients of suspended sediments in such mountain streams. Bedload transport is not regularly measured at the Vallcebre stations, although observations using large net traps during a common event in 1994 suggested that it may not be more than 1% of the total load (Castelltort, 1995).

2. Results and discussion

2.1. Climate and precipitation characteristics

Mean monthly temperatures showed a maximum of about 15°C in July and a minimum of 1°C in January. February and March temperatures were rather warm because of the occurrence of stable weather conditions with high sunshine. Air temperatures below 0°C at 1.5 m above ground occurred a little more than 100 days per year, whereas at the ground surface, about 130 freeze–thaw cycles were recorded on northern aspects and 112 on southern ones. Freezing reached 55 cm below the ground surface on northern exposures and 25 cm on southern ones (Regüés et al., 2000b).

Figure 2.2 shows daily precipitation depths (dots) recorded during 6 years (1998–2003), as well as the relative cumulated trend (grey line). Precipitation is relatively constant throughout the year, except for a marked scarcity in February and

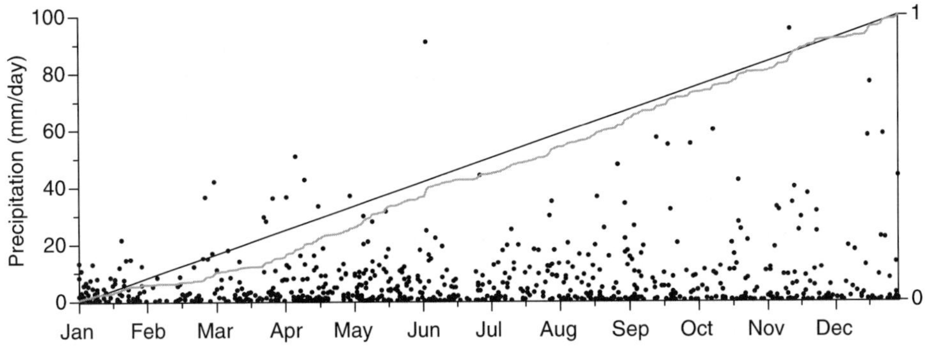

Figure 2.2. Precipitation depths in 24 h (dots) measured during 6 years (1998–2003), and relative cumulative trend (grey line), that represents both the daily amounts and the temporal aggregation of events. The oblique straight black line is shown for comparison of the cumulative trend with a time-constant one.

March and another less pronounced in late June and July. Figure 2.3 shows the highest rainfall intensities in 30 min (dots) along with the relative cumulative trend (grey line). This graph demonstrates the role of summer rainstorms; rainfall intensities are maximum in July–August, with a secondary maximum in May.

2.2. *Physical weathering, runoff generation and erosion rates on badland surfaces*

Early field observations demonstrated the relevant role of frost heaving on the dynamics of badland surfaces (Clotet and Gallart, 1986). The monitoring of regolith temperature, moisture and bulk density during 3 years (Regüés, 1995; Regüés et al., 1995) evidence that regolith bulk density sharply decreases with the occurrence of repeated freezing cycles, leading to the formation of spongy "popcorn" features at the surface. Extreme observed values for bulk density (mean of four measurements) were 0.83 g cm^{-3} (January) and

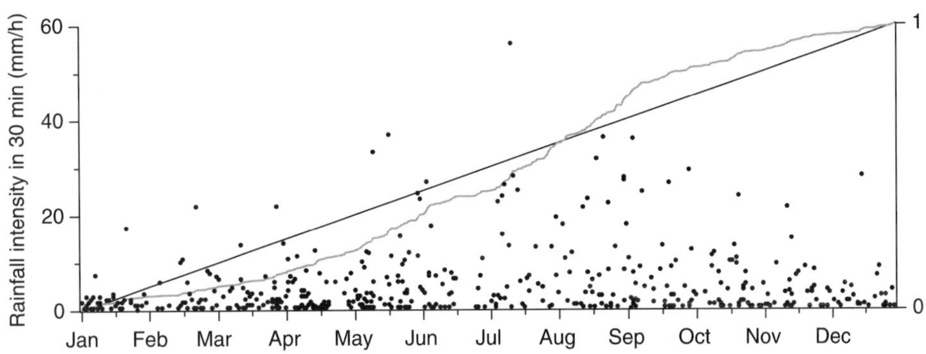

Figure 2.3. Rainfall intensities in 30 min (dots) of the days with precipitation observed during 6 years (1998–2003), and relative cumulative trend (grey line), that represents both the daily amounts and the temporal aggregation of events.

1.46 g cm^{-3} (September). This work showed also that bulk density through the year is positively correlated with temperature ($r = 0.64, p < 0.01$) and negatively with moisture ($r = 0.67, p < 0.01$). Laboratory experiments demonstrated that repeated freeze–thaw cycles lead to the disruption of the soft bedrock and the formation of “popcorn” features, whilst the wetting–drying cycles are much less efficient as weathering agent (Pardini et al., 1995, 1996). The analysis of the amounts of different kinds of energy available for weathering and erosion confirmed the chief role of freezing and its strong difference between sunny and shady aspects, that explain the preferred occurrence of badlands on north-facing slopes. Flora characteristics showed also that vegetation is more stressed by low temperature than by water scarcity (Regüés et al., 2000b).

Rainfall simulation experiments (Regüés, 1995; Regüés and Gallart, 2004) showed that infiltration rates on badland surfaces were relatively high (mean 18 mm h^{-1}) and varying largely throughout the year, with a variation coefficient of 63% and reaching values between 1 and 40 mm h^{-1}. Infiltration rates which ranges between 0.3 and 29 mm are negatively correlated with precipitation before runoff initiation. This behaviour was attributed to the contradictory role of bulk density and moisture of the regolith. Figure 2.4 shows the results of these experiments throughout the year, in terms of maximal runoff rate (for an experimental rainfall intensity of 50 mm h^{-1}), as well as the cumulative trend (grey line). Rates in winter showed the largest variation, due to the role of moisture (Regüés and Gallart, 2004), whereas changes were much smaller along the year with the exception of a cluster of high-runoff rates in October.

Figure 2.5 shows the erosion rates obtained during these rainfall experiments, in terms of effective sediment detachability by rain splash (weight of sediment per unit rainfall kinetic energy). Both the individual experiments (dots) and the cumulative trend (grey line) show the occurrence of highest regolith weakness in October and in the spring.

Monitoring actual erosion rates at the plot (micro-catchment) scale (Castelltort, 1995) demonstrated the grouping of erosion events during the warm part of the year, with 70% of the total annual erosion occurring between May and early August (Fig. 2.6). It appears, therefore, comparing the former figures, that event rainfall intensity was the main driver of erosion events at this scale, overriding the role of regolith behaviour. Yet, the decrease in regolith weakness observed in summer (Fig. 2.5) may be attributed to the outcrop of

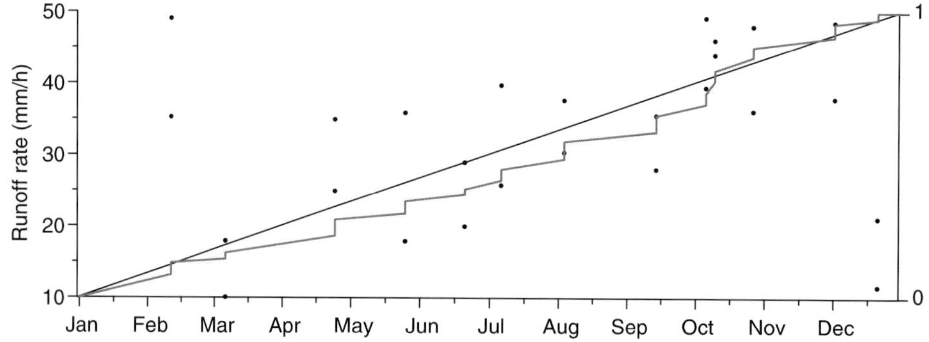

Figure 2.4. Maximal runoff rates (dots) obtained through rainfall experiments on badland surfaces, using an experimental rainfall intensity of 50 mm h^{-1}, along with the relative cumulative trend (grey line).

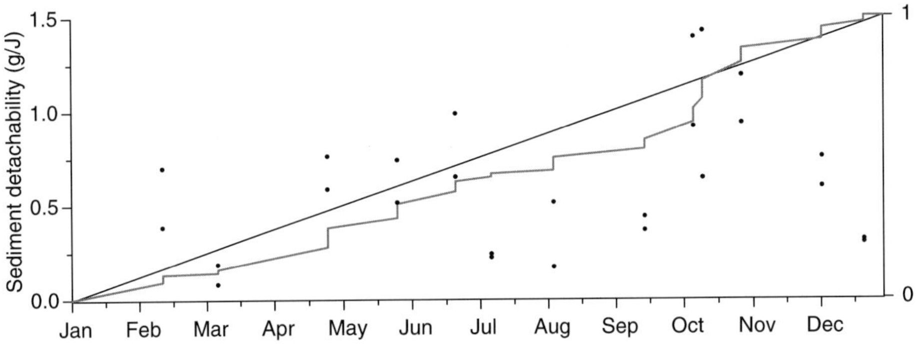

Figure 2.5. Sediment detachability estimates (dots) obtained through rainfall experiments on badland surfaces, in terms of weight of sediment per unit rainfall kinetic energy, along with the relative cumulative trend (grey line).

deeper hard layers of regolith through the erosion of shallow ones. Event erosion rates at this scale may be as high as 6.6 kg m^{-2}, and the annual average was 12 kg m^{-2}. These erosion rates compare very well with early annual estimates of 11–14 kg m^{-2} obtained with plastic bags (Clotet and Gallart, 1986), although are low when compared with rates of 23 kg m^{-2}, obtained with erosion pins (Clotet et al., 1988).

2.3. *Runoff and sediment transport at the small catchment scale*

Figure 2.7 shows the runoff events at the Ca l'Isard gauging station during the period 1995–1999. The main events occurred in the period between November and January, with minor but repeated events in spring and late summer. The comparison with Figures 2.2 and 2.3 evidences that the main runoff response at the catchment scale is driven much more by saturation mechanisms (antecedent conditions and event total precipitation) than by infiltration mechanisms (rainfall intensity). Summer rainstorms produced very active

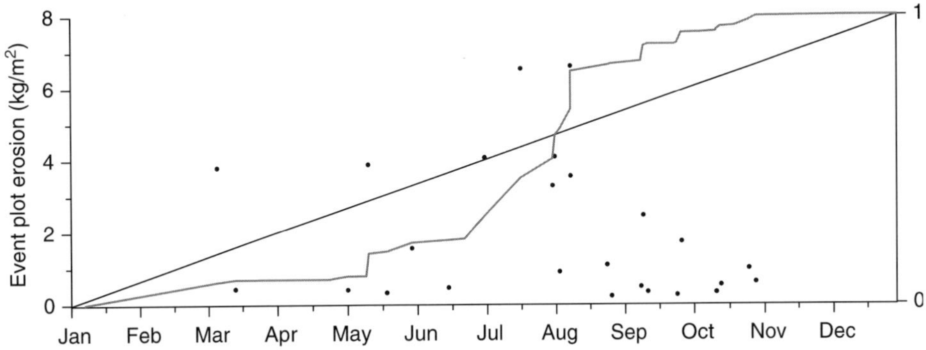

Figure 2.6. Event erosion rates (dots) obtained in an elementary badland catchment (1430 m^2) during 3 years (1991–1993), along with the relative cumulative plot (grey line), that represents both the event amounts and the temporal aggregation of events. Data from Castelltort (1995).

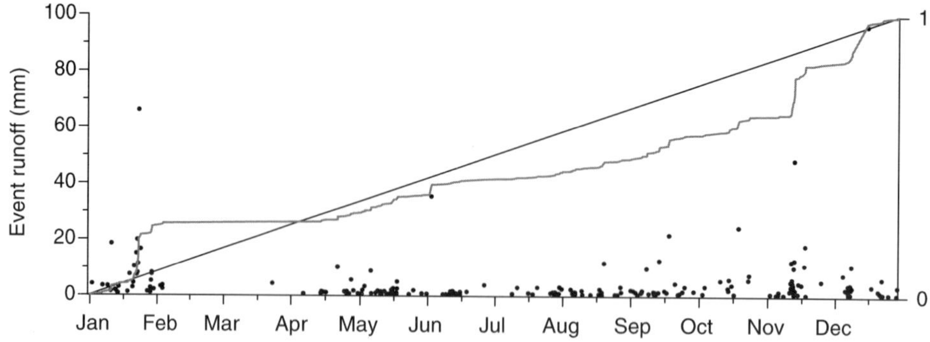

Figure 2.7. Event runoff depths (dots) measured at the Ca l'Isard gauging station (1.32 km^2) during 5 years (1995–1999), along with the relative cumulative plot (grey line), that represents both the event amounts and the temporal aggregation of events.

runoff on badland surfaces, but out of these areas, soils are dry in summer and are able to completely absorb as much as 50 mm of precipitation in 1 day (Llorens and Gallart, 1992). The small runoff events observed in summer are due to the contribution of badlands, while the remaining area did not contribute to runoff (see Gallart et al., 2005 for a longer discussion).

Figure 2.8 shows the suspended sediment transport events at the Ca l'Isard gauging station during the same period as Figure 2.7. As much as 54% of the sediment was transported in November and December, whereas other relevant events occurred in June and September. Some 75% of the transport was performed by 3% of the events, with an average sediment concentration of 41 g l^{-1}.

The study of sediment yield from a nearby small catchment without badlands (Llorens, 1991; Llorens et al., 1997) showed that mean suspended sediment concentrations in stream waters is two to three orders of magnitude smaller (25 mg l^{-1}) as well as sediment yield (4 Mg km^{-2} year^{-1}). This low sediment yield was attributed to the role of both the good

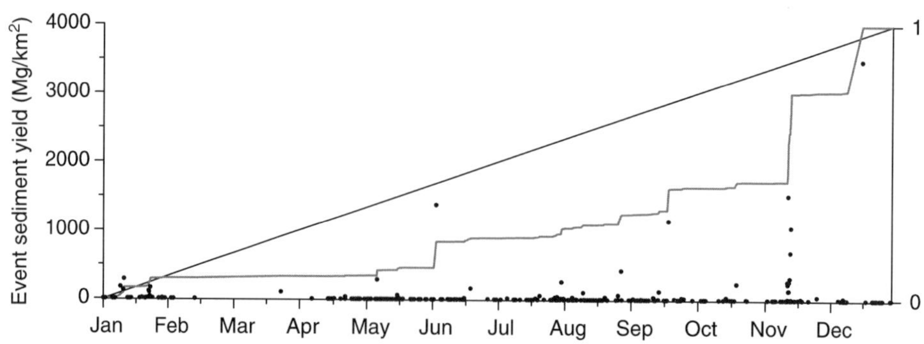

Figure 2.8. Event suspended sediment yields (dots) measured at the Ca l'Isard gauging station (1.32 km^2) during 5 years (1995–1999), along with the relative cumulative plot (grey line), that represents both the event amounts and the temporal aggregation of events.

condition of vegetation cover and the role of the old terraces built for soil conservation when the area was farmed in the past.

Most of the sediment transported at the catchment scale comes, therefore, from the badlands and some other secondary sources like stream bank scars, gullies and small mass movements. Nevertheless, the comparison between Figures 2.6 and 2.8 demonstrates a mismatch between the timing of erosion at the plot scale and the transport at the catchment scale. As stated before, 70% of sediment was eroded from badlands between May and early August, but only 18% of the sediment was transported during the same period at the Ca l'Isard station. Field observations demonstrated that most of the sediments eroded during summer are usually deposited on the stream channels because discharge was too low to transport them. The relationships between stream discharge and sediment concentration during summer floods usually show counter-clockwise hysteresis loops, and stream channels become very muddy after them.

Linear correlation coefficient between event runoff and sediment transport (Regüés et al., 2000a) is higher (0.78) when compared with the correlation coefficients between sediment transport and other relevant event variables as precipitation depth (0.63), peak discharge (0.64) and mean sediment concentration (0.45). The main runoff events are, therefore, also the main sediment transporting events at the catchment scale (Balasch and Castelltort, 1993), as there are large amounts of sediments on the channels available for transportation.

As badlands cover about 4.5% of the Ca l'Isard catchment area, the annual mean erosion rate of 12 kg m^{-2} on the badlands assessed before would represent a mean annual sediment yield of about 500 Mg km^{-2} from the whole catchment. In Figure 2.8, five individual events (during the 5 years) exceeded this figure, with a total sediment yield of 8600 Mg km^{-2}, or 1720 Mg km^{-2} $year^{-1}$. This result suggests first that the largest events would transport sediments deposited in the channels during several years, and that the erosion rates estimated on the badlands are substantially lower than those that may be inferred from sediment yield at the catchment scale.

Figure 2.9 shows annual precipitation, annual sediment yield and averaged multi-annual sediment yield at the Ca l'Isard station for the period 1990–2002. This is a tentative figure because the results were obtained using different methods (the ultra-sonic sensor was installed in 1994) and by different authors (Balasch, 1998; Regüés et al., 2000a), and it is well known that sediment transport measurement is very sensitive to methods and investigators (Walling, 1988). Inter-annual sediment yield from this figure was 1375 Mg km^{-2} $year^{-1}$, which is 2.8 times larger than the sediment yield derived from plot measurements, but only 1.3 times larger than sediment yield derived from erosion pins measurements (23 kg m^{-2} $year^{-1}$ for the badlands or 1035 Mg km^{-2} $year^{-1}$ for the whole catchment). One single event, on 17th December 1997, transported about 3500 Mg km^{-2}, an amount that was exceeded only by 2 years of the 13 in the record, and that corresponds to the sediment yield during 2.5 years of the average. Actually, if 2 years with extreme sediment yield (1997 and 1999) are excluded, the inter-annual average results in 535 Mg km^{-2} $year^{-1}$, an amount very close to the estimate obtained above from plot measurements.

These results raise the question of the accuracy and representativeness of the erosion rates assessments made at the plot and the basin scales. Measurements at the plot scale were made in locations where the collection of sediments was possible, and therefore may

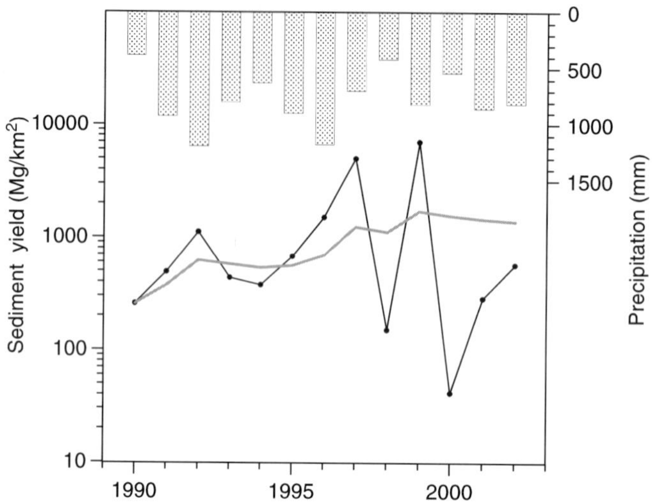

Figure 2.9. Annual precipitation depths (bars), annual sediment yields (dots), and inter-annual average of sediment yield (grey line) at the Ca l'Isard gauging station.

be not representative of very steep conditions, whereas erosion pins were located with more freedom and may be more representative of the effective rates. Furthermore, other secondary sources of sediment, such as bank erosion, mass movements and gully or channel entrenchment have not been taken into account in this assessment.

For the period 1994–1999, the annual suspended sediment transport estimated at Ca l'Isard was 3700 Mg, whereas at the Cal Rodó station, which includes Ca l'Isard the annual sediment transport was only about 2900 Mg (Regüés et al., 2000a). As there are few badland areas in the Cal Rodó catchment not included in Ca l'Isard and those existing have less erosive features, the actual sediment transport at Cal Rodó would be not much greater than at Ca l'Isard. Sediment deposition in the stream bed during this period may not afford an explanation for the difference, because the length of the stream between both gauging stations is only about 800 m, and this would need a conspicuous mean deposition of more than 0.5 m. Therefore, sediment yields from these catchments seem to have been estimated with an error of about 30%. Finally, as the large transport events in 1997 and 1999 largely modify the inter-annual sediment yield, their recurrence intervals should be taken into account for obtaining an adequate estimate of the long-term erosion rate.

3. Conclusions

The sediment dynamics in the Vallcebre catchments is dominated by the occurrence of badlands. There are strong spatial and temporal discontinuities in the conveyance of sediments from the clayey bedrock on badland surfaces to the catchment outlet: the soft bedrock is weathered mainly in winter by the role of repeated freezing–thawing cycles, the silty regolith is eroded in summer from badland surfaces to the stream channels as a consequence of intense rainstorms, and sediments are exported from the basin mainly during the main runoff events that usually occur in autumn. Large floods may transport sediments deposited on the stream beds during several preceding years.

This means that, although nearly all the sediments delivered by the catchments come from badlands, the export of sediments by streams is more controlled by stream dynamics than by processes on badland surfaces. The main reason for this paradox is that the main floods are produced by large precipitation events of moderate intensity occurring in wet periods, when large parts of the catchment contributes to runoff through saturation, whereas in summer, when sediment is eroded from badlands, the catchment is dry and nearly all the stream flow comes just from badlands.

Sediment yield from the catchments showed a very high variation among years, with differences of up to two orders of magnitude. The drier years provided the lesser sediment transport, although the years with the highest sediment transport were not the wetter years, but the years with the larger runoff events. The catchment sediment yield estimated from plot measurements on the badland surfaces was similar to the mean catchment yield measured during years without large events, but substantially smaller than the catchment yield cumulated with all the years.

The representativeness of the erosion rates measured on badland surfaces, the accuracy of sediment transport measurements at the gauging stations, and the effective contribution of the larger floods recorded to the long-term sediment yield are some of the main aspects that deserve further research.

Acknowledgements

This research was possible because of funding from the Agreement between the CSIC and the Spanish Ministry of Environment (RESEL network of the LUCDEME project), as well as the PROHISEM project (REN2001-2268-C01-01/HID) and the PIRIHEROS project (REN2003-08678), both funded by the Spanish Government, and the TempQsim project (EVK1-CT-2002-00112) funded by the European Commission.

The authors are indebted to Maria Sala for her noteworthy support at the beginning of the works, as well as to the memory of Núria Clotet, who led the project until her premature decease in 1990. Data and support provided by P. Llorens, J. Latron, F. Plana, G. Pardini and I. Queralt are acknowledged. The comments made by G. Verstraeten improved the quality of the manuscript.

References

Balasch, J.C., 1998. Resposta hidrològica i sedimentària d'una petita conca de muntanya analitzades a diferent escala temporal. Ph.D. Thesis, Universitat de Barcelona, Spain.

Balasch, J.C., Castelltort, F.X., 1993. Pautas para la modelización con parámetros físicos del transporte de sedimentos en suspensión en una cuenca montañosa, Tomo ll. V Reu. Nal. de Geología Ambiental y Ordenación del Territorio, Murcia, pp. 637–644.

Balasch, J.C., Castelltort, F.X., Llorens, P., Gallart, F., 1992. Hydrological and sediment dynamics network design in a Mediterranean mountainous area subject to gully erosion. In: Bogen, J., Walling, D.E., Day, T. (Eds), Erosion and Sediment Transport Monitoring Programmes in River Basins. IAHS Publication 210, pp. 433–442.

Bonniwell, E.C., Matisoff, G., Whiting, P.J., 1999. Determining the times and distances of particle transit in a mountain stream using fallout radionuclides. Geomorphology 27, 75–92.
Calvo, A., Sirvent, J.M., Palau, E., Romero, M., 1988. Un simulador de lluvia portátil de fácil construcción. In: Sala, M., Gallart, F. (Eds), Métodos y técnicas para la medición en el campo de procesos geomorfológicos, Monografía n° 1. S.E.G., Logroño, pp. 6–19.
Castelltort, F.X., 1995. Erosió, transport i sedimentació fluvial com a integració dels processos geomorfològics d'una conca (conca de Cal Rodó, Alt Llobregat). Ph.D. Thesis, Universitat de Barcelona, Spain.
Castelltort, F.X., Balasch, J.C., 1993. Diseño de una red de monitorización de procesos de erosión y transporte de sedimentos. In: Ortiz, R. (Ed.), Problemática Geoambiental y desarrollo. Actas V Reun. Nal. Geol. Amb. y Ord. del Territorio, Tomo II, pp. 65–672.
Clotet, N., Gallart, F., 1986. Sediment yield in a mountainous basin under high Mediterranean climate. Z. Geomorphol. Supp. Band 60, 205–216.
Clotet, N., Gallart, F., Balasch, J., 1988. Medium term erosion rates in a small scarcely vegetated catchment in the Pyrenees. Catena Suppl. 13, 37–47.
Crosaz, Y., Dinger, F., 1999. Mesure de l'érosion sur ravines élémentaires et essais de végétalisation. Bassin versant expérimental de Draix. In: Mathys, N. (Ed.), Les bassins versants expérimentaux de Draix, laboratoire d'étude de l'érosion en montagne. Cemagref, Antony, pp. 103–118.
Feist, M., Colombo, F., 1983. La limite Crétacé-Tertiaire dans le NE de l'Espagne du point de vue des charophytes. Colloque sur le Sénonien. Géol. Méditerr. Vol. X (3–4), pp. 303–326.
Gallart, F., Solé, A., Puigdefábregas, J., Lázaro, R., 2001. Badland systems in the Mediterranean. In: Bull, N.L., Kirkby, M.J. (Eds), Dryland Rivers: Processes and Management in Mediterranean Climates. Wiley, Chichester, pp. 299–326.
Gallart, F., Llorens, P., Latron, J., Regüés, D., 2002. Hydrological processes and their seasonal controls in a small Mediterranean mountain catchment in the Pyrenees. Hydrol. Earth Syst. Sci. 6 (3), 527–537.
Gallart, F., Latron, J., Llorens, P., 2005. Catchment dynamics in a Mediterranean mountain environment: the Vallcebre research basins (south eastern Pyrenees) I: hydrology. In: Garcia, C., Batalla, R.J. (Eds), Catchment Dynamics and River Processes: Mediterranean and Other Climate Regions. Elsevier, Amsterdam, pp. 1–16.
Llorens, P., 1991. Resposta hidrològica i dinàmica de sediments en una petita conca pertorbada de muntanya mediterrània. Ph.D. Thesis, Universitat de Barcelona, Spain.
Llorens, P., Gallart, F., 1992. Small basin response in a Mediterranean mountainous abandoned farming area: research design and preliminary results. Catena 19, 309–320.
Llorens, P., Queralt, I., Plana, F., Gallart, F., 1997. Studying sediment transfer in a small Mediterranean mountainous basin subject to land abandonment. Earth Surf. Proc. Land. 22, 1027–1035.
Meade, R.H., Dunne, T., Richey, J.E., Santos, U., de, M., Salati, E., 1985. Storage and remobilisation of suspended sediment in the lower Amazon river in Brazil. Science 228, 488–490.
Pardini, G., Pini, R., Barbini, R., Regüés, D., Plana, F., Gallart, F., 1995. Laser elevation measurements of a smectite-rich mudrock following freeze-thawing and wet-drying cycles. Soil Technol. 8 (2), 161–175.
Pardini, G., Vigna Guidi, G., Pini, R., Regüés, D., Gallart, F., 1996. Structural changes of smectite-rich mudrocks experimentally induced by freeze-thawing and wetting–drying cycles. Catena 27, 149–165.
Regüés, D., 1995. Meteorización física en relación con los procesos de producción y transporte de sedimentos en un área acarcavada. Ph.D. Thesis, Universitat de Barcelona, Spain.
Regüés, D., Gallart, F., 2004. Seasonal patterns of runoff and erosion responses to simulated rainfall in a badland area in Mediterranean mountain conditions (Vallcebre, South-eastern Pyrenees). Earth Surf. Proc. Land. 29, 755–767.
Regüés, D., Pardini, G., Gallart, F., 1995. Regolith behaviour and physical weathering of clayey mudrock in a gullied area, as dependent on seasonal weather conditions. Catena 25, 199–212.
Regüés, D., Balasch, J.C., Castelltort, X., Soler, M., Gallart, F., 2000a. Relación entre las tendencias temporales de producción y transporte de sedimentos y las condiciones climáticas en una pequeña cuenca de montaña mediterránea (Vallcebre, Eastern Pyrenees). Cuadernos de Investigación Geográfica 26, 24–41.
Regüés, D., Guàrdia, R., Gallart, F., 2000b. Geomorphic agents versus vegetation spreading as causes of badland occurrence in a Mediterranean subhumid mountainous area. Catena 40 (2), 173–187.
Richard, D., Mathys, S., 1999. Historique, contexte technique et scientifique des BVRE de Draix. Caractéristiques, données disponibles et principaux résultats acquis au cours de dix ans de

suivi. In: Mathys, N. (Ed.), Les bassins versants expérimentaux de Draix, laboratoire d'étude de l'érosion en montagne. Cemagref, Antony, pp. 11–28.

Steegen, A., Govers, G., Nachtergaele, J., Takken, I., Beuselinck, L., Poesen, J., 2000. Sediment export by water from an agricultural catchment in the Loam Belt of central Belgium. Geomorphology 33, 25–36.

Walling, D.E., 1983. The sediment delivery problem. J. Hydrol. 65, 209–237.

Walling, D.E., 1988. Measuring sediment yield from river basins. In: Lal, R. (Ed.), Soil Erosion Research Methods. Soil and Water Conservation Society, Ankeny, IA, pp. 39–73.

Walling, D.E., Owens, P.N., Leeks, G.J.L., 1998. The role of channel and floodplain storage in the suspended sediment budget of the River Ouse, Yorkshire, UK. Geomorphology 22, 225–242.

Catchment Dynamics and River Processes: Mediterranean and Other Climate Regions
Celso Garcia, Ramon J. Batalla, Editors

Chapter 3

Patterns and thresholds of runoff generation and sediment transport on some Mediterranean hillslopes

Adolfo Calvo-Cases[1,*], Carolina Boix-Fayos[2] and Eva Arnau-Rosalen[1]

[1]*Department of Geography, University of Valencia, 46010 Valencia, Spain*
[2]*Department of Soil and Water Conservation, CEBAS-CSIC, Apartado 164, 30100 Murcia, Spain*

Abstract

Runoff and sediment transport data obtained from hillslopes in two limestone areas in southeast Spain are analysed in order to define spatial and temporal thresholds for sediment movement at the patch scale under Mediterranean semiarid and subhumid climatological conditions. The data discussed in this paper include a 7-year series of runoff and sediment collection in open Gerlach plots. The 136 events are analysed in relation to characteristics of rainfall, soil and soil surface components. In both cases slopes behave as a patchwork of runoff and runon areas and the size of the runoff or runon patches being dependent on the climatological conditions. These control the hydrological disconnection between different parts of the slopes. Hortonian and saturation runoff can both be generated and infiltrated downslope. The differences in rainfall characteristics lead to the differences in transport rates between sites. But soil properties and surface components characteristics are also very relevant and do not always behave in the same way. In these conditions runoff rates are similar in both sites, but sediment yield is much higher in the dryer site. The role of the biological factor in controlling runoff and sediment transport is higher in subhumid site. When we move to the drier site, abiotic factors become more important. Some thresholds are defined for the changes in runoff generation mechanisms and rates of sediment transport.

Keywords: Mediterranean hillslopes, erosion, sediment transport, runoff generation, patch scale, soil surface components

1. Introduction

In the 80s and the 90s of the 20th century an intensive effort to derive data and understand processes of runoff generation and sediment movement under Mediterranean climatic conditions was carried out. Thanks to these efforts, a relatively large amount of data on runoff and sediment yields has been gathered and a better understanding of the surface wash processes has been achieved. However, these advances have raised many new questions due to the complexity of the processes of sediment movement in Mediterranean

* Corresponding author. *E-mail address:* adolfo.calvo@uv.es (A. Calvo-Cases).

ecosystems. It was found that arid and semiarid geo-ecosystems tend to be organised in a mosaic-operating pattern of sources and sinks of both water and sediment at different scales (Lavee et al., 1998; Puigdefábregas et al., 1999). Besides, the high spatial variability of the surface properties and the high temporal variability of the rainfall should always be taken into account (Puigdefábregas et al., 1999). So, although runoff generation mechanisms under these environmental conditions are now better understood (Martínez-Mena et al., 1998a,b; Puigdefábregas et al., 1999; Calvo-Cases et al., 2003), these mechanisms and hence sediment movement and delivery, exhibit a complexity that requires further attention (Kirkby et al., 2002; Beven, 2003).

1.1. Advances in the understanding of sediment movement processes on different scales and lithologies

Mediterranean hillslopes look like a patchwork of runoff and runon areas (sources and sinks) that are very much related to the soil, vegetation and rock fragments spatial patterns and the sediment movement along hillslopes.

At the hillslope scale the production of sediment under different climatological conditions on limestone lithology has shown some differences. For example, in subhumid areas total runoff can be higher or similar to the total runoff of semiarid areas. Sediment concentrations and sediment yield are always much lower, however, associated with better structured and less degraded soils (Calvo-Cases et al., 2003).

Research carried out in the arid site of Rambla Honda (Almeria, southeast Spain) on metamorphic rocks, also at the hillslope scale with sparse and patchy vegetation, has shown that a range of positive feedback mechanisms lead to nucleation of resources in the soil beneath plant clumps at the expense of the neighbouring bare ground, thus resulting in the increase of spatial heterogeneity. This spatial heterogeneity is dynamic due to the interaction between plant growth and hillslope fluxes of water and sediments (Puigdefábregas et al., 1999). Vegetation mosaics are organised in a way that maximise the efficiency of water supply to plants and the resistance to sediment movement. When the equilibrium is disturbed because fluxes of water and/or sediment exceed the capacity of the vegetation, redistribution of the plant pattern takes place by differential erosion (Puigdefábregas et al., 1999).

On marls and other types of soft rocks, it has also been shown that when different erosion processes interact, i.e. rill erosion and mass movements, relatively large amounts of sediment are exported from badland areas that are susceptible to mass movement (Harvey and Calvo-Cases, 1991; Calvo-Cases and Harvey, 1996). Bouma and Imeson (2000) explained how, in badland areas under unsaturated topsoil conditions, dynamic soil properties play a major role in the erosion process, whereas under saturation and initiation of overland flow, the increasing importance of runoff hydraulics results in a more complex combination of processes.

At the catchment scale Alexandrov et al. (2003) observed how only 50% of the variance in suspended sediment concentration is explained by variations in discharge, indicating that changes in sediment supply and variations in the importance of source areas are also significant in a semiarid area. In a different type of environment, but with a great interest for the understanding of sediment movement processes at the catchment level,

Trimble (1999) investigated the variation of sources and sinks in a catchment in a long-term study. He argued that sediment yield is poorly related to sediment fluxes within a basin. In the catchment of his study, even though sediment fluxes (from source patches) have changed drastically over the past 140 years, the sediment yield has shown little variation. The change of fluxes was not attributed to a change in climate but to land use changes (Sala and Inbar, 1992; Sala, 2003), controls of sediments by stream-bank structures and geomorphological changes within the catchment. Trimble (1999) stressed how monitoring of sediment yield, especially at plot scale, can lead to erroneous conclusions about erosional processes within a basin.

Concerning the temporal scale, in the Rambla Honda research site, the largest runoff and sediment yields are not recorded during the greatest events but during the intermediate ones. Puigdefábregas et al. (1999) explained this by saying that, firstly large events are also more discontinuous in time, which increases the opportunities of runoff to infiltrate, and secondly that a decline of sediment yield is experienced due to the exhaustion of available sediments during large events.

Conditions for sediment movement tested by Martínez-Mena et al. (2002) have been observed to be detachment-limited during high-intensity storms and transport-limited during medium intensity events in colluvial soils. However, in marls, the erosion was determined by the limited quantity of available sediment. Thresholds of rain intensity of over 15 mm h^{-1} have been considered in these areas as "erosive rainfall" taking into account the total soil loss and transport capacity of the overland flow (Martínez-Mena et al. 2001).

1.2. Erosion and transport rates reported by field studies with a variety of methods and scales

Erosion rates and sediment concentrations reported by field studies established at different scales and using different methods, in Mediterranean semiarid to subhumid climate conditions, show a wide variety of results. These results must be interpreted taking into account the different processes operating at different scales and the artefacts sometimes introduced by the methodology. Erosion rates derived from rainfall simulation experiments carried out in microplots (0.24 m^2) with approximately 55 mm h^{-1} of rainfall intensity on limestones vary from 0 to 6.61 g m^{-2} h^{-1} in Alicante (Spain) under dry soil conditions and from 1.86 to 47.28 g m^{-2} h^{-1} with antecedent soil moisture close to field capacity (Boix-Fayos et al., 1998; Boix-Fayos, 2000). The erosion rates occurring in Murcia with a rainfall intensity varying from 31 to 56 mm h^{-1} on bigger plots (2 m^2) located on Quaternary colluvium show higher average erosion rates (42.2 g m^{-2} h^{-1} with high antecedent soil moisture) (Martínez-Mena et al., 2002). The simulation experiments done on marls in southeast Spain in a wide variety of badland surfaces (Calvo et al., 1992) produced very variable erosion rates. Five groups are distinguished associated with sites, regolith properties and surface cover. The values in each range from: 6–181 g m^{-2} h^{-1}; 24–431 g m^{-2} h^{-1}; 91–624 g m^{-2} h^{-1}; 297–985 g m^{-2} h^{-1} and 964–1900 g m^{-2} h^{-1}, respectively.

The collection of data for several years in closed or open plots under natural rain conditions allows estimating erosion rates per year. Rates between 0 and 8 t ha^{-1} yr^{-1} in closed plots have been found in Alicante on marls (Bautista et al., 1996; Bautista, 1999). In Murcia between 0.012 and 1.84 t ha^{-1} yr^{-1} on limestones (Castillo et al., 1997; Romero Díaz et al., 1998); between 0.006 and 2.4 t ha^{-1} yr^{-1} on marls (Francis, 1986); 2 t ha^{-1} yr^{-1} on marls and sandstones (Romero Díaz et al., 1988).

In open plots under natural conditions two plots (328 and 759 m^2) with a mixed lithology of limestone, dolomites, marls and alluvial deposits in Murcia were monitored for several years, recording soil losses of 0.85 and 2.99 t ha^{-1} yr^{-1} (Martínez-Mena et al., 2001). In an area on slates, after a complete deforestation of the hillslope, Soler and Sala (1995) found rates of sediment removal varying between 3.9 and 45.13 g m^{-2} in 18 months. Using a rainfall simulator, the same area produced very low rates of sediment removal when it was still covered by natural forest: 0.04–0.83 g m^{-2} (Sala and Calvo, 1990).

At the catchment scale, derived from a study of the sedimentation in reservoirs, López Bermúdez and Gutiérrez Escudero (1982) reported 8.8 t ha^{-1} yr^{-1} of erosion rate in nine reservoirs of the Segura river (Murcia). Data reported by Avendaño Salas et al. (1997) show an average of 3.32 t ha^{-1} yr^{-1} of sedimentation rates in reservoirs in Murcia and Alicante, with the exception of the Guadalest reservoir (Alicante) with a much higher sediment accumulation rate (27.03 t ha^{-1} yr^{-1}). Summarising, sedimentation rates between 1.2 and 10.7 t ha^{-1} yr^{-1} were reported by different authors as pointed out by Romero Díaz (2002).

The erosion rates estimated in plots under natural rain are approximately within the same range as the ones reported from the sedimentation in reservoirs. However, the results derived from rainfall simulation experiments must be used as relative values to compare between different soil and rain conditions. As a matter of fact, they are single-event data, and very much depending on factors such as the antecedent soil moisture conditions or the sediment availability on the surfaces at the moment and place of the experiment.

2. Objectives and study area

The objective of this chapter is to define spatial (surface soil conditions) and temporal (rainfall characteristics) thresholds for sediment movement at the patch scale in Mediterranean hillslopes on limestone lithology under two contrasting climatological conditions, semiarid and subhumid.

The study was carried out in the province of Alicante, southeast Spain (Fig. 3.1), in two sites with a strong contrast in climatological conditions due to altitude differences (Mediterranean subhumid and Mediterranean semiarid) and with a contrasting land use history. In both cases, lithology has been taken as a constant (similar types of limestone) as well as other morphological properties such as average slope range or aspect, in order to make the study sites comparable and representative of their conditions.

Cocoll (CC), the study site with the highest altitude (900 m), is located at the northernmost position, with a mean rainfall of 826 mm yr^{-1} and a mean annual temperature of 13.8°C. Under these environmental conditions other Mediterranean areas have a well-developed forest cover and deeper soil profiles. Here, landuse types that affect vegetation and soil prevail, mainly grazing and vegetation removal in the past and forest fires over the last decades (see Boix-Fayos, 1997; 2000; Ramos et al., 2001).

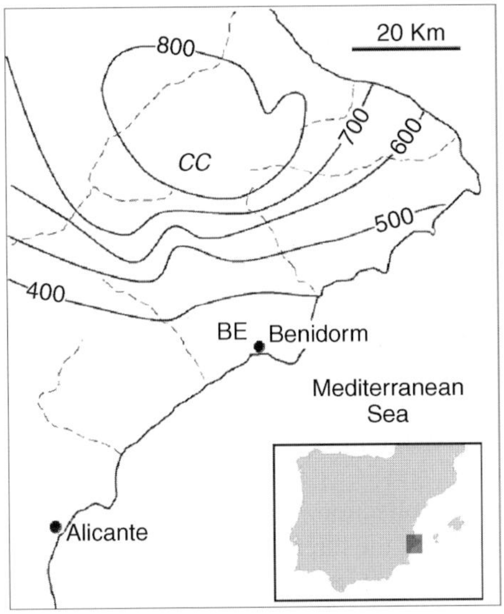

Figure 3.1. Location map of the study areas with expression of the mean annual rainfall. BE = Benidorm, CC = Cocoll.

Benidorm (BE), the lower and drier site, is located 20 km to the south, at 100 m altitude, has a mean rainfall of 387 mm yr^{-1} and a mean annual temperature of 17.9°C. At each study area, a south-facing slope was chosen to carry out the experimental work.

Both hillslopes are located on calcareous bedrock (limestone) from the Upper Cretaceous. Soils are classified as lithic leptosols in the most arid site (BE) and in the highest and wettest site (CC) are mainly lithic leptosols and chromic luvisols (Soriano et al., 1996). Soil aggregate stability has been shown to be dependent on organic matter, clay and calcium carbonate contents (Boix-Fayos, 2000). Interesting relations between soil structure and soil hydrology have appeared, suggesting that small aggregate sizes in the Cocoll area facilitate water infiltration (Boix-Fayos et al., 1998, 2001). Present-day vegetation series are defined as *Rosmarino-Ericion* in Cocoll and *Stipion tenacissima* in Benidorm (Rivas Martínez, 1985).

3. Methods

Previous work carried out on these two hillslopes since 1992 was dedicated to soil properties measurements (Boix-Fayos et al., 1998, 2001; Boix-Fayos, 2000), to the hydrological behaviour and soil water movement, mainly based on rainfall simulation experiments, and relations between soil properties and soil hydrological response (see Boix-Fayos, 2000; Calvo-Cases et al., 2003).

In order to study the soil response under natural rainfall events, erosion plots of Gerlach type (50 cm wide and open) were set up at each site (BE and CC). The plot distribution was done following a criterion of distance to the vegetation clusters and with different proportions of bare-soil and rock-fragment cover in the catchment area of each plot.

Precipitation, total runoff and sediment yields recorded over seven years are presented here (from August 1996 to October 2003). Some of the plots were set up later, in the year 2000 (CC 6 to 9) and two plots in BE (5 and 7) were vandalised and their measurements abandoned in April 2000. In total, full records from August 1996 up through October 2003 are available for 5 plots at each site.

The plot collectors (9 in CC and 7 in BE, Fig. 3.2) were regularly cleaned from sediments. Runoff in the collectors and tanks was measured after each rainfall event. Over half of the measuring period, three plots at each of the sites were continuously monitored with an automatic pressure sensor for water level.

The sediments collected in the traps were dried and separated into three fractions: fine sediments (less than 2 mm in diameter), coarse sediments (more than 2 mm in diameter) and litter (vegetal fragments not considered in the transport rates).

Rainfall was recorded at 10-min intervals (30 min at the beginning of the measurements in 1996) in automatic rain gauges (Cassella 0.2) situated near the plots (see Fig. 3.2) and for each period of sediment-collection, the average, maximum, I30 and I10 intensities were calculated. After removing the intervals without rainfall and with less than 1 mm of rainfall, in order to avoid data with zeros in the calculations, the number of considered events is 59 in BE and 77 in CC. Figure 3.3 represents the time distribution of the measurement periods for each plot.

In order to map the surface component characteristics of the contributing area to each plot, a series of photographs perpendicular to the surface were taken in September 2003. The images, with a resolution of 2 mm per pixel, were georeferenced using the ArcGis 8.1 facility. A digital thematic map was then produced by photointerpretation of a 50×300 cm rectangle upslope from each trap with the following classes: (1) plant cover; (2) litter; (3) >50% of rock-fragment cover; (4) <50% rock-fragment cover; (5) bare soil; (6) bare rock and (7) dead plants (Fig. 3.4). The resulting vector maps were converted into raster of 1 cm^2 resolution and the proportion of each surface component calculated for nested distance intervals of 1 cm of distance increase from the trap was calculated.

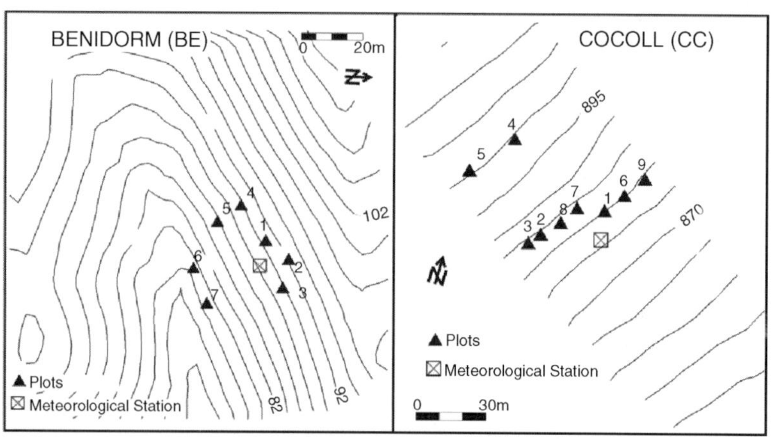

Figure 3.2. Detailed topography of the BE and CC sites and distribution of the plots.

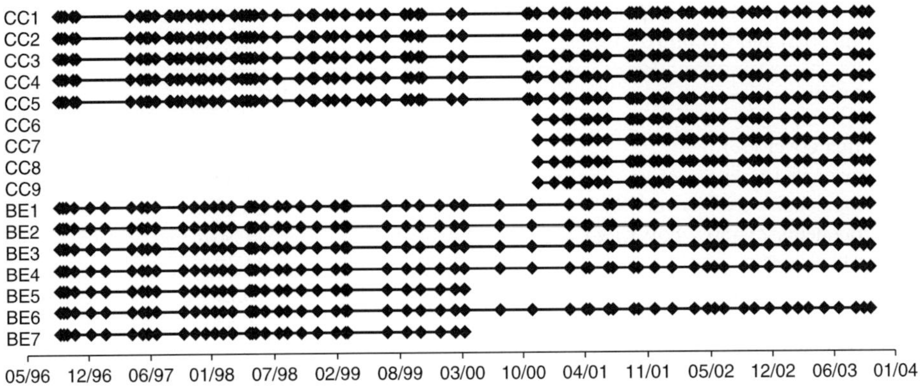

Figure 3.3. Time distribution of the measurement periods (1996–2003) for each plot after removing those with rainfall values of less than 1 mm.

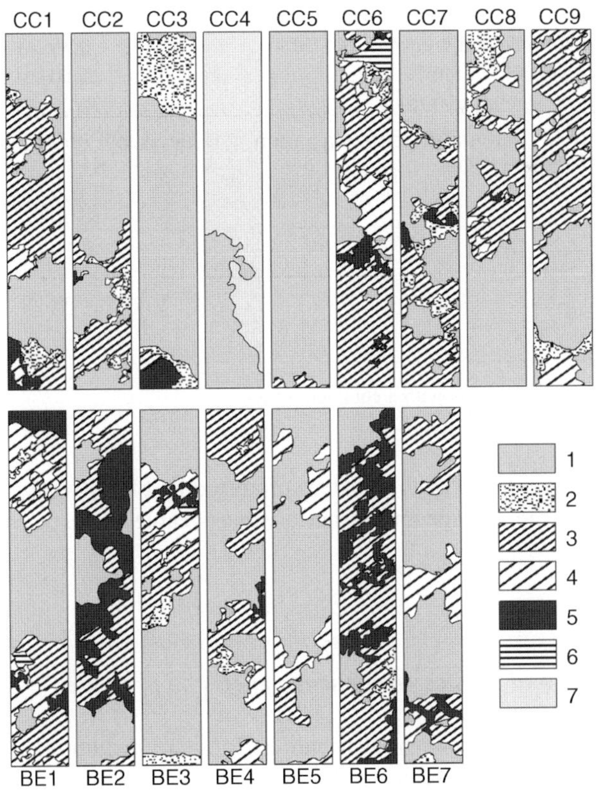

Figure 3.4. Surface components distribution mapped in a 50 × 300 cm rectangle upslope of each plot. (1) Plants cover; (2) litter; (3) more than 50% of rock-fragment cover; (4) less than 25–50%, rock-fragment cover; (5) bare soil; (6) bare rock; and (7) recently dead plants but still covering soil surface.

Table 3.1. Soil, surface components, rainfall, runoff and sediment movement average values for Benidorm (BE) and Cocoll (CC) sites.

Characteristic	BE	CC
Vegetation cover (%)	48.70	63.90
Litter cover (%)	1.86	6.24
Rock fragments >50% cover (%)	26.80	22.49
Rock fragments 25–50% cover (%)	11.30	5.02
Bare soil cover (%)	11.10	1.86
Bare rock cover (%)	0.29	0.49
Aggregates >10 mm (%)	5.55	1.25
Aggregates between 10 and 5 mm (%)	8.85	2.59
Aggregates between 5 and 2 mm (%)	20.87	10.15
Aggregates between 2 and 1 mm (%)	9.51	9.23
Aggregates between 1 and 0.105 mm (%)	13.60	26.58
Aggregates <0.105 mm (%)	3.50	5.30
Stones >10 mm (%)	21.75	18.85
Stones between 10 and 5 mm (%)	11.36	4.74
Stones between 5 and 2 mm (%)	4.98	2.52
Mean weight diameter of aggregates (MWD) (mm)	1.96	0.86
Ratio microaggregation/macroaggregation (RMM)	0.39	1.49
Average distance to vegetation (cm)	25.50	26.00
Maximum distance to vegetation (cm)	70.00	40.00
Total rainfall (mm)	2453.60	5788.60
Average event total rainfall (mm)	41.60	76.20
Average I30 (mm h^{-1})	13.00	14.5
Average I10 (mm h^{-1})	19.70	20.20
Runoff (l m^{-1})[a]	416.00	414.00
Fine sediment (g m^{-1})[a]	987.00	186.00
Coarse sediment (g m^{-1})[a]	1745.00	143.00
Total sediment (g m^{-1})[a]	2733.00	330.00

[a] Total, considering only the plots with a complete series.

Also the distances from the trap to the areas protected by plants were calculated at 1 cm intervals along with the average and variance values (Table 3.1).

4. Results

4.1. Soil properties and surface components: effects on processes

Attending to the surface component distribution (Table 3.1) within the catchment area of the plots (0.5 × 3 m), the area protected by the plants, and the area covered by litter, are larger for CC. On the contrary, rock fragment cover, both with high and low-density fragments, is greater in BE. The bare soil patches are six times bigger for BE. Also the size of individual bare soil patches is bigger in BE and a lot more continuous (Fig. 3.4).

Variables related to soil structure (aggregation indicators and stone content in the soil matrix) show many significant differences between both sites, indicating much larger aggregate sizes and much higher stone content in the soil matrix in BE than in CC (Boix-Fayos et al., 2001).

In BE average distance to plant cover is well correlated with runoff ($r = 0.94$). If plot BE2 is not considered (in which a plant has been growing near the collector) then a good correlation ($r = 0.97$) can also result between the minimum distance to plants and the fine sediment transport rate. In CC fine sediment transport is well correlated with plant distances ($r = 0.88$), but runoff does not show any significant correlation with the distance to plants.

Although average distance from the sediment traps to the vegetation are similar in both sites, the mean values obtained for the maximum distance to the plants in each plot, show that the corridors between plants are more frequent in BE. Therefore, mainly plants and litter protect the CC site soil, although rock fragment cover is important. In the BE site, with an important vegetation cover for the drier conditions, the rock fragments are more important than in CC, as well as the bare soil areas.

4.2. Rainfall characteristics

During the period of measurements the total amount of rainfall collected was 2.4 times greater in CC. Averaging the rainfall amount collected in each period (events) the values were 1.8 times larger in CC. The average rainfall intensity of each period of data collection measured as an average I30 and I10 is slightly higher in CC (Table 3.1) although maximum values are higher in BE for the I10 and higher in CC for the I30.

The number of storms of high magnitude occurs more frequently in CC. For the same time span (1996–2003) the number of events of more than 25 mm was 31 in BE and 52 in CC. For more than 100 mm of rainfall, 5 and 19 events occur in BE and CC, respectively. The maximum event recorded in BE was of 187 mm (169.2 mm in a single shower or period of rainfall without gaps longer than six hours. See Usón and Ramos, 2001) and in CC that magnitude has been exceeded in ten occasions, the highest recorded being 476.2 mm (395.4 mm in a single shower). Also the characteristics of the storms are different. As can be seen in Figure 3.5 that represents the intensity distribution of the two maximum storms in each site, in BE the showers are shorter and higher in intensity.

The rainfall variables (total rainfall, average intensity, I30 and I10) correlate significantly with the response in most of the plots. Especially in BE all the plots except BE5 (high vegetation cover), for coarse sediments, have positive and significant correlation coefficients with runoff and sediment variables (Table 3.2). In contrast, in CC, although many correlations can be found as significant, especially with runoff, values are in general lower and less significant for the sediments (Table 3.2).

Analysing the dataset, it is clear that other factors in combination with rainfall intensity and duration are responsible for runoff generation. In fact, in BE, although a threshold of I30 = 25 and I10 = 40 mm h^{-1} is needed to generate runoff; there are some events that generate runoff under lower intensities. In CC a larger amount of rainfall is needed to generate runoff. Even though some events with I10 = 10 mm h^{-1} can generate runoff if

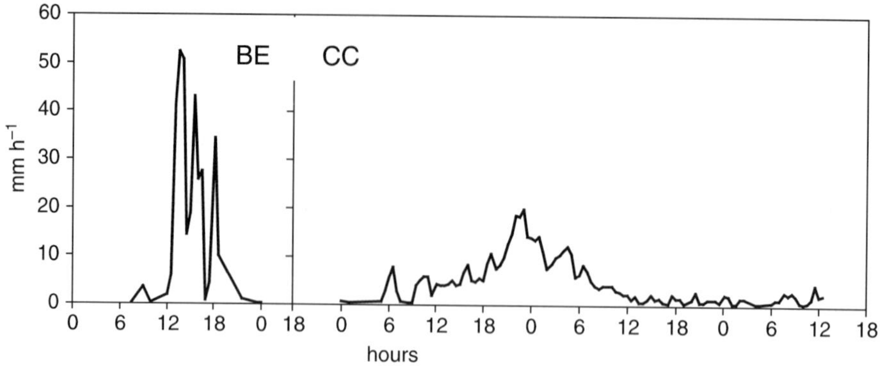

Figure 3.5. Rainfall intensity distribution at 30 min intervals of two extreme events at Benidorn (BE) in October 1997 with a total rainfall of 187 mm and at Cocoll (CC) in April 1997 with a total rainfall of 253.8 mm.

Table 3.2. Pearson correlation coefficients between rainfall properties and runoff (Rf), fine (Fs), coarse (Cs) and total (Ts) sediment.

Plot		Total rainfall	Average rainfall intensity	I30	I10	Plot		Total rainfall	Average rainfall intensity	I30	I10
CC1	Rf	0.58*	0.29*	0.51*	0.45*	BE1	Rf	0.77*	0.45*	0.77*	0.73*
	Fs	0.27*	0.27*	0.64*	0.65*		Fs	0.54*	0.36*	0.69*	0.67*
	Cs	0.17	0.02	0.11	0.08		Cs	0.53*	0.35*	0.68*	0.65*
	Ts	0.24*	0.09	0.29*	0.26*		Ts	0.54*	0.35*	0.68*	0.66*
CC2	Rf	0.64*	0.33*	0.51*	0.48*	BE2	Rf	0.59*	0.34*	0.51*	0.46*
	Fs	0.22	0.21	0.48*	0.50*		Fs	0.52*	0.35*	0.68*	0.66*
	Cs	0.21	0.12	0.30*	0.35*		Cs	0.50*	0.33*	0.61*	0.60*
	Ts	0.28*	0.23*	0.54*	0.58*		Ts	0.50*	0.33*	0.63*	0.62*
CC3	Rf	0.53*	0.03	0.34*	0.27*	BE3	Rf	0.71*	0.50*	0.86*	0.81*
	Fs	0.27*	0.23*	0.73*	0.62*		Fs	0.57*	0.36*	0.74*	0.72*
	Cs	0.29*	0.27*	0.60*	0.58*		Cs	0.54*	0.37*	0.72*	0.69*
	Ts	0.28*	0.25*	0.75*	0.64*		Ts	0.56*	0.37*	0.74*	0.71*
CC4	Rf	0.56*	−0.03	0.26*	0.19	BE4	Rf	0.56*	0.44*	0.85*	0.81*
	Fs	−0.01	0.26*	0.12	0.09		Fs	0.57*	0.39*	0.77*	0.74*
	Cs	0.18	0.13	0.24*	0.23*		Cs	0.58*	0.40*	0.75*	0.72*
	Ts	0.00	0.27*	0.13	0.10		Ts	0.58*	0.39*	0.77*	0.73*
CC5	Rf	0.67*	0.33*	0.56*	0.52*	BE5	Rf	0.80*	0.49*	0.81*	0.76*
	Fs	0.13	0.22	0.47*	0.50*		Fs	0.76*	0.31	0.59*	0.57*
	Cs	0.05	0.25*	0.22	0.29*		Cs	−0.01	−0.01	−0.02	0.05
	Ts	0.11	0.31*	0.43*	0.50*		Ts	0.59*	0.25	0.46*	0.47*
CC6	Rf	0.61*	−0.10	0.44*	0.33	BE6	Rf	0.59*	0.47*	0.66*	0.66*
	Fs	0.64*	0.27	0.72*	0.69*		Fs	0.40*	0.36*	0.71*	0.66*
	Cs	0.22	0.31	0.48*	0.51*		Cs	0.49*	0.29*	0.58*	0.57*

(continued)

Table 3.2. (continued)

Plot		Total rainfall	Average rainfall intensity	I30	I10	Plot		Total rainfall	Average rainfall intensity	I30	I10
	Ts	0.62*	0.29	0.73*	0.71*		Ts	0.52*	0.33*	0.65*	0.64*
CC7	Rf	0.55*	0.06	0.28	0.18	BE7	Rf	0.78*	0.46*	0.76*	0.70*
	Fs	0.34*	0.51*	0.58*	0.66*		Fs	0.78*	0.45*	0.76*	0.72*
	Cs	0.11	0.60*	0.54*	0.71*		Cs	0.74*	0.44*	0.73*	0.69*
	Ts	0.30	0.57*	0.60*	0.72*		Ts	0.76*	0.44*	0.74*	0.70*
CC8	Rf	0.63*	−0.03	0.30	0.22						
	Fs	0.23	0.22	0.16	0.17						
	Cs	0.22	0.16	0.05	0.05						
	Ts	0.24	0.22	0.14	0.15						
CC9	Rf	0.56*	−0.09	0.13	0.07						
	Fs	0.44*	0.46*	0.51*	0.58*						
	Cs	0.09	0.04	0.03	0.04						
	Ts	0.21	0.16	0.17	0.19						

Marked correlations (*) are significant at $p < 0.05000$.

the duration of the event is long enough, a higher intensity ($I10 = 30$ mm h^{-1}) is needed for runoff generation during events of lower magnitude (ca. 20 mm).

4.3. Runoff and sediment movement

At the site scale, runoff and sediment movement rates are well explained by the differences in soil properties, soil surface conditions and the general trends in rainfall characteristics. Considering the plots that have been operating for the entire seven-year period of measurements, the average runoff of the five plots show the same figures in both sites. The higher infiltration rates (1.6 times), measured in CC during rainfall simulation experiments (Calvo-Cases et al., 2003) are not verified by the plot data under natural rainfall conditions, due to the different storm characteristics. In CC, where soils have a greater infiltration capacity, the higher duration and magnitude of the rainfall events produces runoff by saturation excess mechanisms.

In contrast, fine-sediment (<2 mm) movement is five times higher and coarse sediment movement is twelve times higher in BE. The total sediment is eight times greater in BE. Here the higher amount of unprotected soil is significant, but the role of the higher rainfall intensity is also important, especially to explain the large amount of coarse sediments, or even gravels, collected in the traps.

These differences in sediment transport are less pronounced, but still quite different, considering the series of data from all plots in both sites along with plot CC6 where the sediment concentration values are very high (Fig. 3.6). Hydrological response is consistent in both data series.

Analysing average values for individual plots (Fig. 3.6), with the exception of plot CC6, the values are more homogeneous in CC, both in runoff and sediment transport. The higher

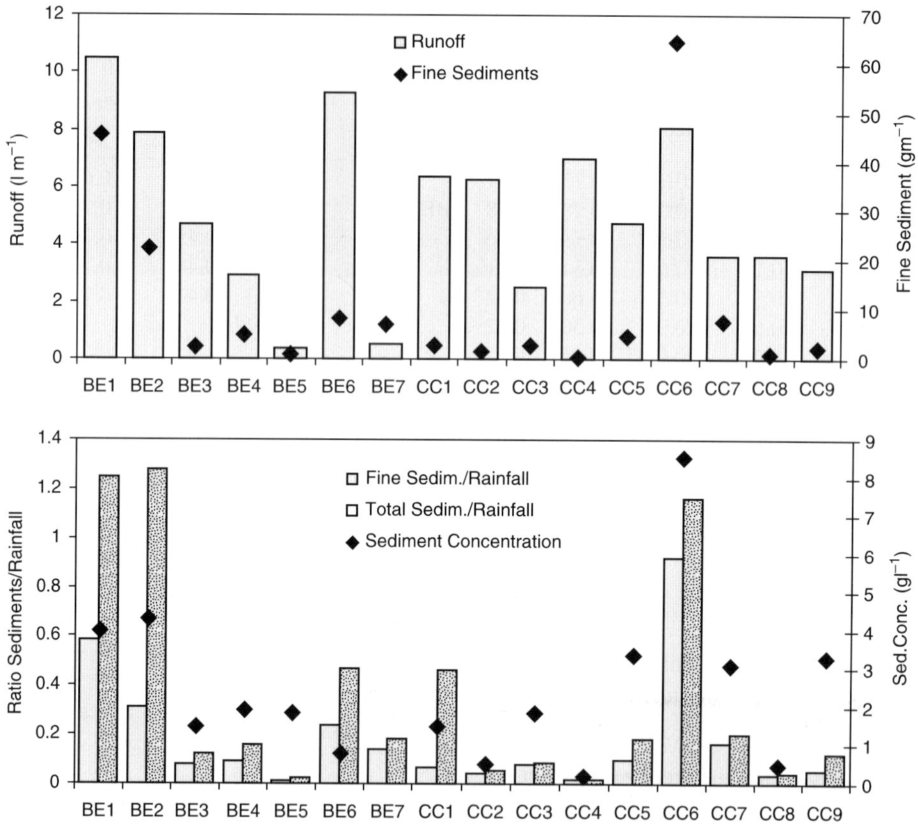

Figure 3.6. Average values during the measuring period (for each plot) of runoff, fine sediment yield, fine sediment/rainfall ratio, total sediment/rainfall ratio and sediment concentration in each plot.

heterogeneity of BE seems to be related with the higher variability in the vegetation cover characteristic of that hillslope.

4.4. The influence of patch size on sediment movement

A series of correlation coefficients have been calculated between the proportions of soil surface cover components, nested at 10 cm-intervals (up to 300 cm upslope of the collector), and the total sediment yield collected from each plot. The purpose is to analyse the influence of the distance (or patch size), together with the soil surface components in the sediment yield at each plot. The correlation coefficient values have been plotted against the distance from the collector (Fig. 3.7) to show how the proportion of each component correlates with the fine, coarse and total sediment, respectively.

Despite the general trends of negative correlations between sediment yield and vegetation/litter and positive correlations between sediment yield and the other three

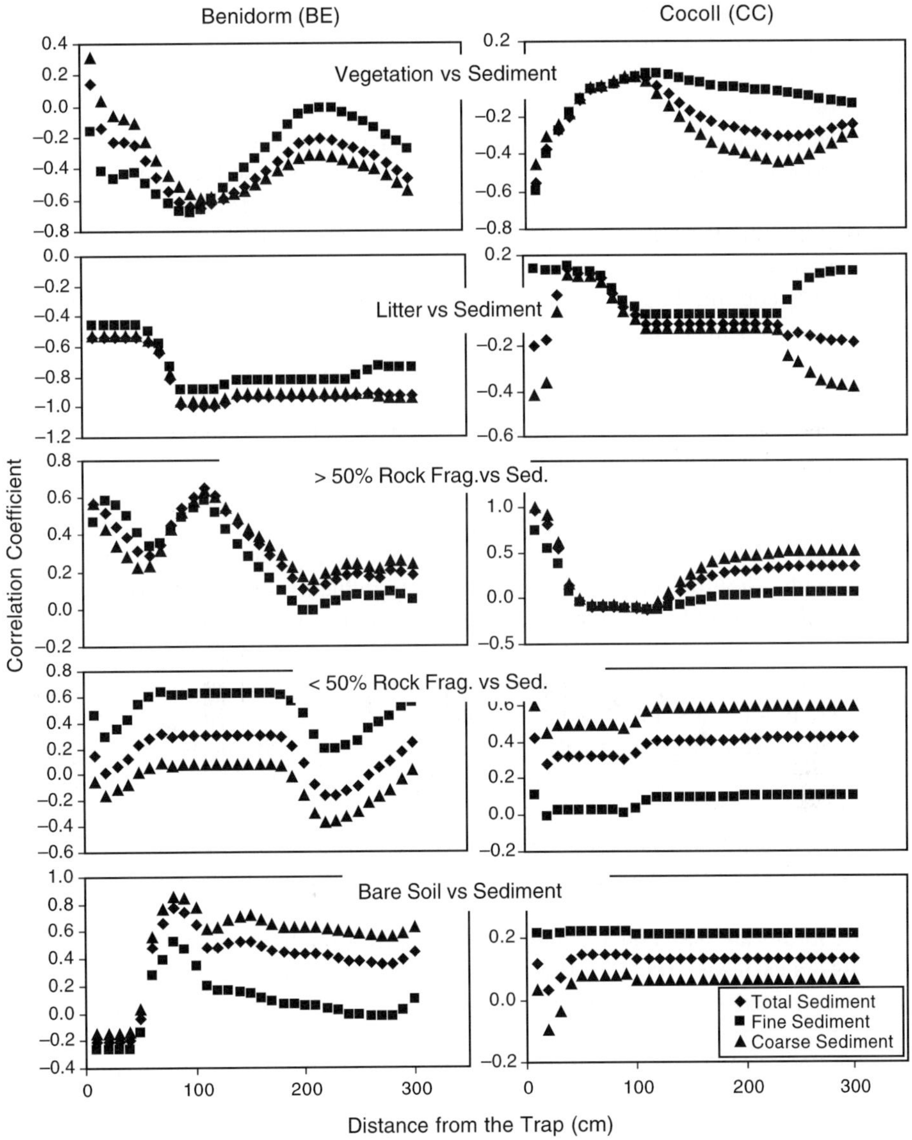

Figure 3.7. Changing correlation coefficients with the distance from the trap and 10 cm intervals between the surface component proportions and fine, coarse and total sediment.

variables of Figure 3.7 (>50% of soil with rock fragments cover, <50% of soil with rock fragments cover and bare soil) the series of correlations show very different curves for both sites when plotted against distance (Fig. 3.7).

In the BE site, the highest correlation values are reached at about one meter of distance with respect to vegetation, litter and >50% rock fragments. Bare soil and <50% rock fragments have the highest correlation values at a distance between 70 and 80 cm. After this distance, the correlations remain high for these two parameters. Percentage of vegetation and percentage of soil with less than 50% rock fragments exhibit better correlations with fine sediments rather than with total and coarse sediments.

Vegetation, litter and, very stony soil surfaces, have an important influence on sediment yield at distances in patches with a length under 1 m, but bare soil and soil surface with <50% rock fragments are associated with higher sediment yields at distances up to 2 m. This indicates that these less protected surfaces can deliver sediment at longer distances and from bigger patches.

In the CC site, the highest correlation coefficients occur with samples taken at very short distances (from 10 to 50 cm), with a maximum at 10 cm and a minimum at 50 cm, between sediment yield and vegetation, litter and percentage of soil with >50% rock fragments. Soil with <50% rock fragments and bare soil, increase their influence in sediment yield from 1 m to 50 cm distance, respectively. They maintain this influence for up to a 3 m distance, but with much lower correlation coefficients, than in the BE site, especially for the fine sediment.

The size of the patches, which influence the sediment yield in the CC site, is much smaller than in BE with respect to vegetation, litter and very stony soil. Bare soil and soil surfaces with a low percentage of rock fragments influence the sediment yield at even larger distances of up to 3 m, but with lower correlation coefficients than in the BE site.

As a consequence, detachment and transport processes are more efficient in the BE conditions because the protective factors (vegetation and litter) become more efficient when present in the first metre from the collector. The high proportion of rock fragment cover is always positively correlated with sediment movement, but becomes significant in the first metre. In the BE conditions, rock fragments are loose on the surface and they move easily not creating a protective component, as can be seen by the high rates of coarse sediment movement (1745 g m^{-1} in average). In contrast, at CC, rock fragments are steadier (143 g m^{-1} in average), with some positive effects in the reduction of transport rates.

These results are based on average values obtained from the longest series of plot data (five plots in each site, Fig. 3.3). But by looking into the individual plot data sets we can make some remarks. At CC, plot CC6 has the highest rates of fine sediment transport and fine sediment concentration of the entire CC site and also has higher rates than any of the BE plots (Fig. 3.6). These differences decrease when sediment transport is considered in relation with rainfall and especially when the ratio includes the coarse sediments. In this case, plots BE1 and BE2 have the maximum transport rates (Fig. 3.6).

4.5. Temporal changes in rainfall-runoff–sediment movement relations

The analysis of the temporal series obtained for each plot (Fig. 3.8a and b) is complex, not only due to the temporal variability of the rainfall, but also because of the changes in the soil properties and the surface components, some of which can be recorded, but others remain ignored and can also affect the results. The maps in Figure 3.4 represent the surface component situation at the end of the series. Unfortunately, this mapping was not made at the beginning, but was done for the first metre upslope from the collector in

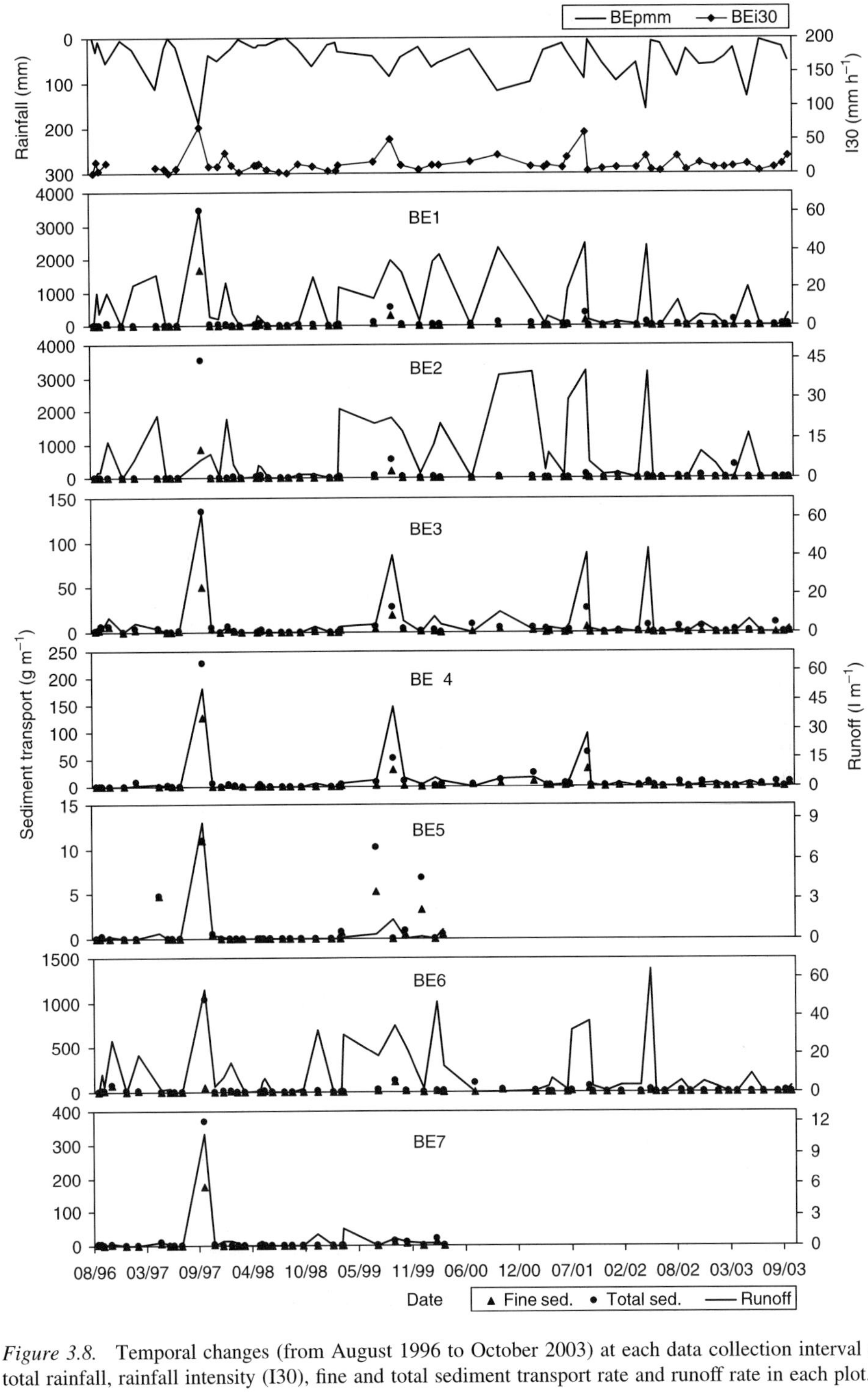

Figure 3.8. Temporal changes (from August 1996 to October 2003) at each data collection interval of total rainfall, rainfall intensity (I30), fine and total sediment transport rate and runoff rate in each plot.

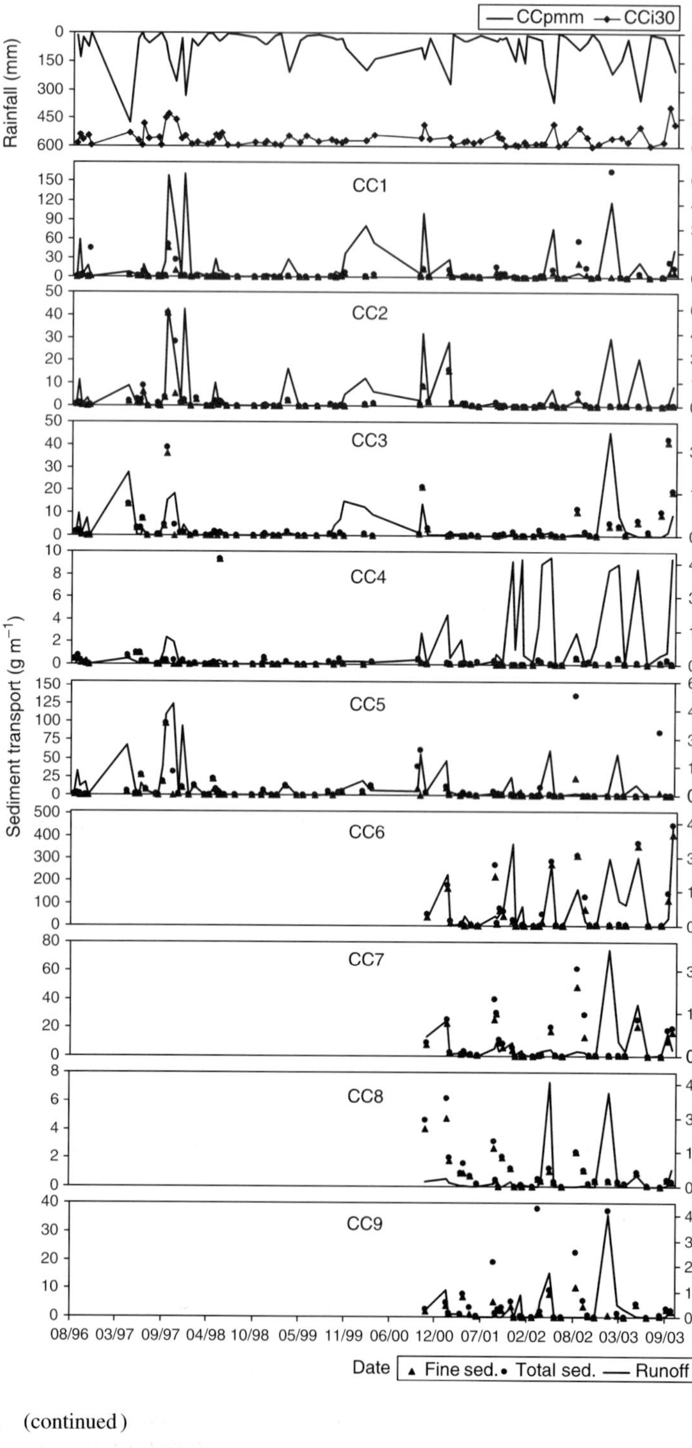

Figure 3.8. (continued)

August 2000. Plot CC4 had undergone one of the main changes affecting runoff response: a high proportion of the plot area was covered by a large plant (*Ulex parviflorus*) and since 2001 that plant had died and, although its dry branches were covering the surface the runoff rate has increased dramatically. Plot CC5 used to have a stony and bare area next to the collector larger than what it has now and because of the growth of the plants that now cover more area, the rates of runoff have been reduced. An increase of coarse sediments in the last events can be attributed to the accumulation of stones next to the collector.

Another change in surface components affecting the results has occurred in plot BE2, where the plant next to the collector was not present at the beginning of the series and in 2000 was two times smaller than now. Although it is still not affecting runoff significantly a decreasing tendency can already be seen in the fine sediment values.

The plots with higher erosion rates (BE1, BE2, BE6 and CC6) exhibit a less-straightforward relationship between runoff and sediments due in part to the exhaustion of the available sediment, during previous very dynamic events. In the BE plots, this is difficult to detect (Fig. 3.8a) because the amount of available sediments is always very high, but in plot CC6 a clear sequence of sediment exhaustion and availability, can be observed (Fig. 3.8b).

In order to distinguish thresholds for runoff and sediment movement, well-vegetated plots are useful as they respond only in extreme events. Plots BE3, BE4 and BE5 (Fig. 3.8a) represent this for the BE site very well. Plot BE3, 10 cm downslope of a big *Stipa* tussock, has only collected a significant amount of runoff four times. In these four events, the time concentration of the rainfall has been very high (more than 50% of total rainfall of the measuring time span in one shower). These four events, generating significant runoff in plot BE3, produced 40% of the runoff collected in all the plots of the slope and 78% of the total sediment moving on the slope.

Plot CC5 has a similar position and role in the CC site. The 16 measurement intervals in which runoff have had significant values, correspond to the events that produced 61% of the total runoff collected in all the plots and 35% of the total sediment. The inversion of these figures between both study areas, especially the lower percentage of total sediment explained by the extreme events in CC, is associated with the exhaustion of available sediment when extreme events occur. In such frequent rainfall conditions of CC, many extreme events occur, proceeded by others that do not generate runoff in protected plots, but mobilise available sediments on bare plots.

In some plots (those with higher erosion rates in CC) the relationship runoff–sediment (fine fraction) shows three types of situations when a certain magnitude is overpassed. On the one hand, few events have a linear increase in runoff and sediments. On the other hand, some events have only a small increase in runoff, but a high increase in sediment movement. Finally, an increase of runoff can appear without a significant increase in sediments. These last two situations are related to sediment availability and exhaustion, respectively, as well as having some influence of the topsoil moistening during longer events. In BE, it is more frequent to have a linear response on the vegetated plots and an exponential response on the less-covered ones. In this case, the development of a mineral crust keeps the sediment removal low until certain thresholds (the series is too short to define them) but when these are exceeded, an exponential increase of sediments occurs.

5. Discussion and conclusions

The results obtained after seven years of measuring runoff and sediment movements are considered representative of the Mediterranean semi-arid conditions (BE site) and Mediterranean sub-humid, but much degraded, conditions (CC site) in limstone areas. They show some clear differences in the process behaviour and the role of the factors controlling hillslope dynamics.

Rainfall is essentially much more scarce in the BE site and characterised by shorter and more intense storms (Maximum I10 higher than in CC). This leads to the higher transport rates obtained in the BE site. But soil properties and surface components characteristics are also very relevant and do not always behave in the same way (i.e. the role of rock fragments cover). In these conditions runoff rates are similar in both sites, but sediment yield is much higher in BE (8.3 times).

At CC, lower sediment concentrations occur despite the occurrence of larger events, the higher percentage of vegetation cover, the higher organic matter availability (Boix-Fayos et al., 1998, 2001), the smaller and more stable aggregates, the higher infiltration rates and a higher efficiency of the rock fragment cover in stabilising the soil. In BE, higher proportions of bare soil, bigger but less stable aggregates, lower infiltration rates and a scarce vegetation cover, allow for higher sediment concentration rates and a quicker response to rainfall.

The mechanisms of runoff generation, analysed in a previous paper (Calvo-Cases et al., 2003), are similar in both sites. But the high frequency in which rainfall thresholds are overpassed in CC, leads to the conclusion that this site is dominated by the *mixed discontinuous runoff model*, described in Calvo-Cases et al. (2003) as a combination of areas with Hortonian overland flow and saturation-excess overland flow. In contrast, in the BE site, rarely does the rainfall exceed the necessary duration to saturate the receiving areas and most of the times a *Hortonian discontinuous runoff model* occurs along the hillsides, with re-infiltration of the runoff generated on the bare and stony patches, under the plants (mainly *Stipa tenacissima*). Plot BE3 is considered as an example of the occurrence of saturation excess overland flow, in some parts of the hillslope. The four times in which the plot has generated runoff are in fact the four maximum events in the rainfall dataset. For BE site conditions, a threshold of 50 mm of rainfall in a continuous and intense shower, seems to be enough to change the runoff generation mechanism. An increase in the spatial continuity of the runoff results in a significant increase of the sediment transport rates as a consequence.

The size of the patches contributing to the sediment production is larger in the semiarid site (BE, about 1 m) than in the subhumid site (CC, 10 cm). This is in agreement with Lavee et al. (1998) who argue that, in the context of Mediterranean environmental conditions, the size of the contribution areas for runoff and sediment, at the slope scale, increase progressively from humid to arid conditions. Moreover, it is found that the role of the biological factor is higher in CC. When we move to the drier site, mineral factors become more important, which is in agreement with the work carried out in Israel (Lavee et al., 1998). As BE site represents a transition between humid and arid conditions, the role of the vegetation and the biological activity in the soil is still relevant, but an increase of the role played by rock fragment cover and mechanical crusts over the operating processes was observed.

In both sites the bare patches (i.e. not covered by plants) are dominated by rock fragments (38.1% in BE and 27.1% in CC). In BE, the rock fragments are very loose over the surface and, in CC, as the soil matrix is less dense, they are slightly embedded in the soil (a few millimetres). Any application of force in BE causes displacement of the rock fragments, whereas in CC this does not occur significantly. In the data set of coarse sediments the amount collected in BE is 12.2 times greater that in CC: a proportion significantly higher for the specific differences in availability.

The development of an armour layer increasing and reducing process activity has been described on several occasions (Kirkby et al., 1998; Kirkby, 2001) but it is usually when the rock fragments are half embedded in the soil that they contribute to an increase in runoff and sediment delivery (Poesen and Lavee, 1994). Runoff generation and sediment movement at the BE site seem to be dependent on soil surface conditions in relation with soil structure, and at the CC site these processes seem to be dependent on organic factors (Boix-Fayos, 2000). However, the specific role played by the stones of the soil surface is not clear yet, and needs further investigation. With the data collected upto now, rock fragments at the soil surface seem to play a double role. In CC with the rock fragments slightly embedded, a positive effect reducing soil detachment can be seen, as it is expressed by low and negative correlations (Fig. 3.7). In contrast at the BE site the rock fragments covering the surface, but in conditions of high mobility, are related positively with the sediment yield, although a positive effect preventing soil detachment can be distinguished. The development of a soil crust and a high stoniness at the soil surface, together with a high-average size of aggregates with low stability in dry soil conditions (Boix-Fayos et al., 2001), produces much higher sediment concentrations in the runoff of the semiarid site of BE compared to the subhumid site of CC.

Acknowledgements

The authors wish to thank the Commission of the European Union for the support provided by the research contract ERMES II (ENV4-CT95-0181). The Spanish Ministries of Environment and of Science and Technology are also acknowledged because of the support through the contract RESEL (Red de Cuencas y Parcelas Experimentales de seguimiento y Evaluación de la Erosión y Desertificación) and project SENSI (AMB99-1246), respectively. A. Basurto, A. Corell, J. Cuenca, J. López, L. Ramirez, A. Reus, and L. Navarro are greatly acknowledged because of their valuable work in the field and in the laboratory and E-Symeonakis for improving the text.

References

Alexandrov, Y., Laronne, J.B., Reid, I., 2003. Suspended sediment concentration and its variation with water discharge in a dryland ephemeral channel, northern Negev, Israel. J. Arid Environ. 53, 73–84.

Avendaño Salas, C., Cobo Rayán, R., Sanz Montero, E., Gómez Montaña, J.L., 1997. Sediment Yield at Spanish Reservoirs and its Relationship with the Drainage Basin Area, Dix-neuvième Congrès des Grands Barajes. Comission Internationale Des Grands Barrages, Florence, Italy, pp. 863–874.

Bautista, S., 1999. Regeneración post-incendio de un pinar (Pinus halepensis, Miller) en ambiente semiárido. Erosión del suelo y medidas de conservación a corto plazo. Ph.D. Thesis, Universidad de Alicante, Spain.

Bautista, S., Bellot, J., Vallejo, R., 1996. Mulching treatment for postfire soil conservation in a semiarid ecosystem. Arid Soil Res. Rehab. 10, 235–242.
Beven, K., 2003. Runoff generation in semi-arid areas. In: Bull, L.J., Kirkby, M.J. (Eds), Dryland Rivers. Hydrology and Geomorphology of Semi-Arid Channels. Wiley, Chichester, pp. 57–105.
Boix-Fayos, C., 1997. The roles of texture and structure in the water retention capacity of burnt Mediterranean soils with varying rainfall. Catena 31, 219–236.
Boix-Fayos, C., 2000. Procesos geomórficos en diferentes condiciones ambientales mediterráneas: el estudio de la agregación y la hidrología de los suelos. Ph.D. Thesis, Universitat de València, Spain.
Boix-Fayos, C., Calvo-Cases, A., Imeson, A.C., Soriano-Soto, M.D., Tiemessen, I.R., 1998. Spatial and short-term temporal variations in runoff, soil aggregation and other soil properties along a Mediterranean climatological gradient. Catena 33, 123–138.
Boix-Fayos, C., Calvo Cases, A., Imeson, A.C., Soriano Soto, M.D., 2001. Influence of soil properties on the aggregation of some Mediterranean soils and the use of aggregate size and stability as land degradation indicators. Catena 44, 47–67.
Bouma, N.A., Imeson, A.C., 2000. Investigation of relationships between measured field indicators and erosion processes on badlands surfaces at Petrer, Spain. Catena 40, 147–171.
Calvo-Cases, A., Harvey, A.M., 1996. Morphology and development of selected badlands in southeast Spain: implications of climatic change. Earth Surf. Processes Landforms 21, 725–735.
Calvo, A., Harvey, A.M., Paya, J., Alexander, R.W., 1992. Response of badlands surfaces in south–east Spain to simulated rainfall. Cuaternario Geomorfol. 5, 3–14.
Calvo-Cases, A., Boix-Fayos, C., Imeson, A.C., 2003. Runoff generation, sediment movement and soil water behaviour on calcareous (limestone) slopes of some Mediterranean environments in southeast Spain. Geomorphology 50, 269–291.
Castillo, V., Martínez-Mena, M., Albaladejo, J., 1997. Runoff and soil loss response to vegetation removal in a semiarid environment. Soil Sci. Am. J. 61, 1116–1121.
Francis, C., 1986. Soil erosion on fallow fields: an example from Murcia. Papeles Geogr. Fís. 11, 21–28.
Harvey, A.M., Calvo-Cases, A., 1991. Process interactions and rill development on badlands and gully slopes. Z. Geomorphol. Suppl. Band 83, 175–194.
Kirkby, M.J., 2001. Modelling the interactions between soil surface properties and water erosion. Catena 46, 89–102.
Kirkby, M.J., Abrahart, R., McMahon, M.D., Shao, J., Thornes, J.B., 1998. MEDALUS soil erosion models for global change. Geomorphology 24, 35–49.
Kirkby, M.J., Bracken, L., Reaney, S., 2002. The influence of land use, soils and topography on the delivery of hillslope runoff to channel in SE Spain. Earth Surf. Processes Landforms 27, 1459–1473.
Lavee, H., Imeson, A.C., Sarah, P., 1998. The impact of climate change on geomorphology and desertification along a Mediterranean arid transect. Land Degrad. Dev. 9, 407–422.
López Bermúdez, F., Gutiérrez Escudero, D., 1982. Estimación de la erosión y aterramientos de embalses en la cuenca hidrográfica del río Segura. Cuadernos Invest. Geogr. 8, 3–18.
Martínez-Mena, M., Albaladejo, J., Castillo, V.M., 1998. Factors influencing surface runoff generation in a Mediterranean semi-arid environment: Chicamo watershed, SE Spain. Hydrol. Processes 12, 741–754.
Martínez-Mena, M., Castillo, V., Albaladejo, J., 2001. Hydrological and erosional response to natural rainfall in a degraded semiarid area of south–east Spain. Hydrol. Processes 15, 557–571.
Martínez-Mena, M., Castillo, V., Albaladejo, J., 2002. Relations between interrill erosion processes and sediment particle size distribution in a semiarid Mediterranean area of SE of Spain. Geomorphology 45, 261–275.
Poesen, J., Lavee, H., 1994. Rock fragments in topsoils: significance and processes. Catena 23, 1–28.
Puigdefábregas, J., Sole, A., Gutierrez, L., Del Barrio, G., Boer, M., 1999. Scales and processes of water and sediment redistribution in drylands: results from the Rambla Honda field site in southeast Spain. Earth Sci. Rev. 48, 39–70.
Ramos, J., Calvo, A., Pérez Badía, R., 2001. Relaciones entre la erosibilidad del suelo y la vegetación en la montaña del Norte de Alicante. Actas III Congreso Forestal Español, Granada, España.
Rivas Martínez, S., 1985. Biogeografía y Vegetación, Real Academia de Ciencias Exactas. Físicas y Naturales, Madrid, p. 103.
Romero Díaz, A., 2002. La erosión en la región de Murcia. Universidad de Murcia, Murcia, p. 339.

Romero Díaz, A., López Bermúdez, F., Thornes, J.B., Francis, C., Fisher, G.C., 1988. Variability of overland flow and erosion rates in a semiárid Mediterranean environment under matorral cover. Murcia, Spain. Catena Suppl. 13, 1–11.
Romero Díaz, A., López Bermúdez, F., Belmonte Serrato, F., 1998. Erosión y escorrentias en el campo experimental "El Ardal". Nueve años de experiencias. Papeles Geogr. 27, 115–130.
Sala, M., 2003. Floods triggered by natural conditions and by human activities in a Mediterranean coastal environment. Geogr. Ann. 85 (A (3–4)), 301–312.
Sala, M., Calvo, A., 1990. Response of four different mediterranean vegetation types to runoff and erosion. In: Thornes, J.B. (Ed.), Vegetation and Erosion. Wiley, Chichester, pp. 347–362.
Sala, M., Inbar, M., 1992. Some hydrologic effects of urbanization in Catalan rivers. Catena 19, 363–379.
Soler, M., Sala, M., 1995. Variabilidad longitudinal de la escorrentía y la erosión en una ladera quemada. Pirineos 145–146, 81–89.
Soriano, M.D., Calvo, A., Boix, C., Pons, V., 1996. Variaciones en las propiedades de los suelos y su agregación en un transecto altitudinal de la provincia de Alicante. Cuaternario Geomorfol. 10, 45–58.
Trimble, S.W., 1999. Decreased rates of alluvial sediment storage in the Coon Creek Basin, Wisconsin, 1975–93. Science 285, 1244–1246.
Usón, A., Ramos, M.C., 2001. An improved rainfall erosivity index obtained from experimental interrill soil losses in soils with a Mediterranean climate. Catena 43, 293–305.

Catchment Dynamics and River Processes: Mediterranean and Other Climate Regions
Celso Garcia, Ramon J. Batalla, Editors

Chapter 4

Repeated patterns of Quaternary discontinuous gullying at El Tormillo, Ebro Basin, Spain

Adrian M. Harvey[1,*] and Mateo Gutiérrez-Elorza[2]

[1]*Department of Geography, The University of Liverpool, Roxby Building, Liverpool, L69 7ZT, UK*

[2]*Seccion de Geomorfología, Departamento de Geología, Facultad de Ciencias, Universidad de Zaragoza, Zaragoza 51009, Spain*

Abstract

Gully systems in the catchment of the Barranco de la Clamor, near El Tormillo, Ebro Basin, northern Spain include midslope, valley floor, and basally induced gully systems. The present system is poorly coupled, with midslope gullies generally petering out without meeting the incised stream network. Even the valley-floor network is poorly integrated, and functions as a serieos of discrete units bounded by major headcuts in resistant sandstone, above which are modern aggradation zones. The modern valley-floor gully pattern is inherited from earlier patterns developed during previous incisional stages of landscape evolution. Through the Holocene the area has switched between incisional and aggradational regimes. Two main fills representing the aggradational phases have been dated to ca. 6000 cal. BP and to the Iron Age, with incision dominating the intervening periods. The regime switches are interpreted to have been climatically induced. The system appears to be vulnerable to regime switching following changes in runoff rates, erosion rates and changes in the coupling characteristics of the system.

Keywords: gully systems, geomorphic coupling, Holocene sequences, El Tormillo, Ebro Basin

1. Introduction

Gullying has often been seen as a response to an environmental change, that modifies hydrological processes, either by an increase in runoff power or by a decrease in the resistance of the surface to erosion. Three main types of gully have generally been recognised on the basis of location (Brice, 1966; Schumm et al., 1984; Campbell, 1997; Harvey, 2003). These are midslope gullies, basally induced hillslope gullies, and valley-floor gullies of the "arroyo" type (Cooke and Reeves, 1976) (Fig. 4.1). Each type may have a different formative mechanism. Midslope gullies relate to conditions on the hillslope itself, and indicate formation in response to either an extreme event or a secular climatic or

---

*Corresponding author. *E-mail address:* amharvey@liv.ac.uk (A.M. Harvey).

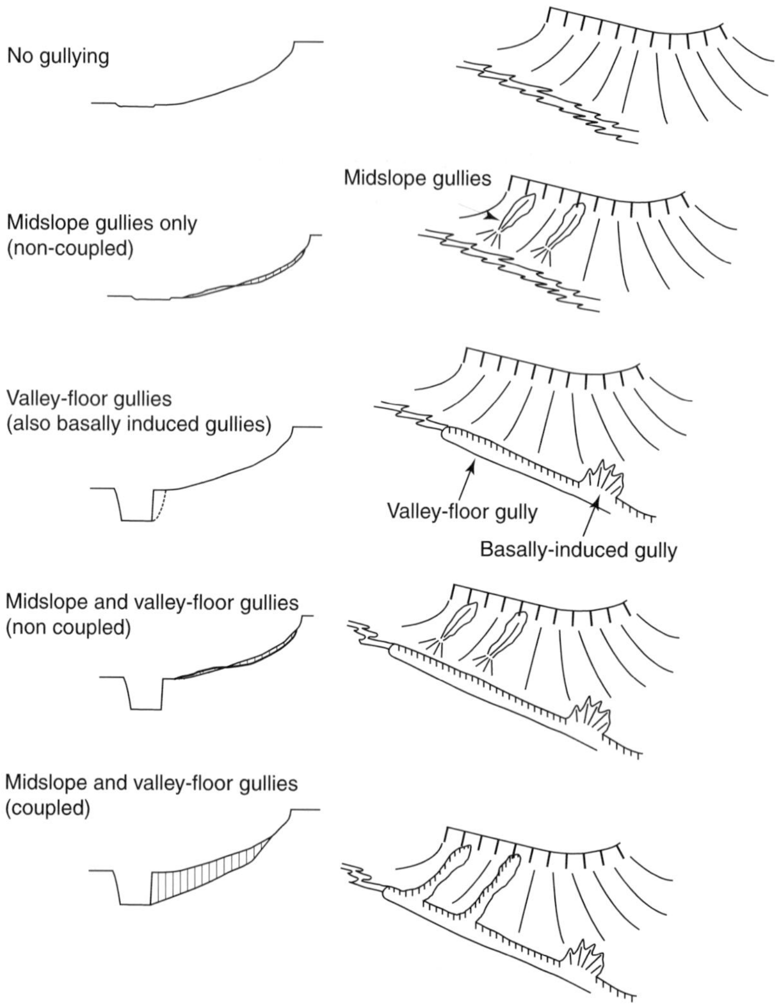

Figure 4.1. Schematic representation of hillslope and valley-floor gullying, indicating the potential range of coupling relationships.

land-cover change, modifying the hillslope hydro-geomorphological processes. Basally induced hillslope gullies have a strong coupling relationship with the stream channel (Harvey, 2001, 2002a) and may be triggered intrinsically (Schumm, 1979) simply by a migrating channel impinging on the valley side or may be triggered by an incising main channel. Valley-floor gullies, both continuous and discontinuous, are essentially incised stream channels, and may owe their origins either to intrinsic controls (Schumm, 1979) or to increased runoff.

All three types of gully have downstream implications. High rates of sediment production by gully erosion may cause sedimentation downstream depending on the coupling characteristics (Harvey, 2002a) and on the downstream critical power relationships

Figure 4.2. View west across the valley of the Barranco de la Clamor, south of the RENFE viaduct. Note valley-floor gullies cut into older fill; partially stabilised midslope gullies cut into the scarp slope below the mesa top.

(Bull, 1979). Non-coupled midslope gullies (Fig. 4.1) may cause colluvial sedimentation at the slope base. Sediment supplied to the channel system, from either coupled midslope or basally induced gullies (Fig. 4.1), will influence within-channel processes. If the result is an increase of critical power (i.e. that required to transport the sediment supplied; Bull, 1979) above the actual power of the stream, alluvial sedimentation on the valley floor will take place. If actual power remains above critical power the sediment will be transported further downstream. Though valley-floor gullies (Fig. 4.1) may initially be the result of excess stream power, the same conditions hold in relation to downstream channel dynamics.

The spatial patterns of erosion and deposition resulting from both active and former phases of gullying are indicative of the system dynamics. This is particularly true for semi-arid Mediterranean regions, where a long history of human settlement and associated pressure on land has combined with a seasonal and storm-dominated climate to produce hillslopes and valley systems near to erosion thresholds.

Hillslopes and valley floors of the Ebro Basin, northern Spain, show complex patterns of modern gullies and bear evidence of previous late Quaternary gullying phases (Fig. 4.2). This paper focuses on the catchment of the Barranco de la Clamor, near El Tormillo, Huesca Province (Fig. 4.3), presenting a description of modern hillslope and valley-floor gullies. It assesses the coupling characteristics of the system, and the extent to which the modern spatial patterns mirror patterns of past phases of gullying. Finally, it considers the chronology and possible causes of the sequence.

2. Study area

The Barranco de la Clamor is a tributary of the Rio Cinca, itself a north-bank tributary of the Ebro. The Clamor drains a catchment cut into Plio-Pleistocene pediment surfaces

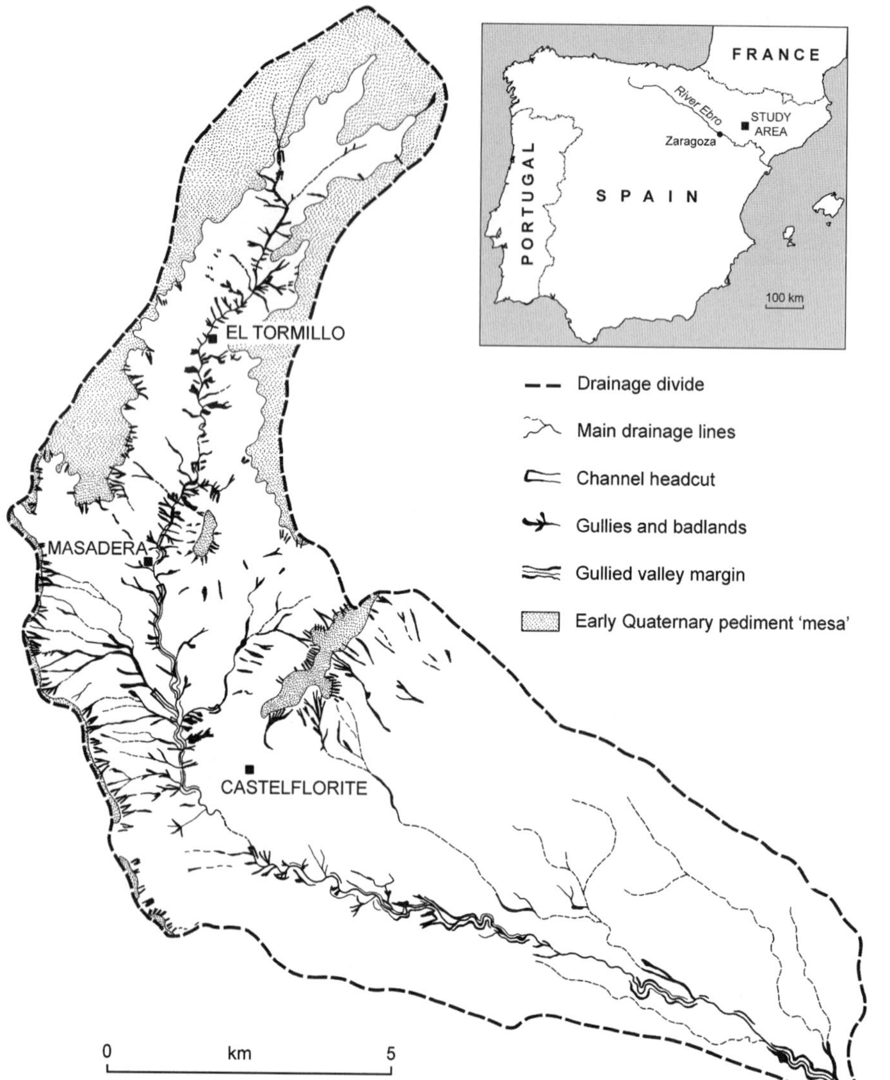

Figure 4.3. Drainage network of the Barranco de la Clamor, El Tormillo area. Inset shows location within northern Spain.

which form the main watershed of the basin. These surfaces are mantled by conglomerates capped by a well-developed calcrete (Sancho and Meléndez, 1992), and form an extensive flat plateau which slopes gently towards the south, and locally forms isolated mesas. The valley has been formed during the Pleistocene by dissection of this surface into the underlying Lower Miocene Sarinena Formation (Gutiérrez-Elorza and Peña-Monné, 1994; Alonso-Zarza et al., 2002; Gutiérrez-Elorza et al., 2002). This comprises a fluviatile sequence, dominantly of weak marls, silts and clays, but which includes localised channel sandstone bodies. These sandstones are resistant to erosion and form low residual relief within the basin, and have inhibited incision by the channel of the barranco.

Some of the residual hills within the basin are capped by Pleistocene terrace gravels (Fig. 4.4), representing stages in the incisional development of the basin. The modern valley

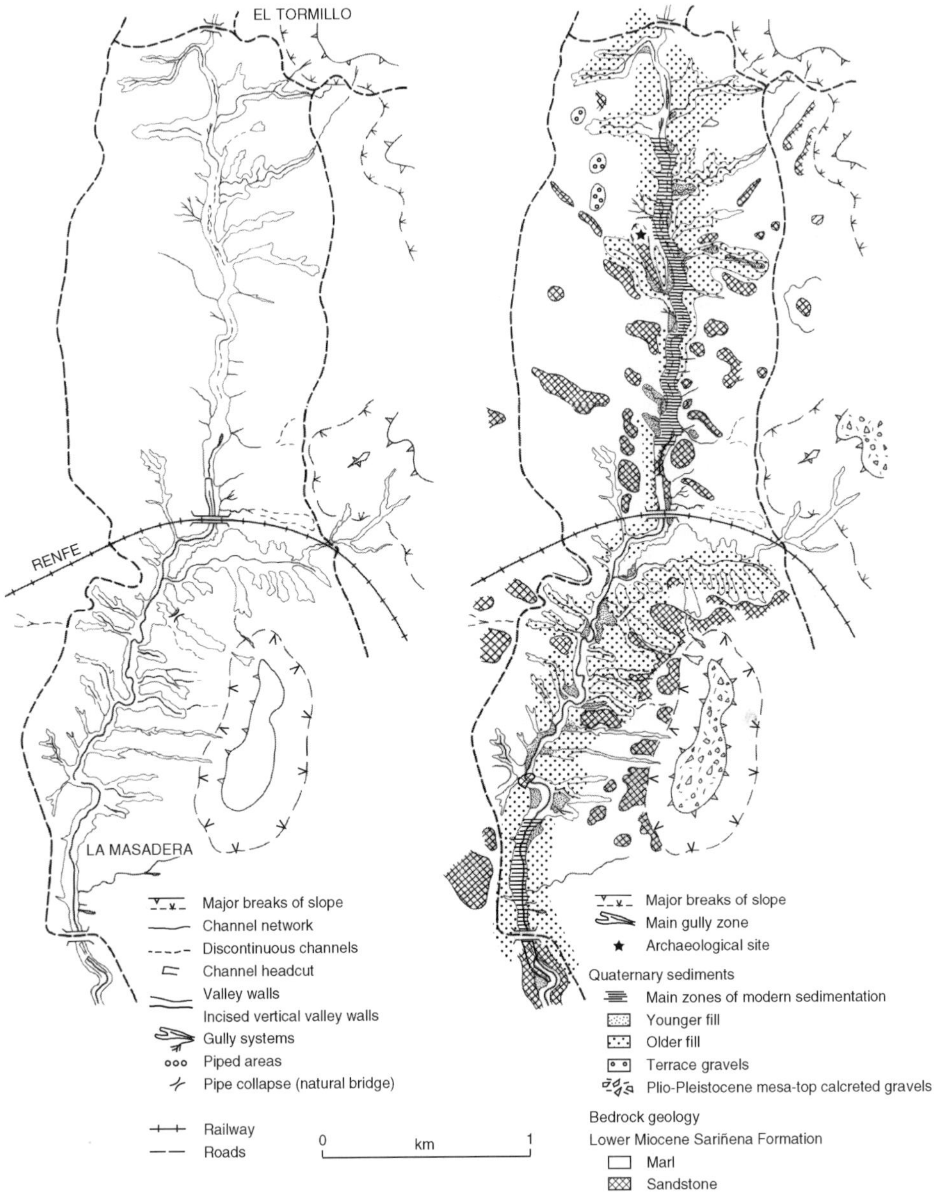

Figure 4.4. Detailed maps of the central part of the study area; left: morphology, drainage and gully networks; right: bedrock and Quaternary geology.

is cut below these terraces, and includes extensive areas of late Quaternary dominantly silty deposits which form at least two distinct fills (see below), similar to infilled valleys elsewhere in the Ebro Basin (Soriano, 1989; Arauzo and Gutiérrez-Elorza, 1994; Gutiérrez-Elorza and Peña-Monné, 1998). The modern channel is incised into these fills, and modern gully systems dissect the valley sides and hillslopes of the basin (see below).

3. Present patterns of gullying

The modern valley floors of the main and many tributary valleys are trenched by linear valley-floor gullies, cut locally into late Quaternary fill deposits and locally into the underlying Miocene rocks. In places the walls of the trenches are dissected by basally induced gullies. In areas of exposed Miocene siltstones these gullies take the form of badlands, but where the trench walls are composed of Quaternary silts, piping and gullies formed by pipe collapse are common. The modern valley-floor trench gullies are discontinuous (Bull, 1997), with three major headcuts in the central part of the main valley between El Tormillo and Castelflorite (Figs 4.4 and 4.5), where the modern channel crosses outcrops of the sandstone units within the Miocene bedrock of the Sarinena Formation. Similar headcuts limit the upstream development of some of the tributary valley-floor gullies. Downstream from each headcut the incised channel is narrow and deeply trenched, but widens to a flat floored valley further downstream, across which the modern stream weaves in a braided pattern. The modern channel carries little sand or gravel and the modern sediments are dominantly of silt. In several locations where the valley floor widens and the gradients are reduced (Figs 4.4 and 4.5), especially immediately upstream of the major headcut at the RENFE railway viaduct (Fig. 4.4), the channel is aggrading. The bowls of trees growing on the floodplain have been buried by > 1.5 m of modern silts.

The steeper slopes of the catchment, especially the scarp slopes below plateau caps, are cut by midslope gullies. The sideslopes of some of these gullies are bare of vegetation and clearly are active under present conditions. Some gully sideslopes are vegetated to varying

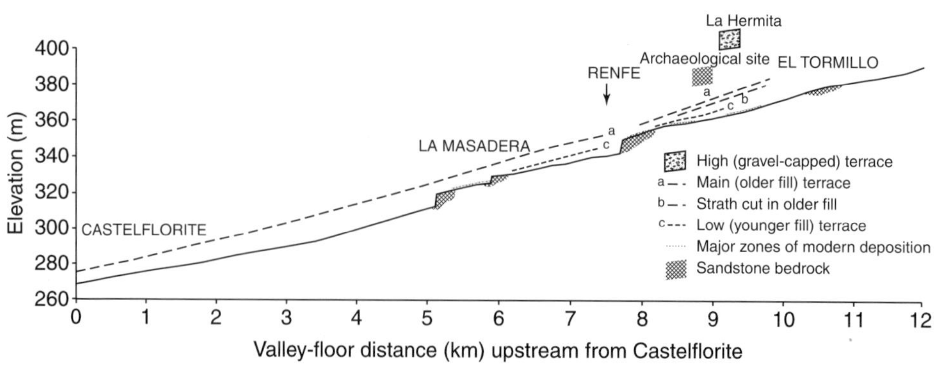

Figure 4.5. Long profile of the Barranco de la Clamor through the central part of the study reach. (Profile between the RENFE viaduct and El Tormillo derived from field survey, that below the viaduct derived from topographic maps modified by field observations.)

degrees and gully erosion processes appear to be less active. Most of these midslope gullies peter out downslope, with only a few showing continuity with the valley-floor gully systems.

The coupling within the present drainage network is far from perfect. Sediment derived from midslope gullying may be deposited on the footslopes and may never reach the main stream channels. Even within the main channels sediment derived either from hillslope sources or from basally induced gully systems may not be discharged through the stream network, but may accumulate within the aggradation zones along the main channel. The headcuts in the resistant sandstone bodies of the Sarinena Formation present local base levels, but do not migrate upstream. The modern system therefore, although dominantly an erosional system, comprises a series of mini sediment source areas and sediment sinks analogous to Newson's (1992) interpretation of Schumm's (1977) classic zonation of the fluvial system. The dynamics of the modern system are therefore not controlled by the regional base level provided by the long-term incision of the Rio Cinca, but must reflect catchment controls by flood discharges and sediment supply, i.e. critical power relationships (Bull, 1979). These controls must in turn reflect climatic and vegetation cover conditions.

4. Multiple phases of landscape evolution

The modern valley is cut largely in Quaternary valley-fill deposits. These are set into an erosional valley system and therefore postdate the incisional phase(s) that followed deposition of the Pleistocene terrace gravels, evident south of El Tormillo (Fig. 4.4). These fills comprise two bodies of dominantly silty sediment, the older and younger fills.

The older fill has an erosional base cut into Miocene marl or sandstone bedrock, normally 1–2 m above the modern stream bed. Thin gravels are common at the base, but the bulk of the fill is composed of silt. Locally aggrading gravels interdigitate laterally with silts (Fig. 4.6), probably representing shallow channel and overbank environments, respectively. In the valley centre the silts are horizontally bedded, rising to a horizontal depositional terrace surface, 12–15 m above the modern channel, here described as the main terrace. Away from the valley centre the silts show less obvious bedding, but where bedding can be identified it suggests a gentle dip towards the valley centre. Here the surfaces also slope towards the valley centre, indicating colluvial rather than alluvial deposition on the footslopes of the catchment.

The older fill buries an erosional topography. The fill thins over sandstone outcrops on the valley floor which now form the headcut zones, and abruptly terminates at sandstone outcrops within the tributary valleys, burying what were former headcut zones. In places where the valley sideslopes are capped by sandstone the slopes themselves are mantled by angular boulders derived from the sandstone ledges above. In these locations a similar boulder-mantled slope is buried beneath the silts of the older fill (Fig. 4.7). Both these phenomena indicate that a topography of deeply trenched linear valley-floor gullies existed prior to burial by the older fill. This is confirmed by considering the spatial patterns occupied by the older fill (Fig. 4.4). The older fill is extensive and occurs throughout the valley system and follows the tributary valley patterns, which themselves have been

Figure 4.6. Section through older and younger fills; Barranco de la Clamor, east bank downstream of archaeological site: m is Miocene marl; o is older fill, y is younger fill, inset into older fill. Note the climbing gravels within the older fill. The eroded top of the section marks the height of a strath terrace; skyline to the left represents the top of the main terrace.

re-excavated to create the present pattern of linear gully systems. It quite clearly represents a major widespread aggradation phase.

The older fill is dissected. It is cut by a strath terrace at about 6 m above the present channel, representing at least a local pause in the general dissection. Below the level of the strath terrace (Figs 4.5 and 4.8) the fill is trenched to about modern stream level. The younger fill is set into this trench (Figs 4.6 and 4.8). It dominantly comprises stratified alluvial silts, but it includes flood-washed sands and fine gravels. Its top surface forms the low terrace 2–3 m above the modern valley floor. It is much less extensive than the older

Figure 4.7. Section in older fill: m is Miocene marls, o is older fill; s is in situ sandstone outcrop of the Miocene Sariñena Formation. Note sandstone boulders on top of the fill, related to modern hillslope topography and sandstone boulders buried by the fill, related to buried topography.

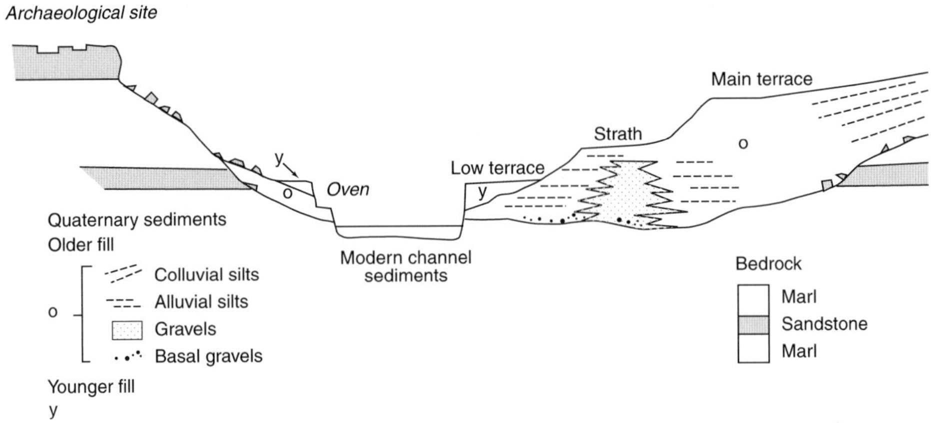

Figure 4.8. Stratigraphic relationships of the deposits in the area near to the archaeological site.

fill (Fig. 4.4), and occurs only as an intermittently preserved depositional terrace within the valley cut into the older fill, but nevertheless represented a switch from an erosionally dominant to an aggrading system.

Following the aggradational phase represented by the younger fill, the modern landscape developed through the dissection of the younger fill and the formation of the modern valley-side gully and pipe systems dissecting both fills. The modern landscape is dominantly erosional but aggradation is occurring on the valley floor between El Tormillo and the headcut above the RENFE viaduct, also in places between there and La Masadera (Fig. 4.4), and again along the valley both upstream and downstream of Castelflorite (locations on Fig. 4.3). Locally, especially between the viaduct and La Masedera, small terraces, younger than the low terrace represent local zones of greater complexity.

The recent evolution of the valley of the Barranco de la Clamor encompasses two totally contrasted landscape styles. These are an aggrading style and a dissectional style, similar to those recognised by Love (1979) for aggradational and dissectional phases in Chaco Canyon, New Mexico. Prior to deposition of the older fill we must envisage a gullied topography with similar spatial characteristics to today's landscape. This pattern developed again during the dissectional phase that followed the older fill, and again since dissection of the younger fill to create today's dominantly erosional landscape. For the aggradation phases we should envisage totally different landscapes, for the older fill a landscape of aggradational colluvial slopes grading towards a non-incised stream in the centre of the valley, and for the younger fill a wide aggrading valley floor set somewhat below the main terrace surface of the older fill, but up to 3 m above the level of the modern valley floor.

The contrasts between the two landscape styles must relate to critical power relationships and involve changes in coupling within the landscape. Under aggradational regimes sediment supply from widespread hillslope erosion must have been high, causing overloading of the system and burial of the previous erosional topography. Coupling from hillslopes to main channels was effective but coupling down the main channels was not effective. Under dissectional regimes relative stream power must have been higher, but with lower rates of hillslope sediment production. Coupling from hillslopes to channels was less effective, but was more effective through the stream system. However, under the current dissectional regime, this coupling is not complete, and local zones of aggradation occur within the channel system. Erosion is focussed along the main channels and on the gullied lower slopes immediately above the trenched channels.

5. Dating the sequence

The ultimate controls of the changes in geomorphic style must relate to runoff and erosion rates and therefore to climate and/or land cover. There are subtle environmental differences between the two main fill episodes exemplified by the fossil content of the two fills (Table 4.1). Gastropod shells are present within both fills, apparently not in life position therefore having been washed in with the silts. All species found in both fills are common within modern dryland Mediterranean regions, therefore do not suggest radically different climatic regimes from those of today. This might indicate Holocene rather than Pleistocene ages for both fills. The assemblages within the two fills do differ, suggesting local environmental differences. The assemblage within the older fill is more diverse, and all species are dryland dwellers, most living on the soil surface. The dominant *Romina decollata* lives within rather than on the soil, suggesting that its shells were deposited with soil eroded from the hillslopes of the catchment. The assemblage within the younger fill is much less diverse, dominated by *Cernuella virgata*, a species characteristic of wet valley floors. The bone of a small rodent was also found bedded within fine gravels near the base of this fill and an unidentified lizard skin and partial skeleton found incorporated within a sandy flood-washed unit both suggest floodwater deposition by the stream on the valley floor. A further indication of a likely Holocene age, at least for the younger part of the sequence, is suggested by the presence of archaeological remains, especially the archaeological site southwest of El Tormillo (Fig. 4.9, location on Fig. 4.4), and its apparent relation to the geomorphic sequence. This archaeological site has undergone multiple occupancy, based on the presence of

Table 4.1. Gastropod fauna within the older and younger fills.

Older fill: species found
Romina decollata[a]
Iberus carthaginiensis
Helix aspersa
Xeroplaxa murcica
Xeroplaxa barceloi
Younger fill: species found
Cernuella virgata[b]
Romina decollata[c]

[a] Overwhelmingly dominant: >90% specimens found.
[b] Overwhelmingly dominant species.
[c] Occasional presence.

Figure 4.9. The archaeological site southwest of El Tormillo. Note the Christian era graves cut into the top of the sandstone outcrop; gully erosion to the left dissects small terraced fields related to the settlement.

pottery within the hillslope deposits below the site, which includes both Iron Age (7–6th century BC) and Musulman (10–11th century AD) and later pottery. Reworked Iron-Age pottery has also been found within the fine gravels within the younger fill, but no pottery has been found within the older fill. A similar approach to dating geomorphic deposits and surfaces has been used elsewhere in northern Spain (Sancho et al., 1988; Soriano, 1989; Gutiérrez-Elorza and Peña-Monné, 1998). The most obvious phase of occupation at the El Tormillo site was during the Christian era (possibly during the 12th century AD), when numerous graves (see Fig. 4.9) were cut into the sandstone capping the hill. To the west of the hill were other structures, including agricultural terraces, now largely degraded by gully erosion. In the valley below the hill, is an oven cut into both older and younger fills (Fig. 4.8), and partly buried by younger colluvium. The position of the oven suggests that its construction took place after dissection of the younger fill. It is now undergoing erosion by the modern stream.

Three Radiocarbon dates have been obtained that allow some dating of the sequence (Table 4.2). Aggradation of the older fill was occurring during the early Holocene, and of the younger fill during the Iron Age. Dissection of the younger fill had taken place by the Christian Era, the date of the carbon ash buried in the floor of the oven. The overall sequence is summarised in Table 4.3.

Table 4.2. Radiocarbon dates.

Site	Material	Lab number	Date (^{14}C years BP)	Calibrated date
Older fill	Shells	Beta-095474	5540 ± 90	BC 4360 (= 6310 BP)
Younger fill	Shells	Beta-095475	2320 ± 70	BC 390 (= 2340 BP)
Oven	Ash/Charcoal	Beta-095473	830 ± 110	AD 1225 (= 725 BP)

Samples taken from Barranco del Clamor 2 km south of El Tormillo (for site stratigraphy see Fig. 4.7).

Table 4.3. Sequence summary.

Late Pleistocene or Pleist-Holocene transition?	Incisional phase Preceding deposition of the older fill	Dissected landscape dominated by large valley-floor gullies
Early Holocene	Aggradational phase. Deposition of the older fill	Aggrading landscape. Hillslopes supplying abundant sediment to the valley system
Late Holocene (post-6000–pre-2500 cal. BP)	Incisional phase. Dissection of the older fill	Dissected landscape dominated by large valley-floor gullies
Post-2500 cal. BP	Aggradational phase. Deposition of the younger fill	Aggrading landscape to flat floored valley set below main terrace surface
Pre-13th century AD	Incisional phase. Dissection of the younger fill	Dissected landscape dominated by large valley-floor gullies
Post-13th century AD	Dominantly incisional phase continues: local valley-floor aggradation and local complexities	Dissected landscape dominated by large-valley gullies: local valley-floor aggradation.

6. Discussion

As elsewhere in the Mediterranean region (Macklin and Passmore, 1995; Fuller et al., 1998) the geomorphic system of the Barranco de la Clamor underwent major threshold-controlled changes between incisional and aggradational regimes during the late Quaternary. Climatic change, both directly through rainfall characteristics and indirectly through changes in vegetation cover, and human-induced land-use changes, especially in the later part of the period, are possible underlying causes. There is evidence in northern Spain for periods during the Holocene of climates cooler and wetter than today and of intervening warmer drier climates (Gutiérrez-Elorza and Peña-Monné, 1998). Wetter climates might stimulate greater vegetation growth, thus inhibiting erosion, but whether flood runoff is higher under wetter or drier climates is uncertain. Irrespective of the underlying causes of change, the mechanisms are not entirely straightforward. The response of both hillslopes and valley floors and their coupling are important. The aggradation phases identified in the Barranco de la Clamor appear to have resulted from excess sediment generation on the hillslopes, but the phases of incision appear to have resulted from higher flood discharges within the valley system. Neither is explicable simply by a climatic change to either cool wet or warm dry conditions. As the older fill pre-dates significant human impact, human disturbance cannot account for all the changes observed.

Other studies of late Quaternary aggradation/incision sequences within the Ebro basin (Peña et al., 1983; Soriano, 1989; Arauzo and Gutiérrez-Elorza, 1994; Macklin et al., 1994; Macklin and Passmore, 1995; Fuller et al., 1998; Gutiérrez-Elorza and Peña-Monné,

1998; Lopez-Aviles et al., 1998) indicate that such regime changes are both time-transgressive and regionally variable, and almost certainly influenced by local threshold conditions.

There is no clear regional picture of the style of geomorphic activity in the valleys of the Ebro Basin during the late Pleistocene, though Macklin and Passmore (1995) and Fuller et al. (1998) suggest a period of alluvial aggradation. In southern Spain aggradation at that time appears to have been widespread (Harvey, 2002b), however, climatic and vegetation conditions may have been very different in southern than in northern Spain. In the Barranco de la Clamor a major incision phase pre-dates the older fill; its age is uncertain. If our radiocarbon dates are reliable, aggradation of the older fill was occurring during the early Holocene and continued at least until 6000 cal. BP. When it was initiated is not clear and it could well have begun during the late Pleistocene. All that we can say in this context is that at some stage in the late Pleistocene or the early Holocene there was a major climatically triggered switch from an incisional to an aggradational system.

A clearer regional picture can be identified for the later Holocene. Gutiérrez-Elorza and Peña-Monné (1998) argue for a climatic rather than a human control of the late Holocene sequences, and identify basin-wide aggradation phases during cooler wetter phases of the late Holocene, during the Iron Age and again during the Little Ice Age (see also Macklin et al., 1994). The Iron Age phase is co-incident with our younger fill. In the intervening periods of warmer, drier climates incision appears to have been dominant, in the case of the Barranco de la Clamor during the period prior to deposition of the younger fill and during the period following it, as the present landscape developed. The influence of the Little Ice Age is uncertain, but it may account for some of the minor complexities in the central part of the study area.

The modern landscape exhibits a range of gully types, including mostly midslope gullies, valley-floor gullies, and basally induced gullies. The midslope gullies tend to be non-coupled in relation to the channel network, however during aggradation phases sediment movement from the hillslopes to the valley system must have taken place. The valley-floor gullies include the incised main channel itself and its tributaries. The main channel forms a series of discrete sediment systems, bounded by headcuts in resistant sandstone, above which are modern valley-floor aggradation zones (Bull, 1997). There is a lack of effective coupling between the reaches. The tributary valleys similarly end in sandstone headcuts, and in general do not show continuity with the hillslopes above. The present pattern of valley-floor gullies has been inherited from previous incisional phases by excavation of earlier fill deposits. In addition, there are extensive valley-side basally induced gullies, including badlands cut into Miocene marls and linear and piped gully systems cut into Quaternary fills.

Although under the present regime, the potential supply of sediment to the stream system is high, it is insufficient to overwhelm the system and switch it from incision to overall aggradation. Even with today's potential for human-enhanced erosion rates the Barranco de la Clamor is not in an aggradational regime. This itself is further evidence to support a primary climatic rather than a human control of the sequence.

However, the intrinsic properties of the system, a semi-arid climate, a variable vegetation cover, highly erodible substrates, and a system characterised by marked spatial discontinuities, suggest that the threshold between incisional and aggradational regimes could easily be crossed. The system could be vulnerable to dramatic changes resulting from changes in runoff rates, erosion rates from any of the three gullied zones, and changes in the coupling characteristics of the system.

Acknowledgements

We wish to thank the Department of Geography of the University of Liverpool and the Departamento de Geologia, Universidad de Zaragoza for funding towards the costs of field work and the costs of the Radiocarbon dates. We thank the Cartographics section, particularly Sandra Mather, of the Department of Geography, University of Liverpool, for producing the illustrations. We also wish to thank Professor Chris Paul, formerly of the University of Liverpool for identifying the gastropods and for comments on their ecology, and Dr. Francisco Burillo of the University of Zaragoza for discussions relating to the archaeology of the El Tormillo site.

References

Alonso-Zarza, A., Armenteros, I., Braga, J.C., Munoz, A., Pujalte, V., Ramos, E., Aguirre, E., Alonso-Gavilan, G., Arenas, C., Baceta, J.I., Carballiera, J., Calvo, J.P., Corrochano, A., Fornos, J.F., Gonzalez, A., Luzon, A., Martin, J.M., Pardo, G., Payros, A., Perez, A., Pomar, L., Rodriguez, J.M., Villena, J., 2002. Tertiary. In: Gibbons, W., Moreno, T. (Eds), The Geology of Spain. Geological Society, London, pp. 292–334, Ch. 13.

Arauzo, T., Gutiérrez-Elorza, M., 1994. Evolución de los valles de fondo plano del centro de la Depresión del Ebro. In: Arnaez, J., Garcia-Ruiz, J.M., Gomez-Villar, A. (Eds), Geomorfología en España. III Reunion Nacional de Geomorfologia, Vol. I, pp. 277–290.

Brice, J.C., 1966. Erosion and Deposition in the Loess Mantled Great Plains, Medicine Creek, Nebraska, Professional Paper 352H. U.S. Geological Survey, pp. 254–335.

Bull, W.B., 1979. Threshold of critical power in streams. Geol. Soc. Am. Bull. 90, 453–464.

Bull, W.B., 1997. Discontinuous ephemeral streams. Geomorphology 19, 227–276.

Campbell, I.A., 1997. Badlands and badland gullies. In: Thomas, D.S.G. (Ed.), Arid-Zone Geomorphology, 2nd edn. Wiley, Chichester, pp. 261–291.

Cooke, R.U., Reeves, R.W., 1976. Arroyos and Environmental Change in the American South-West. Oxford University Press, Oxford, 213 pp.

Fuller, I.C., Macklin, M.G., Lewin, J., Passmore, D.G., Wintle, A.G., 1998. River response to high-frequency climate oscillations in southern Europe over the last 200 k.y. Geology 26 (3), 275–278.

Gutiérrez-Elorza, M., Peña-Monné, J.L., 1994. Depresion del Ebro. In: Gutierrez-Elorza, M. (Ed.), Geomorfologia de Espana. Rueda, Madrid, pp. 305–342.

Gutiérrez-Elorza, M., Peña-Monné, J.L., 1998. Geomorphology and late Holocene climatic change in Northeastern Spain. Geomorphology 23, 205–217.

Gutiérrez-Elorza, M., Garcia-Ruiz, J.-M., Goy, J.-L., Gracia, F.J., Gutiérrez-Santolalla, F., Marti, C., Martín-Serrano, A., Pérez-Gonzalez, A., Zazo, C., Aguirre, E., 2002. Quaternary. In: Gibbons, W., Moreno, T. (Eds), The Geology of Spain. Geological Society, London, pp. 335–366, Ch. 14.

Harvey, A.M., 2001. Coupling between hillslopes and channels in upland fluvial systems: implications for landscape sensitivity, illustrated from the Howgill Fells, northwest England. Catena 42, 225–250.

Harvey, A.M., 2002a. Effective timescales of coupling within fluvial systems. Geomorphology 44, 175–201.

Harvey, A.M., 2002b. Uplift, dissection and landform evolution: the Quaternary. In: Mather, A.E., Martín, J.M., Harvey, A.M., Braga, J.C. (Eds), A Field Guide to the Neogene Sedimentary Basins of the Almeria Province, South-East Spain. Blackwell, Oxford, pp. 225–322.

Harvey, A.M., 2003. Badlands. In: Goudie, A. (Ed.), Encyclopedia of Geomorphology. Routledge, London.

Lopez-Aviles, A., Ashworth, P.J., Macklin, M.G., 1998. Floods and Quaternary sedimentary style in a bedrock-controlled reach of the Bergantes River, Ebro Basin, northeast Spain. In: Benito, G., Baker, V.R., Gregory, K.G. (Eds), Palaeohydrology and Environmental Change. Wiley, Chichester, pp. 181–196.

Love, D.W., 1979. Quaternary fluvial adjustments in Chaco Canyon, New Mexico. In: Rhodes, D.D., Williams, G.P. (Eds), Adjustments of the Fluvial System. Kendall/Hunt, Dubuque, IA, pp. 277–308.

Macklin, M.G., Passmore, D.G., 1995. Pleistocene environmental change in the Guadalope basin, northeast Spain: fluvial and archaeological records. In: Lewin, J., Macklin, M.G., Woodward, J.C. (Eds), Mediterranean Quaternary River Environments. Balkema, Rotterdam, pp. 103–113.

Macklin, M.G., Passmore, D.G., Stevenson, A.C., Davis, B.A., Benavente, J.A., 1994. Responses of rivers and lakes to Holocene environmental change in the Alcaniz region, Teruel, north-east Spain. In: Millington, A.C., Pye, K. (Eds), Environmental Change in Drylands: Biogeographical and Geomorphological Perspectives. Wiley, Chichester, pp. 113–130.

Newson, M.D., 1992. Geomorphic thresholds in gravel-bed rivers–refinement for an era of environmental change. In: Billi, P., Hey, R.D., Thorne, C.R., Tacconi, P. (Eds), Dynamics of Gravel-Bed Rivers. Wiley, Chichester, pp. 3–20.

Peña, J.L., Echevarria, M.T., Petit-Maire, N., Lafont, R., 1983. Cronología e interpretación de las acumulaciones holocenas de la Val de las Lenas (Depresion del Ebro, Zaragoza). Geographicalia 30, 321–332.

Sancho, C., Meléndez, A., 1992. Genesis y significado ambiental de los caliches pleistocenos de la region de Cinca (Depresion de Ebro). Rev. Soc. Geol. Espana 5, 81–93.

Sancho, C., Gutiérrez, M., Peña, J.L., Burillo, F., 1988. A quantitative approach to scarp retreat starting from triangular slope facets, central Ebro Basin, Spain. Catena Suppl. 13, 139–146.

Schumm, S.A., 1977. The Fluvial System. Wiley, New York, 338 pp.

Schumm, S.A., 1979. Geomorphic thresholds: the concept and its applications. Trans. Inst. Br. Geogr. New Ser. 4, 485–515.

Schumm, S.A., Harvey, M.D., Watson, C.C., 1984. Incised Channels. Morphology, Dynamics and Control. Water Resources Publications, Littleton, CO, 200 pp.

Soriano, M.A., 1989. Infilled valleys in the central Ebro Basin (Spain). Catena 16, 357–367.

Chapter 5

The influence of soil saturation on the stability of abandoned agricultural hillslope terraces under Mediterranean climatic conditions

Ali Zgaier and Moshe Inbar*

Department of Geography, University of Haifa, Haifa 31999, Israel

Abstract

The strength of two residual soils, terra rossa and rendzina, which constitute most of the abandoned agricultural hillslope terraces near Yirka village, Western Galilee, Israel, was determined by two methods: shear by a basal direct field shear apparatus and probing by a dynamic penetrometer. The first apparatus was used to determine the cohesion and the angle of internal friction of the top soil and the second to inspect the change of the undrained shear strength of the above soils with depth. The strength parameters with midsummer water content (16.91% for terra rossa and 11.07% for rendzina) were found to be: terra rossa: $c \approx 10.57$ *kN/m*^{2}*,* $\phi = 41.24°$*; rendzina:* $c \approx 10.72$ *kN/m*^{2}*,* $\phi = 42.34°$*. The strength parameters of the same soils with midwinter water content were: terra rossa:* $c' \approx 2.18$ *kN/m*^{2}*,* $\phi' = 40.75°$*; rendzina:* $c' \approx 5.24$ *kN/m*^{2}*,* $\phi' = 39.03°$*. Stability calculations confirmed by field investigations showed that the drop of cohesion of both soils in consequence of their saturation during the rainy season can cause terrace landsliding unless the terraces are retained by sufficiently strong enough retaining walls.*

Keywords: erosion, agricultural terraces, landslides, shear box, Mediterranean mountains, Galilee

1. Introduction

Mountain semiarid habitats are most common in the Mediterranean regions, and their ecosystems are characterized by intensive erosive processes. Agricultural land use is a major factor affecting erosion processes, and in mountainous areas the conservation or abandonment of terraces is a crucial factor in soil loss. For more than two millennia soil accumulated in the human-made terraces, which were the economic basis for flourishing cultures. In the Eastern Mediterranean countries terrace construction began in ancient periods, at least since the 15th century BCE (Ron, 1966; Aizenberg, 1994). In the Iberian countries early terraces are probably from the Roman period. Other researchers suggest older ages and speculate that terracing developed earlier than 5000 year B.P. (Sandor, 1998).

---

*Corresponding author. *E-mail address:* inbar@geo.haifa.ac.il (M. Inbar).

Abandonment of agricultural land is widespread in the mountainous area of the Mediterranean countries of Europe. Erosion processes increased due to the abandonment of terrace fields, as in Spain (García-Ruiz, 1989; López-Bermudez, 1990) Greece (Lehman, 1993), and in old land terrace systems in Israel (Naveh and Dan, 1973). An extensive bibliography on the terraces in the Mediterranean mountain region covering about 120 items was published by Rodríguez-Aizpeolea and Lasanta-Martínez (1992).

Landsliding is one of the most important mass-wasting processes that affect agricultural hillslope terraces retained by dry-built stone walls in Mediterranean regions. Field investigations in areas of abandoned agricultural terraces near Yirka, Western Galilee, Israel, showed that a terrace begins to erode only after it has been affected by a landslide (Zgaier, 2000). Terrace landslides occur in winter during or immediately after heavy rains (Salas-Pinto and Vásques-Villanueva, 1987; Gallart and Clotet-Perarnau, 1988; Grimalt-Gelabert et al., 1992; Lehmann, 1993, 1994; Pallares-Bou, 1994; Zgaier, 2000). The aim of this study was to check the influence of soil saturation on the strength of the terrace soils, and the effect of the changes of soil strength on the stability of these terraces against landsliding.

2. Soils and methods

2.1. Site and soils

Field tests for the determinations of soil strength were carried out on abandoned agricultural terraces near Yirka, Western Galilee, Israel. Most of the terraces in this region are made of terra rossa and rendzina and in terms of land use they may be divided into two groups, cultivated and abandoned. Old olive trees grow on most of the cultivated terraces. The abandoned terraces were used prior to their abandonment, during the late 1950s, for growing cereals and legumes. Average annual precipitation at the study area is 710 mm. The rainy season begins usually in September and ends in May. Most of annual precipitation ($\sim$77%) falls from November to February. The rainiest month is January, followed by February. Mean annual temperature at the study area is 20°C. Mean summer (August) temperature is 25°C, and mean winter temperature (January) is 10°C.

The terra rossa here falls within the range of the inorganic clays of high plasticity – inorganic silts–organic clays of high plasticity, whereas the rendzina falls within the range of the inorganic clays of medium plasticity – inorganic silts–organic clays of slight to low plasticity on the chart of the Unified Soil Classification System.

Soils are shallow and their depths vary from a few cm to 50–80 cm. Their occurrence is usually in pockets. In the terraces soil accumulation depends on the height of the retaining wall. Some of the physical properties of the soils at the study area are given in Table 5.1.

2.2. The direct field shear apparatus

The apparatus used for the determination of the strength components of the inspected soils was a basal direct shear apparatus constructed specially for this purpose generally in accordance with the apparatus designed by Chandler et al. (1981). This apparatus is

Table 5.1. Some physical properties of the soil layers (60 cm depth) at the sites of shear tests.

Soil	Average water content in midsummer dry state (%)	Average water content in midwinter saturated state (%)	Average dry unit weight (kN/m^3)	Average saturated unit weight (kN/m^3)	Average porosity	Average void ratio	Average shrinkage limit (%)	Average plastic limit (%)	Average liquid limit (%)	Plasticity index
Terra rossa	16.91	36.33	16.572	18.533	0.5	1.12	9.56	28.43	57.85	29.42
Rendzina	11.0	31.35	16.18	19.024	0.48	0.93	17.94	21.94	37.89	15.95

composed of three main elements: a shear box, a winch, which provides the force needed for shear, and an electronic dynamometer for the measurement of the force needed for shear. The natural normal stress on the soil sample is simulated by the means of concrete blocks.

- The shear box was built of heavy 9.5 mm thick steel plates and reinforced with angle braces to minimize box deformation (Fig. 5.1). Its basal shear area is 300 × 300 mm. Two steel loops were mounted on its front plate. A sturdy synthetic rope was inserted through these loops to attach the shear box to the winch.
- Each block had a mass of about 30 kg, a base of about 400 × 200 mm, and a thickness of about 200 mm.

Figure 5.1. The direct field shear box.

- The shear force is provided by a 5-t capacity winch attached to the shear box by a synthetic rope inserted into its hook. Another similar synthetic rope, the anchor rope, was inserted into the anchor shackle of the winch. During the shear tests the anchor rope was wound around the trunk of a nearby olive or fig tree.
- The force needed for shear was measured by a 3.0 t capacity Ron 2300 electronic dynamometer.
- The shear tests were conducted in August 1996 on dry and artificially saturated soil samples. The strength parameters obtained from the results of the tests on dry soil samples represent the strength of the soil in the midsummer dry state. The first step of a sample preparation was the removal of the top 10–15 cm thick soil layer from the spot chosen for testing. Then a 20 mm thick rectangular steel plate with an area equal to that of the base of the shear box (0.09 m^2) was laid on that spot. Next, 15 cm deep trenches were dug around it (Fig. 5.2). The resulting soil monolith was finally trimmed with a steel knife so that its dimensions were equal to the inside of the shear box ($30 \times 30 \times 15$ cm^3). The surface of the soil monolith was also trimmed so that its inclination became zero.

For the preparation of saturated soil samples, 30 l of water were poured on each spot ($\sim 30 \times 30$ cm^2) chosen for shear. Sample preparation in these cases proceeded according to the steps mentioned above, after all the water had vanished from the soil surface (after $\sim$1.5 h). The strength of such saturated soil samples is effective because of rapid soil drainage (Selby, 1993). During each test, on dry or saturated soil, a soil sample was taken for laboratory soil moisture measurement.

- The shear box was placed over the soil monolith, and then a synthetic rope was inserted through its two loops and also through one of the two shackles of electronic dynamometer sensor, and tied. The other shackle of the sensor was directly attached to the winch hook. Through the same shackle of the winch another rope was inserted, then it was wound around the trunk of a nearby tree in alignment with the rope, and tied. Finally the handle of the winch was turned carefully till shear occurred, then the force needed for shear was read off the dynamometer screen.

Figure 5.2. Sample preparation for an *in situ* shear test.

2.3. The dynamic penetrometer

Dynamic probing methods for the determination of soil strength are widely used for soil engineering problems (Card and Roche, 1988; Nixon, 1988; Tonks and Whyte, 1988; Bathurst et al., 1997).

Tests for the determination of the undrained shear strength (C_u) of the soil were carried out on three terraces, one of terra rossa and two of rendzina. They were conducted on dry soil (during July and August 1996). The apparatus used was a SRS 15 dynamic penetrometer (Fig. 5.3), whose specifications are given by Tonks and Whyte (1988) (Table 5.2).

During each test the number of blows needed to drive the rod into the soil were recorded for each 10 cm increment. Tests on terraces lying on hard rocks (limestone and nari) were stopped only when the rod of the penetrometer hit the bedrock. Tests on terraces lying on soft rocks (marl) were stopped when the rod had been driven 10–30 cm into the bedrock.

Figure 5.3. The SRS 15 dynamic penetrometer.

Table 5.2. Specifications of the SRS 15 Dynamic Penetrometer (after Tonks and White, 1988).

Mass (kg)	Drop (m)	Free fall energy of hammer (kJ)	Projected area of tip (mm^2)	Specific energy (kJ/m^2)
50	0.5	0.245	1500	163

The undrained shear strength was computed from the results of the dynamic penetration tests by the following equation (Tonks and Whyte, 1988):

$$C_u \approx K_1 N_{100}$$

where:

N_{100} = the average number of blows needed to drive the rod of the penetrometer 10 cm between depths 20 and 30 cm (the depth at which shear by the direct field shear apparatus took place).
K_1 = empirical coefficient. Its value was taken as 7, the average of all values of this coefficient for the cohesive soils mentioned by Tonks and Whyte (1988).

The undrained shear strength from the results of the direct field shear tests were computed by the following equation (Carson and Kirkby, 1972; Abramson et al., 1996):

$$C_u = 0.5 q_u = c\,\mathrm{tg}(45 + \phi/2)$$

where q_u is the unconfined compressive strength, c the soil cohesion and ϕ is the angle of internal friction.

3. Results and discussion

3.1. Shear strength

The failure envelopes of field shear tests of terra rossa and rendzina in dry and saturated states were as follows (Figs 5.4–5.7):

Terra rossa

Dry state (average water content: 16.91%)
$S = 10.569 + \sigma_n \mathrm{tg} 41.24°$
Saturated state (average water content: 36.33%)
$S = 2.1829 + \sigma_n \mathrm{tg} 40.75°$.

Rendzina

Dry state (average water content: 11.07%)
$S = 10.722 + \sigma_n \mathrm{tg} 42.34°$
Saturated state (average water content: 31.35%)
$S = 5.2419 + \sigma_n \mathrm{tg} 39.03°$
where S is the shear strength and σ_n the normal stress.

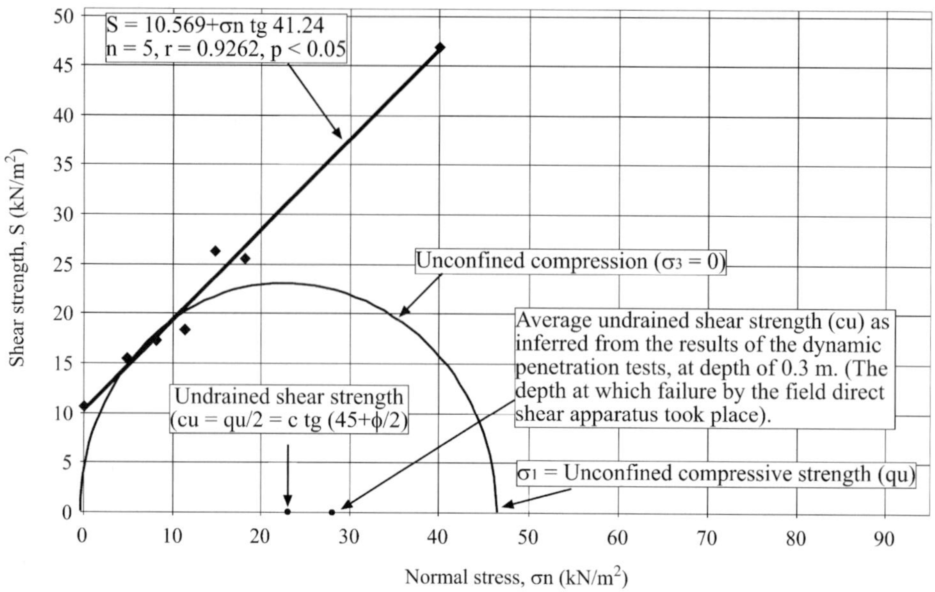

Figure 5.4. Failure envelope of consistency tests on dry terra rossa.

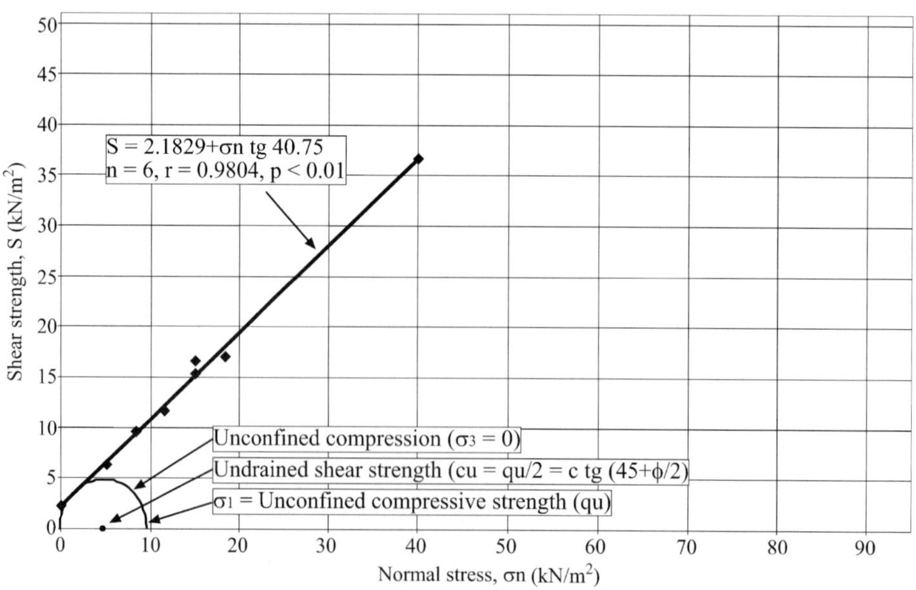

Figure 5.5. Failure envelope of consistency tests on saturated terra rossa.

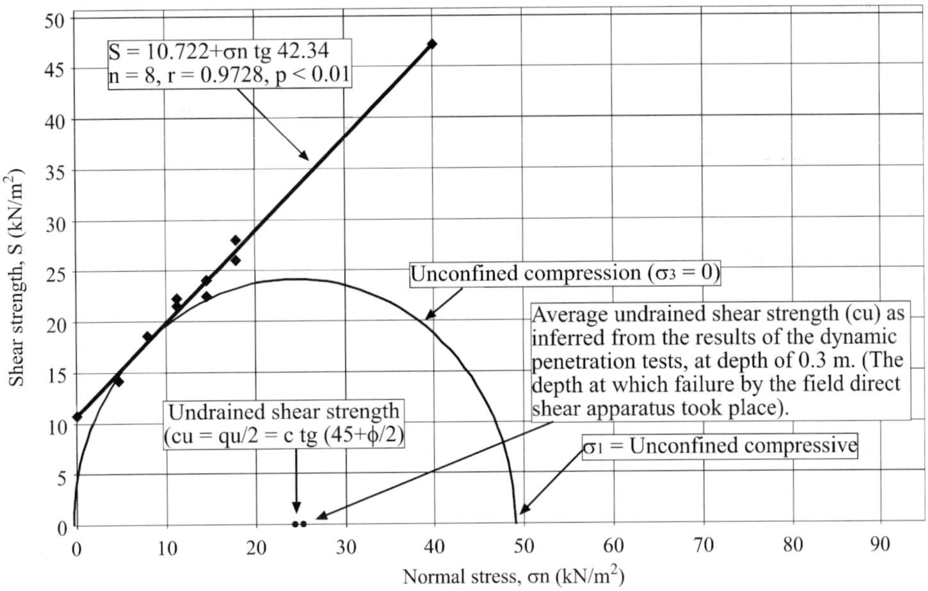

Figure 5.6. Failure envelope of consistency tests on dry rendzina.

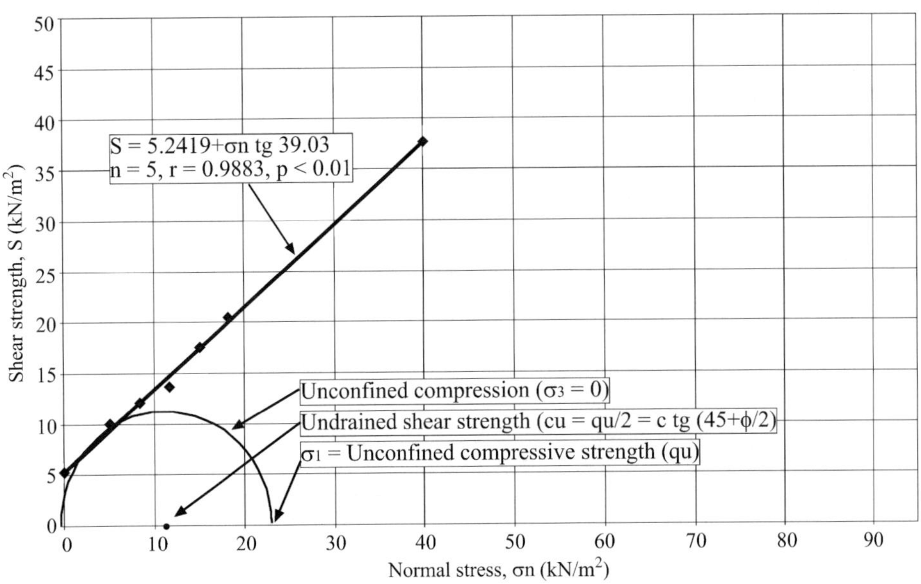

Figure 5.7. Failure envelope of consistency tests on saturated rendzina.

From the above equations three main facts were evident: the cohesion of both soils was small; their angles of internal friction were large; and saturation caused a significant drop in their cohesion and only a slight decrease in their angles of internal friction. The large drop in soil cohesion after saturation is attributed to a drop in soil moisture suction.

Two factors may be responsible for the small values of cohesion of both soils:

(i) The abundance of stones, especially in terra rossa.
(ii) The fact that both soils had never been over-consolidated in the past.

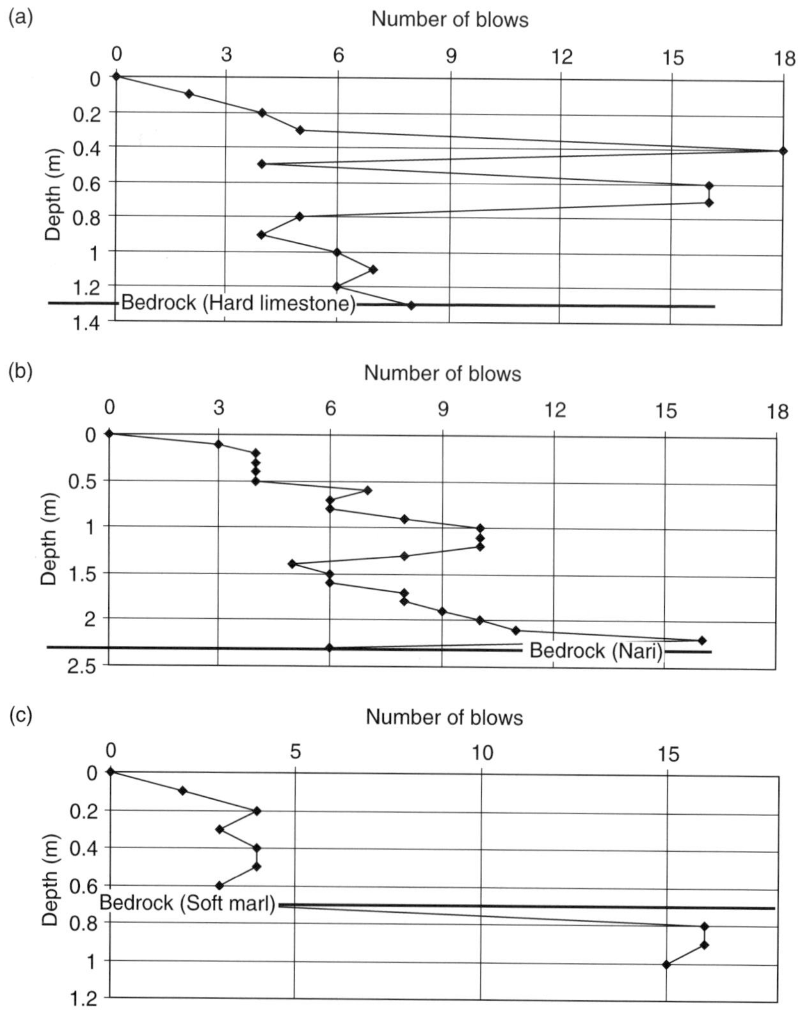

Figure 5.8. Typical dynamic probe logs for terra rossa on hard limestone (a), rendzina on nari (b) and rendzina on soft marl (c).

The large values of the angles of internal friction and the slight drop of these values after saturation were attributed to the high stone content of both soils.

3.2. Dynamic penetration tests

Three typical penetration logs of the inspected soils are depicted in Figure 5.8. To check the match of the results obtained by the dynamic penetrometer and by the field shear apparatus, a comparison was made between the undrained shear strength of both soils terra rossa and rendzina, at the depth of 30 cm (the depth at which shear by the direct field shear apparatus took place) obtained by both methods (Table 5.3).

The undrained shear strength of both soils, as computed from the results of the dynamic penetration tests, was higher than that computed from the results of the field shear tests, and the difference was higher for terra rossa than for rendzina.

One or both of two factors was responsible for the high values of the undrained shear strength computed from the results of the dynamic penetration tests:

- The extra blows needed to crush the stones found in the soil profile, perpendicular to the path of the penetrometer's head. Such stones had to be crushed before the penetrometer could be driven into the soil. Stone crushings were displayed in the penetration log as prominent deviations from its general trend.
- The use of two different methods for the determination of soil strength. Different methods frequently give different results (Zimbone et al., 1996).

The three penetration logs displayed in Figure 5.8 show that the change of the undrained shear strength with depth is not significant. We can, therefore, consider the shear strength parameters determined by the direct field shear tests as representing the shear strength of the whole soil profile.

The influence of soil shear strength on the stability of terraces in the study area, with midsummer dry water content and midwinter saturated water content, is depicted in Figures 5.9–5.12. The following variables and their change with soil depth are shown in these figures for dry and saturated rendzina and terra rossa: (a) maximal depth of soil at the site of field shear experiment; (b) depth of tension zone; (c) lateral earth pressure;

Table 5.3. Values of the undrained shear strength of terra rossa and rendzina in dry states as computed from the results of the dynamic penetration test and the field shear tests.

Soil	The average number of blows needed to drive the rod of the penetrometer 10 cm between the depths 20 and 30 cm	Undrained shear strength (C_u) (kN/m^2)	
		According to the penetration tests	According to the field shear tests
Terra rossa	3.25	28.00	23.32
Rendzina	3.60	25.20	24.28

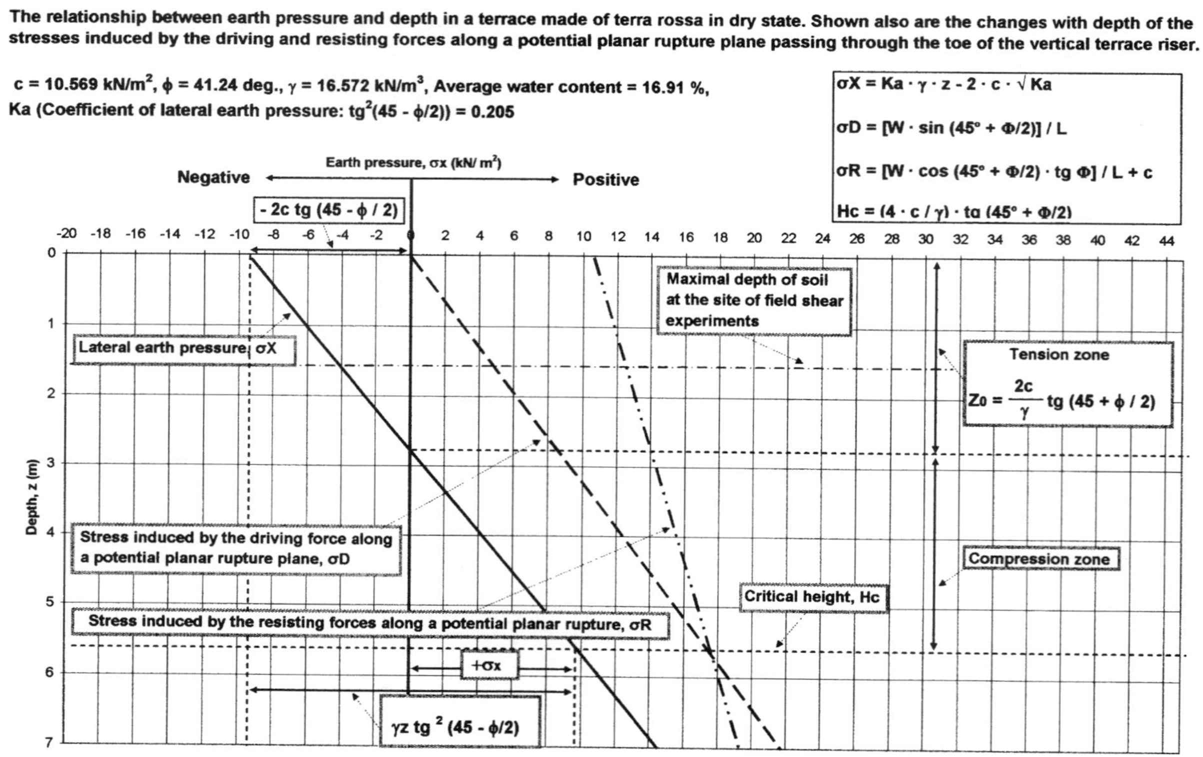

Figure 5.9. Change of earth pressure with depth in a terrace of dry terra rossa. Shown also are the changes with depth of the stresses induced by the driving and resisting forces along a potential planar rapture plane passing through the vertical terrace riser.

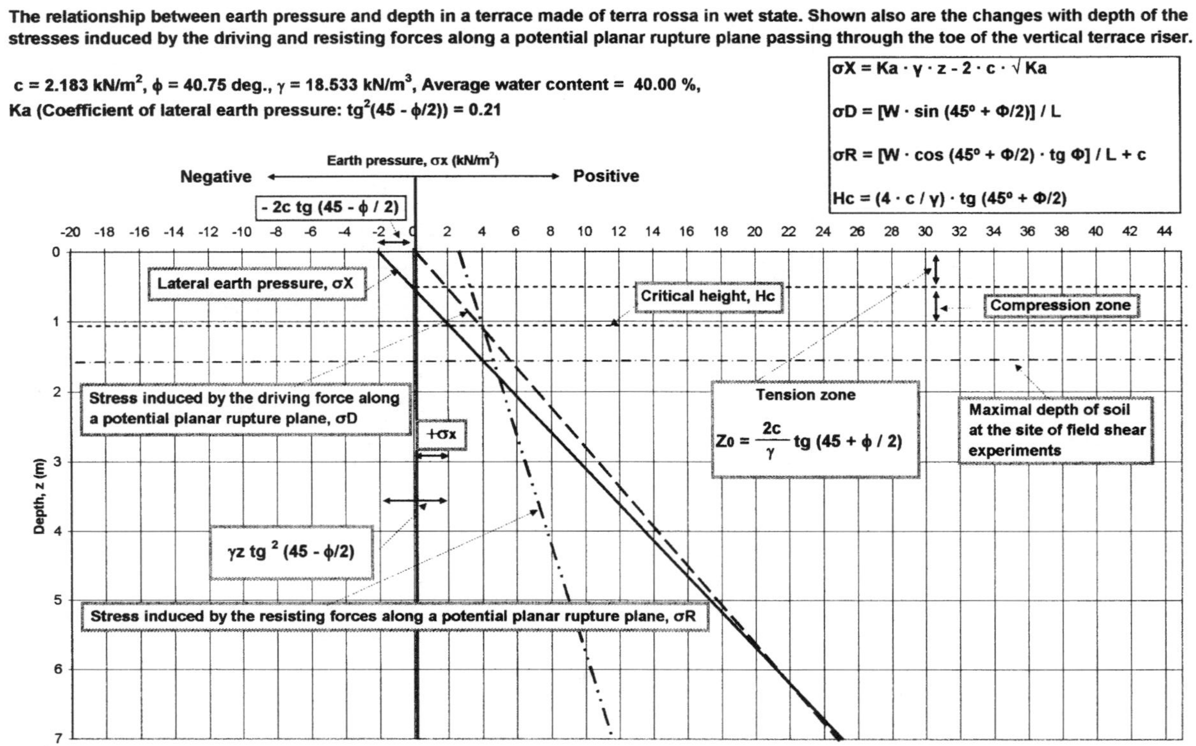

Figure 5.10. Change of earth pressure with depth in a terrace of saturated terra rossa. Shown also are the changes with depth of the stresses induced by the driving and resisting forces along a potential planar rapture plane passing through the vertical terrace riser.

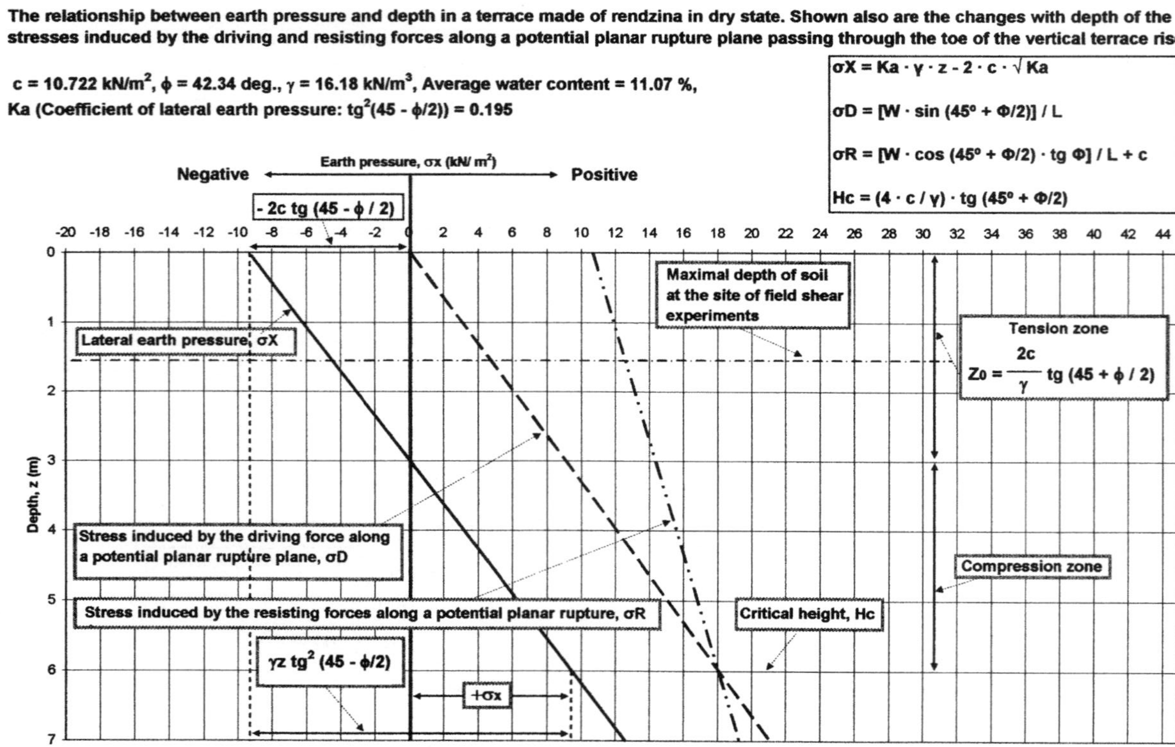

Figure 5.11. Change of earth pressure with depth in a terrace of dry rendzina. Shown also are the changes with depth of the stresses induced by the driving and resisting forces along a potential planar rapture plane passing through the vertical terrace riser.

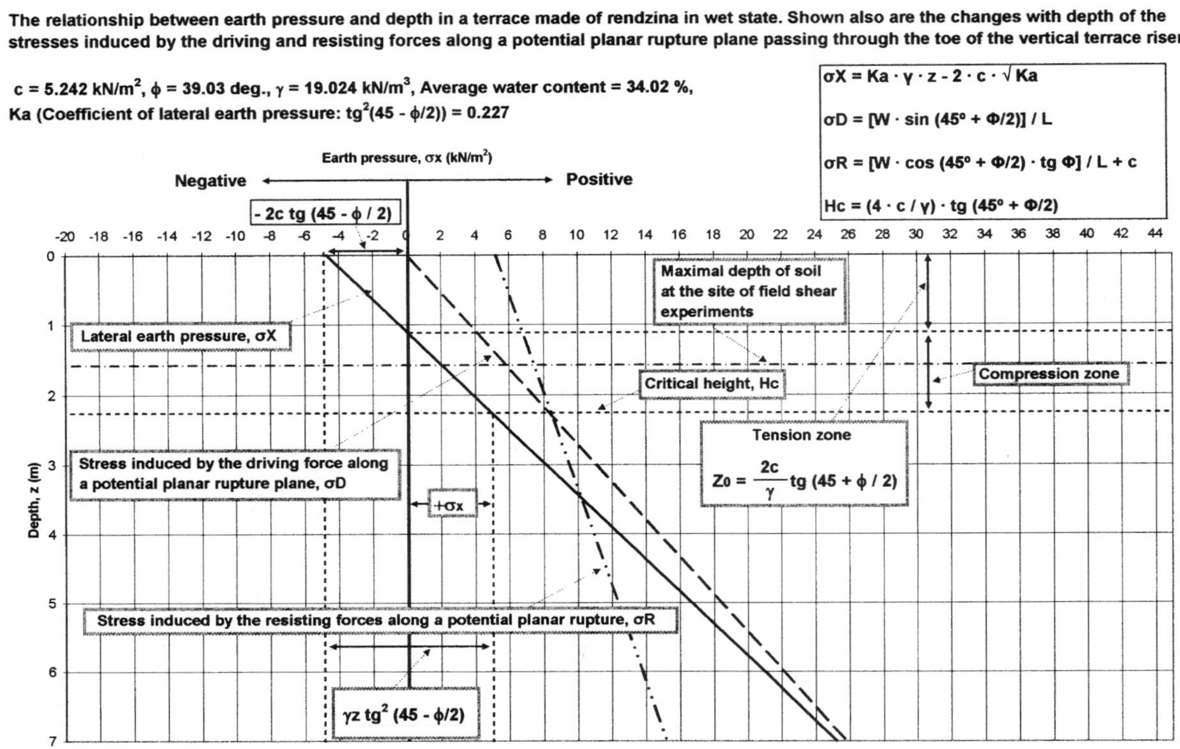

Figure 5.12. Change of earth pressure with depth in a terrace of saturated rendzina. Shown also are the changes with depth of the stresses induced by the driving and resisting forces along a potential planar rapture plane passing through the vertical terrace riser.

Table 5.4. Critical heights of the riser of terraces of terra rossa and rendzina in dry saturated states.

Soil type	Critical height of terrace riser in midsummer dry state (m)	Critical height of terrace riser in midwinter saturated state (m)
Terra rossa	5.63	1.03
Rendzina	6.00	2.31

(d) critical height of the terrace riser; (e) depth of the theoretical compression zone; (f) driving force along a potential planar rupture plane; and (g) resisting force along a potential planar rupture plane.

The critical height (H_c) of the terrace riser, which controls the stability of the terrace against landsliding, was calculated according to the Culmann equation (Selby, 1993).

$$H_c = \frac{4c}{\gamma}\frac{\sin\alpha\cos\phi}{[1-\cos(\alpha-\phi)]}$$

In the case of a vertical terrace riser ($\alpha = 90°$), the case for which the calculations were done, this reduces to:

$$H_c = \frac{4c}{\gamma}\mathrm{tg}\left(45+\frac{\phi}{2}\right)$$

where c is the soil cohesion, α the angle of terrace riser, ϕ the angle of internal friction, and γ the soil unit weight.

The critical heights of unretained terrace risers for both types of soils are given in Table 5.4. The small critical heights of risers of saturated terraces (~1 m for terra rossa terraces and ~2.3 m for rendzina terraces) emphasize the importance of retaining walls in preventing the collapse of the risers. Unretained terraces and terraces with partly destroyed

Figure 5.13. Bulge of the retaining wall of an agricultural terrace.

Figure 5.14. Typical landslide of an agricultural terrace.

retaining walls are prone to collapse if their risers are higher than the critical height. Field investigations in the study area showed that even strong well-built retaining walls, under the effect of the lateral earth pressure and the drop of cohesion due to soil saturation, with time develop bulges at the level of one third of the retaining wall's height from its base (Fig. 5.13). Sooner or later such bulges lead to the collapse of retaining wall or even to landslides, (Fig. 5.14) which involve both the retaining wall and the retained soil. The lack of maintenance of most of terraces in the study area contributed greatly to terrace destruction and degradation.

4. Conclusion

Saturation of terra rossa and rendzina, the soils of which the terraces in the study area are composed, caused a large drop in their cohesion and a slight drop in their angles of internal friction. The drop in the cohesion of terra rossa was larger than that of rendzina. The drop of cohesion due to saturation can lead to terrace landslides. The direct basal field shear apparatus and the dynamic penetrometer were found to be useful in the determination of soil strength.

Acknowledgements

The research was supported under grant C 12-006, U.S.–Israel Cooperation Development Research Program, Office of the Science Advisor, U.S. Agency for International Development. We wish to thank the comments by R.P.C. Morgan and an anonymous reviewer for the constructive suggestions for the improvement of the manuscript.

References

Abramson, L.W., Lee, T.S., Sharma, S., Boyce, G.M., 1996. Slope Stability and Stabilization Methods. Wiley, New York.

Aizenberg, E., 1994. Nahal Refaim – a bronze era village southwest of Jerusalem. Kadmonyot 103–104, 82–95, in Hebrew.

Bathurst, J.C., Burton, A., Kilsby, C.G., Sheffield, J., 1997. Large basin modeling and landslide erosion modeling. Mediterranean Desertification and Land Use, Second Annual Report, pp. 225–239.

Card, G.B., Roche, D.P., 1988. The use of continuous dynamic probing in ground investigation, Proceedings of the Geotechnology Conference, Institution of Civil Engineers, Birmingham, UK, pp. 119–122.

Carson, M.A., Kirkby, M.J., 1972. Hillslope Form and Process. Cambridge University Press, Cambridge.

Chandler, M.P., Parker, D.C., Selby, M.J., 1981. An Open-Sided Field Direct Shear Box, British Geomorphological Research Group. Technical Bulletin 27.

Gallart, F., Clotet-Perarnau, N., 1988. Some aspects of the geomorphic processes triggered by an extreme rainfall event: the November 1982 flood in the Eastern Pyrenees. Catena Suppl. 13, 79–95.

García Ruiz, J.M., 1989. Erosion processes in abandoned fields. A case study in Central Spanish Pyrenees. Geooko, Plus 2. Darmstadt.

Grimalt Gelabert, M., Blázquez Salom, M., Rodríguez Gomila, R., 1992. Physical factors, distribution and present land-use of terraces in the Tramuntana mountain range. Pirineos 139, 15–25.

Lehmann, R., 1993. Terrace degradation and soil erosion on Naxos island, Greece. In: Wicherek, S. (Ed.), Farm Land Erosion in Temperate Plains Environment and Hills. Elsevier, Amsterdam, pp. 429–450.

Lehmann, R., 1994. Landschaftsdegradierung Bodenerosion und -Konservierung auf der Kykladeninsel Naxox, Griechenland. Physiogeographica, Basler Beiträge zur Physiogeographie Band 21.

López Bermúdez, F., 1990. Soil erosion by water on the desertification of a semi-arid Mediterranean fluvial system: the Segura basin, Spain. Agr. Ecosyst. Environ. 33, 129–145.

Naveh, Z., Dan, J., 1973. The human degradation of Mediterranean landscapes in Israel. In: Di Castri, F., Mooney, H. (Eds). Mediterranean Type Ecosystems. Springer, New York, pp. 373–390.

Nixon, I.K., 1988. Introduction to Papers 10–13, Proceedings of the Geotechnology Conference Penetration Testing in the UK. Institution of Civil Engineers, Birmingham, UK, pp. 105–111.

Pallares-Bou, J., 1994. Procesos que conducen a la rotura de muros en terrazas de cultivo (Norte Castellón). Cuaternario y Geomorfología 8, 23–36.

Rodríguez Aizpolea, J., Lasanta Martínez, T., 1992. Los bancales en la agricultura de la montaña Mediterránea: una revisión bibliográfica. Pirineos 139, 118–123.

Ron, Z., 1966. Agricultural terraces in the Judean mountains. Israel Explor. J. 16, 33–49.

Salas-Pinto, D.F., Vásques-Villanueva, A., 1987. Andenes: Parámetros, Operación y Mantenimiento. Universidad Nacional Agraria. La Molina, Lima-Peru.

Sandor, J.A., 1998. Steps toward soil care: ancient agricultural terraces and soils, Proceedings of the 16th World Congress of Soil Science, Montpellier, France.

Selby, M.J., 1993. Hillslope Materials and Processes. Oxford University Press, Oxford.

Tonks, D.M., Whyte, I.L., 1988. Dynamic soundings in site investigations: some observations and correlations, Proceedings of the Geotechnology Conference Penetration Testing in the UK. Institution of Civil Engineers, Birmingham, UK, pp. 113–117.

Zgaier, A., 2000. Erosional processes on abandoned agricultural hillslope terraces under Mediterranean climatic conditions, Yirka, Western Galilee. Ph.D. Thesis, University of Haifa, Israel.

Zimbone, S.M., Vickers, A., Morgan, R.P.C., Vella, P., 1996. Field investigations of different techniques for measuring surface soil shear strength. Soil Technol., 101–111.

Catchment Dynamics and River Processes: Mediterranean and Other Climate Regions
Celso Garcia, Ramon J. Batalla, Editors

Chapter 6

The long-term effects on soil properties from a forest fire of varying intensity in a Mediterranean environment

Xavier Úbeda[1,*], Sara Bernia[1] and Elisabeth Simelton[2]

[1]*Department of Physical Geography, University of Barcelona, 08028 Barcelona, Spain*
[2]*Earth Sciences Centre, Physical Geography, University of Göteborg, PO Box 460, Göteborg, 405 30 Sweden*

Abstract

Forest fires are frequent in the Mediterranean area and the intensity of these fires determines their impact on soil properties, runoff and erosion rates. Furthermore, Mediterranean plant species are affected differently by forest-fire intensity. Different soil and vegetation scenarios, therefore, occur in an area following a forest fire depending on the intensity of the fire and, consequently, soil recuperation times vary. The aim of this chapter is to test the values of a number of soil parameters in an area subject to different fire intensities in 1994. The sampling was carried out in 2001, 7 years after the fire. Comparison of a low intensity burnt area (LIB) and a high intensity burnt area (HIB) with a control forest (CF) shows that soil parameters have yet to recover the control values. Recovery has been slowest in the HIB area. For example, in 2001, soil pH in the CF was 5.94 and 6.37 in the LIB area and 6.49 in the HIB area. The soil organic carbon was 6.01% in the CF, 3.29% in the LIB area and 1.67% in the HIB area. The total soil nitrogen was 0.44% in the CF, 0.23% in the LIB area and 0.15% in the HIB area. The concentration of cations (Ca^{2+}, Mg^{2+}, Na^{+} and K^{+}) in the soil was higher in the CF, with the exception of potassium where values in the LIB area and CF were similar in both plots. The best correlation between bases was found in CF. Interestingly, no correlation was found between sodium concentrations and the other cations, while calcium concentrations presented a weaker correlation with soil organic carbon and nitrogen in the HIB area than they did in the LIB area, where, in fact, these parameters showed a strong correlation. This analysis has demonstrated that after 7 years there are still significant differences between the three areas of different burnt intensity in all the properties except sodium.

Keywords: Mediterranean environment, forest fires, fire intensity, soil properties, long-term effects

1. Introduction

Each year some 50,000 fires sweep through 700,000–1,000,000 ha of Mediterranean forest at a considerable ecological and socio-economic cost (Vélez, 1997). The area affected by fire tripled between 1960 and 1985, growing from 200,000 to 600,000 ha, while the number of fires increased from 1300 fires/year to 9000 (Conacher and Sala, 1998). In Catalonia, for

*Corresponding author. *E-mail address:* xubeda@ub.edu (X. Úbeda).

example, between 1973 and 1998 wildfires affected an area of land equivalent to 14% of the region (Panareda and Arola, 1999). One of the 1217 fires in Catalonia during 1994 was the Llagostera fire (see this study), which burned a total area of 76,625 ha, of which 55 ha was forest.

Various studies have been undertaken on the short-term effects of heating on soil chemical quality (Raison et al., 1993; Giovannini, 1997), on soil hydrology and physical erosion (Sala et al., 1993; Cerdà, 1998; Úbeda, 1999), on water repellence (Singer and Le Bissonais, 1998; Rochinbaud and Hungerford, 2000), and on the recovery of vegetation (Cerdà, 1998; Turner et al., 1999) and fauna.

The ways in which fires modify soil nutrients are complex. The main processes have been stated as: (1) transformation of nutrients from organic to inorganic forms, (2) volatilization, (3) erosion of ash and nutrient-rich surface soil, (4) alteration of nitrogen-fixing systems and (5) modification of decomposition rates of litter and soil organic matter (Raison et al., 1993; Panareda and Arola, 1999).

The interrelationships between these processes and the duration of these modifications are not easily monitored in terms of causes and effects. Indeed, the duration of the impact of fire on soils can vary from minutes to several years (Raison et al., 1993; Driscoll et al., 1999). In this context, Luis-Calabuig and Tarrega (1993) claim that the correlation between time and soil variables is always negative, but that within 2 years few differences remain in these properties in areas affected by different fire intensities. Alcañiz et al. (1996) concluded that commonly the indirect effects of fire have consequences that are more serious for the soil than the damage caused by the fire.

Although soils return to their initial temperature within a few hours of a fire dying out (Dimitrakopoulos and Martin, 1994), the intensity of that fire will influence the post-fire processes (Giovannini, 1997). For example, a higher rate of erosion is recorded due to the presence of fire-induced hydrophobic layers and the absence of vegetation. After a year, soil erosion (kg/ha) in an area affected by severe fire has been found to be ten times greater than that in an area affected by light burning, and 50 times greater than that in the control plot (Giovannini and Luchessi, 1993). After 2.5 years, erosion rates were still found to be fluctuating at a level slightly higher than those of the control forest (Úbeda, 1998). Furthermore, different combustibility rates have been reported as influencing soil chemical composition at least 10 years after a fire (Carballas et al., 1993).

Soil pH may fluctuate for 5 and 50 years as the crude ash remains incompletely incorporated within the soil (Giovannini, 1994), cations are being leached, or new humus is being formed (Carballas, 1997). Carballas et al. (1993) report that carbon and inorganic nitrogen concentration in burnt and unburnt forests in NW Spain recovered to similar pre-fire levels in 2 and 5 years respectively after a fire. They found that soil enzymatic activities began to increase slowly after 2 years, while after 5 years there was a considerable increase and after 10 years the levels were similar to those of the undisturbed forest. However, the study did not examine fire intensity. Soil N availability would seem to fall as the result of a fire (Alcañiz et al., 1996). In Canada, the cover and the type of post-fire succession has been found to influence soil N availability for at least 140 years (Driscoll et al., 1999).

Studies of forest fires and post-fire soil processes have typically examined short-term effects. Few studies of the effects of wildfires on topsoils have a duration of more than

3 years, even though the effects on soil and vegetation may be noted for between 10 and 50 years (Giovannini, 1994), and still be detected after 140 years (Driscoll et al., 1999).

The aim of this study is to extend up previous research conducted in an area affected by a wildfire to determine the effects of two different fire intensities on the soil (Úbeda, 1998). The changes studied in 1994 after the wildfire showed differences between a control plot and low and high fire intensities. At that moment, the soil properties changed because the fire intensity, the ash addition and volatilization of some elements. After some years the soil may also be different in every plot due to erosion processes. This study examines several common soil properties to determine the effect of post-fire processes 7 years after the 1994 wildfire in Llagostera (Catalonia).

2. Study area

The results reported here refer to soils in the Cadiretes massif, in the northernmost part of the Catalan Coastal Ranges, north east Spain (Fig. 6.1), at an altitude of between 190 and 250 masl.

The fire started on July 5, 1994 and burnt 55 ha. The burnt area lies roughly on a facies limit (N–S direction) with a coarse-grain size to the east and a medium-grain size to the west. Further west, towards Llagostera, lies a region of granodiorite, biotitic granites and Ordovician metamorphosed quartz-feldspar schist, slates and sandstones. The NW–SE faults are from the Hercynian orogenesis and were reactivated during Alpine folding. The Llagostera basin is primarily of Pliocene eroded granitic sand and clay, and quaternary alluvial sand and clay.

Erosion processes have largely shaped the granite area of the massif, although in some areas metamorphic features underlie this relief erosion. The massif is covered by dense

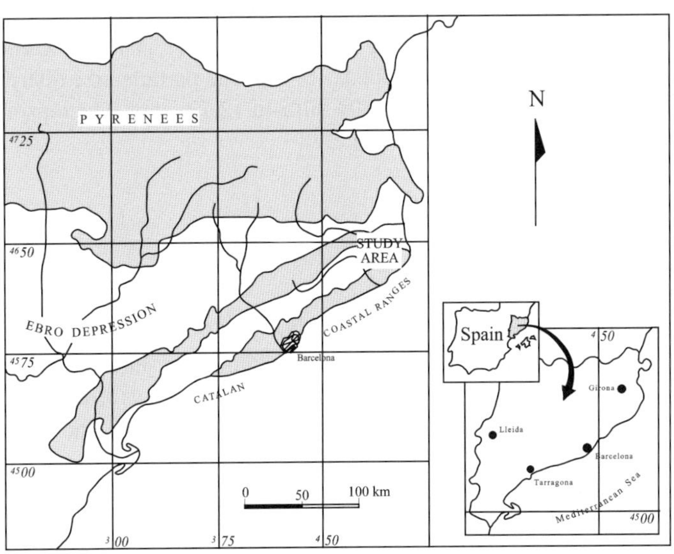

Figure 6.1. Location of the study area.

Table 6.1. Topsoil description.

Soil horizon	Thickness	Control forest (CF)	Thickness	High intensity burnt (HIB)	Thickness	Low intensity burnt (LIB)
O	0–2 cm	Pine needles on top and fresh humus. Sandy loamy sand	None	None	0–2 cm	Half-burnt pine needles on top and fresh humus. Sandy loamy
A	2–7 cm	Sandy loamy with few rock fragments Structure weak medium angular blocky; non-sticky, non-platic; common fine pores, common medium roots	0–3 cm	Sandy loam with few rock fragments; weak fine granular structure; slightly sticky, slightly plastic; few fine pores; few fine roots	2–6 cm	Sandy loam with few rock fragments; weak very fine Sub-angular blocky; non-sticky, non-plastic; common fine pores; common fine roots

Mediterranean vegetation, e.g. *Quercus suber*, *Arbutus unedo*, *Erica arborea*, and in some places, *Pinus* plantations occur. Mean annual rainfall ranges between 700 and 800 mm, with a maximum in autumn (37% of the total), followed by winter (26%), spring (26%) and summer (13%). Autumn is the season showing the highest rainfall intensity with values frequently in excess of 15 mm/h. Summer temperatures often exceed 25°C; whereas winter temperatures are generally mild and almost never fall below 0°C. Evapotranspiration exceeds precipitation from June to August.

The soils in the control forest and the low intensity burnt area (LIB) are *typic Xerochrepts*, whereas the high intensity burnt area (HIB) is *lithic Xerochrept* (Soil Taxonomy). Slope aspects are 9–10° (Úbeda, 1998). The particle-size distribution of the sandy loam soil texture is 74.8% sand, 12.4% silt and 12.7% clay. The topsoil description is as follows in Table 6.1 (Simelton, 2001).

3. Methodology

3.1. Delimitation of the burnt plots

Fire intensities were determined immediately in the field following the forest fire in 1994. Three areas were distinguished according to tree remains, the state and number of leaves and branches, soil surface and the quantity and colour of ashes: black-moderate, grey-medium and white-strong combustion (Moreno and Oechel, 1989). The characteristics of each unit were as follows:

Low intensity burnt area. The tallest trees retained some leaves (although not all were green, and many would eventually fall), and a large number of branches, including smaller ones. A considerable amount of litter was deposited just after the fire, and a 2-cm layer of burnt black humus was still present two years after the fire. All the

Quercus suber survived as well as some *Pinus*. This area corresponds to the lower part of the slope adjacent to a stream.

High intensity burnt area. The tallest trees lost all their leaves and branches and only the trunks remained. *Arbutus unedo* also disappeared completely. The soil surface was left totally uncovered, but after the fire and before the first rains, a large horizon of grey and white ashes was present.

3.2. Sampling

This follow-up study was carried out between May and June 2001 (7 years after the fire). The three days immediately preceding sampling were dry and sunny with day temperatures of 22–23°C. The total amount of precipitation in the 10-day period before sampling was approximately 30 mm.

A total of 90 samples (approx. 500 g) from the upper 3 cm of the topsoil were collected in approximately 100 m^2-sized central areas of the same control forest (CF), and the high and low intensities areas as delimited by Úbeda (1998). Thirty samples were taken from each area (HIB, LIB and control forest).

3.3. Laboratory methods

To enable comparison with the earlier study undertaken between 1994 and 1996 (Úbeda, 1998), the same analytical methods were employed here. The pH was calculated in a water solution of 1:2.5 (soil:water). Soil organic carbon (OC) was calculated as the percentage lost on ignition (Bascomb, 1974). The elemental analysis was performed using the NA 2100 protein Carlo Erba Instrument for Total Nitrogen. The Na^+, Ca^{2+}, Mg^{2+} and K^+ content was estimated using ammonium acetate extraction. The Na^+ and K^+ was analysed by atomic absorption AAS (accuracy: 0–20 ppm, 1/20 dissolution), whereas in the case of Ca^{2+} and Mg^{2+} atomic emission was used (0–20 ppm, 1/50 dissolution).

3.4. Data treatment

The data treatment was performed using the SPSS 11.0 and Statgraphics 5.0 programs. All the data analyses were designed so as to detect any relationship or pattern of conduct between the variables. The first analysis of the data involved calculation of their summary statistics: minimum, maximum, mean, standard deviation and variance. The results are shown in Table 6.2 and are presented in accordance with Webster's (2001) specifications. The second analysis examines the relationships between the variables according to a Pearson Correlation test (Table 6.3). Some of the relationships are described in scatter diagrams. Finally, a contrast analysis was performed on the data to determine whether any of the differences between the variable values recorded at the two fire affected sites and those recorded from the control plot were statistically significant. The application of this test requires that three conditions must first be met (Mateos-Aparicio and Martín, 2003): (1) Normality, which was verified by performing the Kolmogorov–Smirnov test; (2) Independence of the observations, which was verified by examining the scatter plots of the residuals compared with the values observed for each variable and parameter (observations are considered to be independent when no

Table 6.2. Summary statistics.

Fire intensity		N	Minimum	Maximum	Mean	Standard deviation
Control						
	pH	30	4.93	6.68	5.94	0.34
	OC (%)	30	3.94	8.37	6.01	1.00
	total N (%)	30	0.190	1.060	0.44	0.18
	C/N	30	6.7	37.5	15.9	6.9
	Ca (ppm)	30	2225.0	8883.0	4451.1	1482.6
	Mg (ppm)	30	423.00	1713.00	877.07	310.31
	Na (ppm)	29	71.00	1962.00	474.03	378.52
	K (ppm)	30	183.00	477.00	311.47	67.53
High (HIB)						
	pH	30	6.14	6.75	6.49	0.15
	OC (%)	30	0.53	7.50	1.67	1.31
	Total N (%)	30	0.070	0.520	0.15	0.09
	C/N	30	6.8	14.4	10.4	1.7
	Ca (ppm)	30	1604.0	4233.0	2618.6	669.5
	Mg (ppm)	30	346.00	1233.0	589.17	214.40
	Na (ppm)	30	187.00	1466.00	566.33	313.49
	K (ppm)	30	182.00	373.00	268.37	56.50
Low (LIB)						
	pH	30	5.56	7.14	6.37	0.30
	OC (%)	30	0.78	12.78	3.29	2.30
	Total N (%)	30	0.080	0.720	0.23	0.13
	C/N	30	9.6	18.6	13.9	2.6
	Ca (ppm)	30	1251.0	5397.0	2909.3	910.7
	Mg (ppm)	30	278.00	1227.00	691.10	229.30
	Na (ppm)	30	182.00	1834.00	620.73	445.30
	K (ppm)	30	124.00	601.00	339.57	106.45

underlying trends are observed in the scatter plots); and (3) Homocedasticity, this means that the variances of the population are homogeneous, which was checked by applying Levene's variance equality test.

After ensuring that these requirements were met, we performed the ANOVA analysis for the parametric variables (see Table 6.4) and the Kruskal–Wallis chi-squared test for the non-parametric variables (see Table 6.5). The hypotheses for the two contrast tests applied were:

- H_0: (null hypothesis, $P_{value} > 0.05$) the significance values in the comparison of the three different types of management (both fire intensities and control forest) for each parameter are not statistically significant and can be explained by chance.
- H_A: (alternative hypothesis, $P_{value} < 0.05$) there are statistically significant differences between the three different types of management for each parameter analysed and they cannot be explained by chance.

Table 6.3. Pearson's correlation indexes

Fire Intensity			OC (%)	Total N (%)	C/N	Ca (ppm)	Mg (ppm)	Na (ppm)	K (ppm)
Control	pH	*R*	0.012	0.014	−0.064	0.168	−0.008	−0.044	−0.002
		Significance	0.951	0.943	0.736	0.376	0.966	0.819	0.990
		n	30	30	30	30	30	29	30
	OC (%)	*R*		0.010	0.469**	0.039	0.018	0.121	0.126
		Significance		0.958	0.009	0.839	0.927	0.532	0.506
		n		30	30	30	30	29	30
	Total N (%)	*R*			−0.780**	0.815**	0.627**	0.337	0.151
		Significance			0.000	0.000	0.000	0.074	0.426
		n			30	30	30	29	30
	C/N	*R*				−0.670**	−0.496	−0.181	−0.143
		Significance				0.000	0.005	0.346	0.451
		n				30	30	29	30
	Ca (ppm)	*R*					0.696**	0.527**	0.306
		Significance					0.000	0.003	0.100
		n					30	29	30
	Mg (ppm)	*R*						0.572**	0.290
		Significance						0.001	0.120
		n						29	30
	Na (ppm)	*R*							0.520**
		Significance							0.004
		n							29
High (HIB)	pH	*R*	0.078	0.112	0.207	0.069	0.113	−0.023	0.161
		Significance	0.683	0.557	0.273	0.716	0.552	0.902	0.394
		N	30	30	30	30	30	30	30
	OC (%)	*R*		0.986**	0.691**	0.405**	0.078	0.080	0.293
		Significance		0.000	0.000	0.026	0.682	0.673	0.116
		n		30	30	30	30	30	30
	Total N (%)	*R*			0.612**	0.337	0.050	0.053	0.273
		Significance			0.000	0.068	0.792	0.783	0.144
		n			30	30	30	30	30
	C/N	*R*				0.527**	0.181	0.032	0.259
		Significance				0.003	0.339	0.868	0.167
		n				30	30	30	30

Continued

Table 6.3. continued

Fire Intensity			OC (%)	Total N (%)	C/N	Ca (ppm)	Mg (ppm)	Na (ppm)	K (ppm)
	Ca (ppm)	*R*					0.797**	0.123	0.297
		Significance					0.000	0.518	0.111
		n					30	30	30
	Mg (ppm)	*R*						0.219	0.255
		Significance						0.246	0.173
		n						30	30
	Na (ppm)	*R*							0.303
		Significance							0.103
		n							30
Low (LIB)	pH	*R*	−0.032	0.043	−0.283	0.195	−0.261	0.119	0.326
		Significance	0.865	0.822	0.130	0.301	0.164	0.529	0.079
		n	30	30	30	30	30	30	30
	OC (%)	*R*		0.976**	0.560**	0.684**	0.462*	−0.010	0.369*
		Significance		0.000	0.001	0.000	0.010	0.957	0.045
		n		30	30	30	30	30	30
	Total N (%)	*R*			0.412*	0.703**	0.450*	0.036	0.450*
		Significance			0.024	0.000	0.013	0.851	0.012
		n			30	30	30	30	30
	C/N	*R*				0.397*	0.449*	−0.115	0.148
		Significance				0.030	0.013	0.545	0.435
		n				30	30	30	30
	Ca (ppm)	*R*					0.743*	0.270	0.661**
		Significance					0.000	0.149	0.000
		n					30	30	30
	Mg (ppm)	*R*						0.131	0.556**
		Significance						0.489	0.001
		n						30	30
	Na (ppm)	*R*							0.348
		Significance							0.059
		n							30

*Significance correlation at level 0.05.

**Significance correlation at level 0.01.

Table 6.4. ANOVA results of the parametric variables (2001 data).

	Squared sum	df	Squared average	F	Significance
Mg (ppm)					
Inter-groups	1,278,604.156	2	639,302.078	9.844	0.000
Intra-groups	5,650,296.733	87	64,945.939		
Total	6,928,900.889	89			
Na (ppm)					
Inter-groups	323,571.939	2	161,785.970	1.103	0.336
Intra-groups	12612177.499	86	146,653.227		
Total	12935749.438	88			

Table 6.5. Kruskal–Wallis Chi-squared test for the non-parametric variables (2001 data).

	pH	OC (%)	Total (N) (%)	C/N	Ca (ppm)	K (ppm)
Chi-Squared	40.196	55.116	46.847	23.247	33.409	10.158
df	2	2	2	2	2	2
P value	0.000	0.000	0.000	0.000	0.000	0.006

4. Results

4.1. Vegetation recovery following the fire

The number and type of vegetation species that resprouted following the fire differed in each area. In the LIB area 16 species were identified, the same number as in the control plot. By contrast, in the HIB area only 10 species were identified. More significantly, however, was the fact that in the latter area 55% of the vegetation corresponded to the same two species: *Cistus monspeliensis* and *Cistus salviifolius*. In the LIB area, these two species accounted for just 9.4% of the vegetation, while in the CF they represented 2.1%. This result is significant in understanding the results described below.

4.2. Comparison of plots (7 years after the fire)

The soil pH (Table 6.2) was still higher in both burnt plots (LIB: 6.37 and HIB: 6.49) than in the control (5.94) after 7 years. Particularly noteworthy is that the range of results was much greater in the control plot than in the burnt plots (Fig. 6.2), with a standard deviation of pH of 0.34 compared with values of 0.30 (LIB) and 0.15 (HIB) in the burnt plots.

Most soil parameters (5 of 8) were highest in the control plot. The high intensity plot recorded lowest values in OC, Total N, C/N, Ca^{2+} and Mg^{2+} (Table 6.2). Potassium concentrations recorded very similar values at the three sites: 339 (LIB) and 268 ppm (HIB) in the burnt plots and 311 ppm in the control.

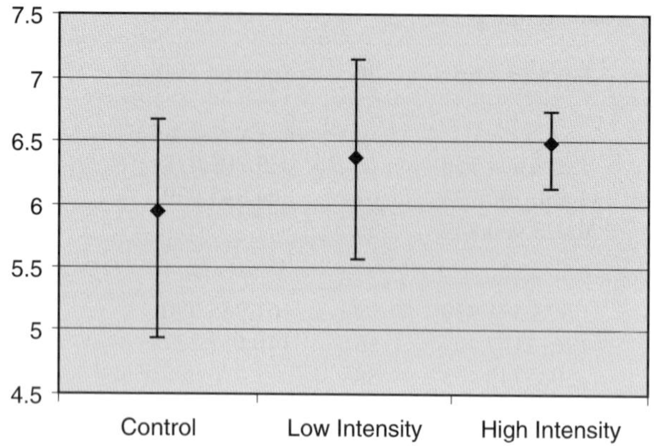

Figure 6.2. Mean, minimum and maximum pH values in each fire intensity (2001 data).

Of significance is the fact that the standard deviation was lowest for 7 of the 8 parameters in the HIB area. This was particularly true of the nitrogen concentrations and soil OC in the HIB area, where the values recorded in all 30 samples were extremely similar.

4.3. Correlations between soil parameters

The Pearson's correlation coefficients between parameters in each studied area are shown in Table 6.3. The best correlation between bases was found in CF. Interestingly, no correlation was found between sodium concentrations and the other cations in burnt areas, while calcium concentrations presented a weaker correlation with soil OC and nitrogen in the HIB area than they did in the LIB area, where, in fact, these parameters showed a strong correlation. The potassium is just well correlated with calcium ($r = 0.661^{**}$) and magnesium ($r = 0.556^{**}$) in the LIB area. The best correlation between OC vs N were found in LIB: $r = 0.976^{**}$, and HIB: $r = 0.986^{**}$. In contrast, in CF the correlation only reached $r = 0.010$. The pH is not correlated in any studied area with any parameter with a significant correlation at level <0.05.

4.4. Contrasts between fire intensities

The Kolmogorov–Smirnoff test showed that variables were normally distributed. The results testing the homogeneity of the variance are shown in Table 6.4. A parametric contrast test (ANOVA) could only be run on the magnesium and sodium concentrations as they presented a significance value greater than 0.05. Magnesium concentrations presented a significant difference between the three plots with a P value <0.01 while, in the case of sodium no significant difference was found P value >0.01 (Table 6.4).

The Kruskal–Wallis chi-squared test was run on the non-parametric variables. The remaining parameters presented significant differences between the plots because all the P values were lower than 0.01 (Table 6.5).

4.5. Comparison of 1994 and 2001 soil parameters in the burnt forest areas

A similar evolution was seen in the soil parameters at the two burnt plots in 1994 and 2001 (Table 6.6). The soil pH of the undisturbed plot is around 6, but whereas the pH has fallen over time in the other two plots it has not yet returned to its pre-fire level. The organic carbon and nitrogen levels have also fallen over time despite the fact that the first sampling showed values that were higher than those of the control. By contrast, the cation concentrations have increased and are now similar to those recorded in the control plot.

The non-parametric and parametric contrast statistics treatment (Tables 6.7 and 6.8) show how the Na^+ (P value $= 0.060$) is the only parameter which does not have significant differences between the values of 1994 and 2001, because the significance is higher than 0.05. All the other studied soil properties have changed significantly between 1994 and 2001.

Table 6.6. Comparison between the values in 1994 and 2001 in the low and high intensity and in the control forest.

	Low intensity		Control forest	High intensity	
	1994 (Úbeda, 1998)	2001	(2001)	1994 (Úbeda, 1998)	2001
PH	6.72	6.37	5.94	6.98	6.49
OC (%)	13.6	3.29	6.01	9.8	1.67
N (%)	0.64	0.23	0.44	0.44	0.15
Ca (ppm)	1074	2909	4451	1369	2619
Mg (ppm)	213	691	877	281	589
Na (ppm)	380	621	474	330	566
K (ppm)	308	340	311	277	268

Table 6.7. ANOVA results of the parametric variables (1994–2001 comparison).

K (ppm)	Squared sum	df	Squared average	F	Significance
Inter-groups	84,915.798	4	21,228.949	3.479	0.010
Inter-groups	640,675.465	105	6,101.671		
Total	725,591.263	109			

Table 6.8. Kruskal–Wallis Chi-squared test for the non-parametric variables (1994–2001 comparison).

	pH	OC (%)	Total (N) (%)	C/N	Ca (ppm)	Mg (ppm)	Na (ppm)
Chi-squared	75.460	83.931	68.423	55.495	69.453	59.952	9.041
df	4	4	4	4	4	4	4
P value	0.000	0.000	0.000	0.000	0.000	0.000	0.060

5. Discussion

An increase in pH after forest fires has been reported in both field (Giovannini, 1994) and laboratory experiments (Fernández et al., 1997). These increases have been attributed to the increased availability of cations and the consumption of organic acids during the oxidation of litter and soil organic matter (Fisher and Binkley, 2000).

This variation in pH can affect vegetal nutrition (Guerrero, 2003). Here, after 7 years the pH values in both burnt plots were still higher than in the CF while the vegetation was markedly different in the plot suffering high intensity burning. Similar findings have been reported by Trabaud (1992) and Alonso et al. (1996). The recovery of vegetation types also varied so that while in the LIB area 16 species were recorded, only 10 appeared in the HIB area. This would seem to be a further consequence of fire intensity.

A number of studies have been conducted to determine how long it takes for soil pH to return to previous levels. Mataix-Solera (1999) reported a 4-month lag and concluded that this was determined by the effects of the wind that eroded the ashes during the first few months following the fire. Antos et al. (2003) reported an increase in pH of 0.4, one year after a fire and still reported the same levels 3 years later. A number of studies have reported changes in pH that have persisted for 50 years (Khanna and Raison, 1986; Etiégni and Campbell, 1991). To understand this varying behaviour, many factors need to be taken into consideration, including climate, soil type, ash accumulation, erosional processes and, above all, vegetation regrowth. Antos et al. (2003) reported that soil pH would remain high in their particular area of study until the vegetation – conifers – regained dominance at the site.

Levels of total soil N and organic C have been reported as increasing, decreasing and remaining unchanged after a fire (Choromanska and DeLuca, 2001; Wan et al., 2001). However, any interpretation of soil C and N must be tempered by a consideration of the size of the soil fraction and layer being examined. The 2–6 mm fraction has been reported as containing a considerable proportion of soil N (Cromack et al., 1999), while the effects of a disturbance can be manifested at a greater depth than that sampled here (Little and Klock, 1985). According to Corti et al. (1998), a more complete understanding of soil nutrient dynamics requires the examination of coarser fractions and deeper soil layers.

Here, total soil N and organic C presented lower values after 7 years, and this was most marked in the HIB area. This seems to be a clear indication of the effects of fire intensity on these soil parameters. Some studies (e.g. Marcos et al., 1999) have found that 2 years after the fire, soil N and C levels were the same as those of the control. Fluctuations in total N availability seem to be linked to the growth of perennial herbaceous vegetation (Lillis, 1993; Marcos et al., 1999). The rapid drop in this parameter might be attributed to leaching, microbial immobilization, and plant uptake (Antos et al., 2003). In the latter study, a significant negative correlation was noted between N and total plant biomass. Furthermore, this relationship became stronger over time, suggesting that plant uptake may become an increasingly important factor in controlling the concentration of mineral N in forests. The rapid decline in N availability may be responsible for the decline, or loss, of many weed-like species that characterize the early stages of post-fire succession in forest systems (Dyrness, 1973; Schoonmaker and McKee, 1988).

Decreasing standard deviation values in the soils of burnt forests have been reported elsewhere (Antos, et al., 2003), but the latter claim that the scale of the plot needs to be taken into consideration. On occasions, the standard deviation after fire has been found to be similar to the standard deviation in agricultural soils (McNabb et al., 1986). Beckett and Webster (1971) concluded that although variability increases with the area sampled, up to half of the variance within an agricultural field may be present within any square metre. This would also seem to be the case in a burnt forest soil.

Antos et al. (2003) concluded that additional research is needed to refine our understanding of the spatial and temporal scales over which plants respond to, and in turn mediate, soil resource availability after large-scale disturbances. The duration of the indirect effects attributable to the recovery of the vegetation and ash levels depend principally on the quantity of ash and the effects of the erosion processes that occur after the fire (Mataix-Solera, 1999). Here, we saw that the recovery rates of all the soil parameters in the HIB area were not as great as those recorded in the LIB area. Similar findings have been also reported (Andreu et al., 1994). The impact of the fire on the soil and the latter's unprotected state explain why recovery after a high intensity fire is slower.

It is not unsurprising that after 7 years the soil conditions should differ in each of the three plots because runoff production and the erosion generated also differed. In the first year, runoff production was recorded as $14.2\ l\ m^{-2}$ in CF, $1.37\ l\ m^{-2}$ in the LIB area and $24.8\ l\ m^{-2}$ in the HIB area. These rates of runoff produced erosion levels of $0.16\ kg\ m^{-2}$ in CF, $0.1\ kg\ m^{-2}$ in the LIB area and $3.1\ kg\ m^{-2}$ in the HIB area (Úbeda and Sala, 1998).

The characteristics of the topsoil, in comparison with a control plot, are not only different in term of changes due to the fire and erosion processes; moreover, the topsoil sampled in the burnt areas may correspond to a deeper soil in the control plot. Lavé and Burbank (2004) reported erosion rates between 0.9 and 1.6 mm/year in areas greatly enhanced by recurrent fires in southern California. These erosion processes may explain that the correlation results are different in each burnt area. The first top-centimetres of the soil in the burnt areas have disappeared as a consequence of these processes.

The ion concentration in the runoff varied for each plot (Úbeda and Sala, 2001). In the first year after the fire, the ionic denudation concentration levels were $120.5\ mg\ l^{-1}$ in the CF, $566.2\ mg\ l^{-1}$ in the LIB area and $59.3\ mg\ l^{-1}$ in the HIB area. The total ionic denudation ($mg\ m^{-2}$) was higher in the HIB area as the total runoff was also higher ($24.8\ l\ m^{-2}$ compared with $1.37\ l\ m^{-2}$ in the LIB area).

6. Conclusions

Seven years after the fire, the soil parameters in the burnt areas of the forest have yet to return to their pre-fire levels. The concentrations of cations in the soil are now roughly similar to their undisturbed values, but C and N percentages still differ considerably. Similarly, soil pH has not recovered its pre-fire levels.

In conclusion, there appears to be a marked sequence to the effects of fire intensity on soil parameters. Immediately after the fire the intensity has an impact on soil characteristics. These changes affect vegetation regrowth and distribution, which in turn affects the soils in terms of protection, humus formation and nutrient recuperation. This sequence is crucial in understanding the long-term effects on soil parameters following a forest fire. The largely similar results recorded here in the area affected by high intensity fire would seem to be attributable in part to the influence of the more homogeneous vegetation type which undergoes regrowth after the fire.

Acknowledgements

This work was possible thanks to an exchange between the University of Göteborg and the University of Barcelona. We would like to thank John A. Moody for reviewing the first manuscript. This work is part of the "RESEL" project from the Spanish Ministry of Environment.

References

Alcañiz, J.M., Serrasolsas, I., Vallejo, R., 1996. Efectes dels incendis forestals sobre el sòl. In: Terradas, J. (Ed.), Ecologia del Foc. Proa, Barcelona, pp. 111–130.

Alonso, M., Vega, J.A., Bará, S., 1996. Biochemical parameters to indicate fire damages in *Pinus pinaster* stand, Fire Ecology and the European Biota, Toledo, 7.

Andreu, V., Rubio, J.L., Forteza, J., Cerni, R., 1994. In: Sala, M., Rubio, J.L. (Eds), Long term effects of forest fires on soil erosion and nutrient losses, Soil Erosion as a Consequence of Forest Fires. Geoforma ediciones, Logroño, pp. 79–89.

Antos, J.A., Halpern, C.B., Miller, R.E., Cromack, K., Halaj, M.G., 2003. Temporal and spatial changes in soil carbon and nitrogen after clearcutting and burning of an old-growth Douglas-Fir forest, USDA, Research Paper, 555, Pacific Northwest Research Station.

Bascomb, C.L., 1974. Physical and chemical analysis of <2 mm samples. In: Avery, B.W., Bascomb, C.L. (Eds), Soil Survey Laboratory Method, 6. Harpenden, pp. 14–15.

Beckett, C.M., Webster, R., 1971. Soil variability: a review. Soils Fertil. 34, 1–15.

Carballas, M., 1997. Effects of fire on soil quality, biochemical aspects, Proceedings of the European School of Climatology and Natural Hazards Course. European Commission, Luxembourg, pp. 249–261.

Carballas, M., Acea, M.J., Cabaneiro, A., 1993. Organic matter, nitrogen, phosphorus and microbial population evolution in forest humiferous acid soils after wildfires. In: Trabaud, L., Prodon, R. (Eds), Fire in Mediterranean Ecosystems. Commission of European Communities, Banyuls-Sur-Mer, pp. 379–385.

Cerdà, A., 1998. Post-fire dynamics of erosional processes under Mediterranean climatic conditions. Zeistchrift für Geomorphologie 42 (3), 373–398.

Conacher, A.J., Sala, M., 1998. Land Degradation in Mediterranean Environments of the World. Nature and Extent, Causes and Solutions. Wiley, England.

Choromanska, U., DeLuca, T.H., 2001. Prescribed fire alters the impact of wildfire on soil biochemical properties in a ponderosa pine forest. Soil Sci Soc Am J 65, 232–238.

Corti, G., Ugolini, F.C., Agnelli, A., 1998. Classing the soil skeleton (greater than two millimeters): proposed approach and procedure. Soil Sci Soc Am J 62, 1620–1629.

Cromack, K., Miller, R.E., Hellgerson, O.T., 1999. Soil carbon and nutrient in a coastal Oregon Douglas-fir plantation with red alder. Soil Sci. Soc. Am. J. 63, 232–239.

Dimitrakopoulos, A.P., Martin, R.E., 1994. Effect of moisture content on soil heating during simulated wildland fire conditions. In: Sala, M., Rubio, J.L. (Eds), Soil erosion as a consequence of forest fires. Geoforma ediciones, Logroño, pp. 207–216.

Driscoll, K.G., Arocena, J.M., Massicotte, H.B., 1999. Post-fire soil nitrogen content and vegetation composition in Sub-Boreal Spruce forests of British Columbia's central interior, Canada. Forest Ecol. Manag. 121 (3), 227–237.

Dyrness, C.T., 1973. Early stages of plant succession following logging and burning in the western Cascades of Oregon. Ecology 54, 57–68.

Etiégni, L., Campbell, A.G., 1991. Physical and chemical characteristics of wood ash. Bioresour. Technol. 37, 173–178.

Fernandez, I., Cabaneiro, A., Carballas, T., 1997. Organic matter changes immediately after a wildfire in an Atlantic forest soil and comparison with laboratory soil heating. Soil Biology and Biochemistry, 29, 1–11.

Fischer, R.F., Binkley, D., 2000. Ecology and Management of Forest Soils, 3rd ed. John Wiley, New York.

Giovannini, G., 1994. Effect of fire on soil quality. In: Sala, M., Rubio, J.L. (Eds), Soil Erosion as a Consequence of Forest Fires. Geoforma ediciones, Logroño, pp. 15–27.

Giovannini, G., 1997. The effect of fire on soil quality-physical and chemical aspects, Proceedings of the European School of Climatology and Natural Hazards Course. European Commission, Luxembourg, pp. 217–248.

Giovannini, G., Luchessi, S., 1993. Effects of fire on soil physico-chemical characteristics and erosion dynamics. In: Trabaud, L., Prodon, R. (Eds), Fire in Mediterranean Ecosystems. Commission of European Communities, Banyuls-Sur-Mer, pp. 403–412.

Guerrero, C., 2003. Uso de diferentes residuos orgánicos en la restauración de suelos forestales quemados. Ph.D. Thesis, Universidad Miguel Hernández, Spain.

Khanna, P.K., Raison, R.J., 1986. Effect of fire intensity on solution chemistry of surface soil under a Eucalyptus pauciflora forest. Aust. J. Soil Res. 24, 423–434.

Lavé, J., Burbank, D., 2004. Denudation processes and rates in the Transeverse Ranges, southern California: Erosional response of a transitional landscape to external and anthropogenic forcing. J. Geophys. Res. 109, F01006, doi: 10.1029/2003JF000023.

Lillis, M., 1993. Use of water and nutrient by mediterranean resprouters and reseeders in post-fire succession. In: Trabaud, L., Prodon, R. (Eds), Fire in Mediterranean Ecosystems. Commission of European Communities, Banyuls-Sur-Mer, pp. 365–377.

Little, S.N., Klock, G.O., 1985. The influence of residue removal and prescribed fire on distributions of forest nutrients, Research Paper PNW-338, Portland, OR: USDA, Forest Service, Pacific Northwest Research Station.

Luis-Calabuig, E., Tárrega, R., 1993. Studies of post-fire regeneration in *Quercus pirenaica* ecosystems in Leon Province (NW Spain). In: Trabaud, L., Prodon, R. (Eds), Fire in Mediterranean Ecosystems. Commission of European Communities, Banyuls-Sur-Mer, pp. 69–86.

Marcos, E., Tárrega, R., Luis-Calabuig, E., 1999. Alteraciones producidas por un incendio forestal en el suelo de una repoblación de *Pinus radiata*. Edafología 6, 27–35.

Mataix-Solera, J. 1999. Alteraciones físicas, químicas y biológicas en suelos afectados por incendios forestales. Contribución a su conservación y regeneración. Ph.D. Thesis, Universidad de Alicante, Spain.

Mateos-Aparicio, G., Martín, M., 2003. Análisis de la varianza y la covarianza. In: Lévy Mangin, J.P., Varela Mallou, J. (Eds), Análisis Multivariante para las Ciencias Sociales. Prentice Hall, Madrid, pp. 147–213.

McNabb, D.H., Cromack, K., Fredriksen, S.L., 1986. Variability of nitrogen and carbon in surface soils of six forest types in the Oregon Cascades. Soil Sci. Soc. Am. J. 50, 1037–1041.

Moreno, J.M., Oechel, W.C., 1989. A single method for estimating fire intensity after a burn in California chaparral. Acta Oecologica 10/1, 57–68.

Panareda, J.M., Arola, J., 1999. Els incendis forestals. Eumo Editorial, Vic, Barcelona.

Raison, R.J., O'Connell, A.M., Khanna, P.K., Keith, H., 1993. Effects of repeated fires on nitrogen and phosphorus budgets and cycling processes in forest ecosystems. In: Trabaud, L., Prodon, R. (Eds), Fire in Mediterranean Ecosystems. Commission of European Communities, Banyuls-Sur-Mer, pp. 347–363.

Rochinbaud, P.R., Hungerford, R.D., 2000. Water repellency by laboratory burning of four northern Rocky Mountain forest soils. J. Hydrol. 231–232, 207–219.

Sala, M., Soler, M., Pradas, M., Schrader, K., Lussi, A., 1993. Post-fire soil and vegetation dynamics in natural and afforested areas in Southern Europe: The role of fire intensity, Annual Report, University of Barcelona. Barcelona.

Schoonmaker, P., McKee, A., 1988. Species composition and diversity during secondary succession of coniferous forests in the western Cascade Mountains of Oregon. Forest Sci. 34, 960–979.

Simelton, E. 2001. Texture and nutrient status in the topsoil seven years after low, and high intensity wildfires, NE Catalonia, Spain. M.Sc. Thesis, Göteborg University, Sweden.

Singer, M.J., Le Bissonais, Y., 1998. Importance of surface sealing in the erosion of some soils from a Mediterranean climate. Geomorphology 24, 79–85.

Trabaud, L., 1992. From the cell to the atmosphere: an introduction to interactions between fire and vegetation, Fire in Mediterranean Ecosystems. Ecosystems Research Report 5, 13–21.

Turner, M.G., Romme, W.H., Gardner, R.H., 1999. Prefire heterogenity, fire severity and early postfire plant reestablishment in subalpine forests of Yellowstone National Park, Wyoming. Int. J. Wildland Fire 9, 21–36.

Úbeda, X. 1998. Efectes de les diferents intensitats de foc, durant els incendis forestals, en els paràmetres físico-químics del sòl i en l'increment de l'escolament i l'erosió. Ph.D. Thesis, Universitat de Barcelona, Spain.

Úbeda, X., 1999. Canvis en la capacitat d'infiltració d'un sòl després d'un incendi forestal al massís de Cadiretes. Importància de la intensitat del foc en les taxes d'infiltració. Scientia Gerundensis 24, 63–78.

Úbeda, X., Sala, M., 1998. Variations in runoff and erosion in three areas with different fire intensities. Geoödynamik XIX, 179–188.

Úbeda, X., Sala, M., 2001. Chemical concentrations in overland flow from different forested areas in a Mediterranean Environment: burned forest at different fire intensity and unpaved road. Zeistchrift für Geomorphologie 45 (2), 225–238.

Vélez, R., 1997. Recent history of forest fires in Mediterranean area, Proceedings of the European School of Climatology and Natural Hazards Course. European Commission, Luxembourg, pp. 15–26.

Wan, S., Hui, D., Luo, Y., 2001. Fire effects on nitrogen pools and dynamics in terrestrial ecosystems: a meta-analysis. Ecol. Appl. 11, 1349–1365.

Webster, R., 2001. Statistics to support soil research and their presentation. Eur. J. Soil Sci. 52, 331–340.

Catchment Dynamics and River Processes: Mediterranean and Other Climate Regions
Celso Garcia, Ramon J. Batalla, Editors

Chapter 7

Landscape disturbance and organic carbon in alluvium bordering steepland rivers, East Coast Continental Margin, New Zealand

Basil Gomez[1,*] and Noel A. Trustrum[2]

[1]*Geomorphology Laboratory, Indiana State University, Terre Haute, IN 47809 USA*
[2]*Institute of Geological and Nuclear Sciences, PO Box 30 368, Lower Hutt, New Zealand*

Abstract

In the past ~150 years, forest clearances have accelerated upland erosion and carbon losses from terrestrial ecosystems. High rates of overbank deposition on the Waipaoa River flood plain make it possible to distinguish the contribution of individual events to stratigraphy, and the fine-grained alluvium on McPhail's bend retains the impact of deforestation. When hillslope erosion intensified in the late 1920s following deforestation of the headwaters (1880–1920), the amount of organic carbon associated with the overbank sediments increased by ~50%. However, the amount of carbon involved is low (<1–2% by weight), and transmission losses to flood plains bordering the lower reaches of all three major rivers draining the East Coast Continental Margin of New Zealand are small (between 3 and 11% of each river's annual organic carbon flux to the Pacific ocean). From the perspective of organic carbon transferred by geomorphological processes and associated with fluvial sediments, the low storage potential of these flood plains accrues both from their small size (which amounts to <2% of total basin area) and the magnitude–frequency characteristics of event sediment yields.

Keywords: flood plains, New Zealand, organic carbon sequestration, steepland rivers

1. Introduction

Only a small proportion of the sediment mobilized on hillslopes is delivered to basin outlets, and fluvial sediment fluxes, which are modulated by transfers into and out of storage, may be strongly influenced by exchanges between rivers and their flood plains (Walling et al., 1996). Soil and regolith removed from hillslopes are among the primary sources of sediment stored in flood plains, and flood plain alluvium is an easily accessible, often high-resolution, sediment archive that contains chronologically ordered evidence of changing conditions in the catchment environment (Collins et al., 1997; Alto et al., 2003). During the last ~150 years, the predominant influence on flood plain sedimentation in many parts of

---

*Corresponding author. *E-mail address:* bgomez@indstate.edu (B. Gomez).

the world has been a succession of human impacts, beginning with the clearance of indigenous forests (Walling, 1996). Deforestation not only increases upland erosion and sediment deposition downstream, it also affects the cycling of carbon and other soil-derived biogenic elements (Lal, 1995; Ver et al., 1999). Since 1850, there has been a net release of carbon from terrestrial ecosystems, a large proportion of which occurred early in the past century, as a result of the conversion of native forest to agricultural land in areas outside the tropics (Houghton, 1995).

Some of the carbon released by forest clearances may be sequestered on flood plains as alluvial sediments accumulate (Stallard, 1998). The impact of minor sinks, such as flood plains, on the behavior of the biogeochemical cycles of elements such as carbon is not nearly as significant as that of the large reservoirs of the ocean and terrestrial biosphere. Nevertheless, some knowledge of background values and the degree to which they have been modified by human interference is required before a quantitative assessment can be made of the potential of any reservoir to act as a sink for an anthropogenically enhanced flux. However, it remains that few data are available on carbon storage in flood plain alluvium (Morozova and Smith, 2003). Thus, it has proved difficult to assess the impact of anthropogenic activity on the river carbon budget, and estimates of the effect of flood plain storage on the export of carbon typically are based on presumed pre-agricultural conditions (Stallard, 1998). Efforts to elucidate the mechanisms that influence the transport of organic carbon along the aquatic continuum and on flood plains also have focused on large, low-gradient rivers (Junk, 1985; Ittekkot et al., 1986; Richey et al., 1990; Depetris and Kempe, 1993; Cole and Caraco, 2001). This belies the importance of turbid steepland rivers (Milliman and Syvitski, 1992; Leithold and Blair, 2001; Gomez et al., 2003). Many rivers in the latter class have small ($<10 \times 10^4$ km^2) drainage basins that were deforested in the 19th and 20th centuries. Individually, small steepland rivers are not as important to biogeochemical cycling as their large, low-gradient counterparts (cf. Smith and Alsdorf, 1998), but collectively they export a significant amount of organic carbon to the Earth's oceans (Kao and Liu, 1996; Lyons et al., 2002).

In this chapter, we document the impact of anthropogenic activity on the transfer and deposition of organic carbon associated with the alluvial sediments of the Waipaoa River flood plain during the last 150 years, during which time the indigenous forest was almost eradicated by European colonists. We also derive comparable first-order estimates of the annual loss of organic carbon to flood plain storage for the other two major rivers (the Waiapu and Uawa) draining the East Coast Continental Margin (ECCM), North Island, New Zealand. Collectively, these three rivers generate $\sim$55 Mt of suspended sediment annually (Hicks and Shankar, 2003). This amounts to $\sim$0.3% of the total global suspended sediment input to the ocean (from $\sim$0.0033%, or 4942 km^2, of the Earth's total land area). Erosion-related carbon fluxes from the pastoral steeplands on the ECCM are equally significant (Page et al., 2004). The data presented here are of importance, not only because they furnish a historical perspective on the sequestration of organic carbon associated with flood plain sediments in the lower reaches of three prototypical small, steepland rivers with extremely high suspended sediment loads, but also because they provide a baseline against which the magnitude of the impact associated with deforestation may be evaluated.

2. Study area

The Waipaoa River drains the eastern flanks of the axial Raukumara Range (Fig. 7.1A), which is located adjacent to the boundary of the Australian and Pacific lithospheric plates, within the zone of active deformation associated with the Hikurangi subduction margin (Lewis and Pettinga, 1993). The 2205 km^2 river basin is underlain by: (1) plates of thrust Late Cretaceous and Early Tertiary mudstone and argillite; (2) Early Cretaceous greywacke; and (3) a cover sequence of poorly consolidated Neogene marine sedimentary rocks (Mazengarb et al., 1991; Mazengarb and Speden, 2000). Polynesian settlers (Maori)

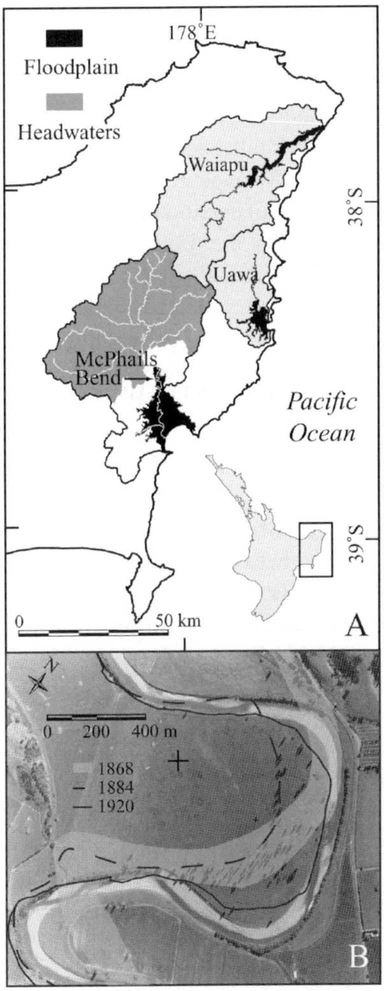

Figure 7.1. A: Location map, showing McPhail's bend and the headwater areas (shaded) in the upper Waipaoa River basin impacted by deforestation after 1875. Black shading delimits the entire area of the lowland flood plains bordering the Waipaoa (unlabelled), Uawa and Waiapu rivers. B: Configuration of McPhail's bend (1868–1988) and location of core site (cross).

disturbed the lowland forests (Wilmshurst et al., 1999), but the activities of European colonists impacted the entire basin. Wholesale clearance of the indigenous forest began in the late 1820s (Mackay, 1982), and most of the lowlands had been deforested by 1875 (Murton, 1968). Clearances in the headwaters accelerated in the period from 1890 to 1910 and continued until 1920. Today only 2.5% of the basin remains under native forest. The 2073 km^2 Waiapu and 664 km^2 Uawa River basins have similar geologic settings and deforestation histories to the Waipaoa River basin (Pullar, 1962; Rijkse and Pullar, 1978; Rijkse, 1980; Mazengarb et al., 1991). The lower reaches of all three rivers are bordered by well-drained, low-relief flood plains ("flats") that are intensively farmed or cultivated, and were regularly impacted by flooding until the 1950s, by which time an extensive system of artificial levées had began to be constructed. Along the lower reaches of the Waipaoa River these levées (stopbanks in local parlance) reduced, by 70%, the flood prone area of the Poverty Bay Flats (the Waipaoa River flood plain).

The ECCM experiences a maritime climate that, periodically, is perturbed by cyclonic storms of tropical origin. Sediment is generated mainly by gully erosion and shallow landsliding, which occurs when rainfall exceeds ~200 mm in <72 h (Reid and Page, 2002). Sediment production in the Waiapu River basin is contingent on gully erosion, which generates sediment at all scales of events and shows no threshold effects (Hicks et al., 2004). In contrast, most sediment in the Uawa River basin is generated by shallow landsliding (Hicks et al., 2004). The annual suspended sediment load of the Waiapu and Uawa rivers is ~35 and ~5×10^6 t, respectively (Hicks and Shankar, 2003). In the Waipaoa River basin, although the majority of the suspended sediment is produced by gully erosion, sediment production is augmented by shallow landsliding which typically generates 10 to 20% (and as much as 60%) of the suspended sediment during intense rainstorms (Page et al., 1999; Trustrum et al., 1999; Hicks et al., 2000; Reid and Page, 2002), and the river annually delivers ~15×10^6 t of suspended sediment to the Pacific ocean. Flows $\geq 1800\ m^3\ s^{-1}$ (which inundate the flood plain) transport 24% of the mean annual suspended sediment load, whereas, 50% of the load is transported by flows $<500\ m^3\ s^{-1}$, and 83% by flows between the mean flow ($34.7\ m^3\ s^{-1}$) and the mean annual flood ($1346\ m^3\ s^{-1}$). The bulk of the suspended sediment in transport is in the silt range, and suspended sediment concentrations in flood flows are very high ($>30\ g\ l^{-1}$) (Hicks et al., 2000; Gomez et al., 2003). Most floods last for only a few tens of hours, but sediment accumulates rapidly on the flood plain. The average (post-1850) rate of vertical accretion at three sites on the flood plain is ~$60\ mm\ yr^{-1}$, and contemporary (average) rates of vertical accretion on the flood plain are $14–18\ mm\ h^{-1}$ (Gomez et al., 1998, 1999).

Overbank sedimentation on the flood plains of the major rivers draining the ECCM is driven by the supply of material from headwater tributaries and accelerated in the historic period following deforestation (Pullar and Penhale, 1970; Pullar and Rijkse, 1977). In the Waipaoa River basin, the most intensive phase of deforestation occurred between 1890 and 1910 (Hamilton and Kelman, 1952). Following the conversions to pasture, extensive shallow landsliding was first observed during the winters of 1893 and 1894. The incidence of shallow landsliding increased during the first decade of the 20th century, and by the end of the second decade it had become a pervasive erosional process throughout the headwaters (Henderson and Ongley, 1920). Gully erosion in the headwaters is associated with terrain underlain by argillite, and many gullies may have been initiated during the unusually wet winters of 1916, 1917, and 1918 (Hamilton and Kelman, 1952). By the early

1930s, the effects of accelerated hillslope erosion had permeated the headwater regions and had begun to have a discernable impact on the flood plain (Pullar, 1962).

McPhail's bend is located towards the upstream end of the main Waipaoa River flood plain (Fig. 7.1B), and has never been tilled. Consequently, the stratigraphic record for the historic period is exceptionally detailed (Fig. 7.2A). Most depositional units have been tied to specific events, and the high rates of fine-sediment deposition make it easy to distinguish the contribution of individual events identified from historical records to flood plain stratigraphy (Gomez et al., 1999). Under the present hydrologic regime, overbank sedimentation occurs on McPhail's bend at discharges $\geq 1500\ m^3\ s^{-1}$, and floods in excess of this magnitude commonly are associated with storms that exceed the threshold

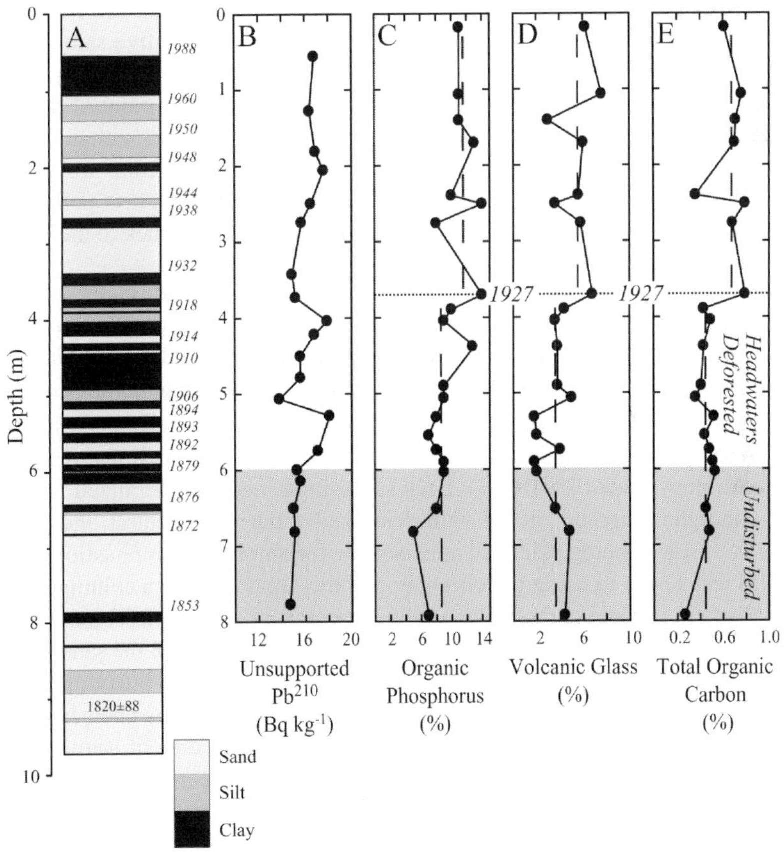

Figure 7.2. A: Generalized stratigraphy and probable dates associated with major flood units identified in the sediment core from McPhail's bend (after Gomez et al., 1999). The sediment samples discussed in this chapter were obtained solely from silt units, and wood fragments near the base of the core yielded the radiocarbon date. B: Concentration of unsupported ^{210}Pb in the sediment core. C: Concentration of the organic phosphorus-bearing fraction as a percentage of total phosphorus. D: Percentage of volcanic glass in the sand fraction. E: Percentage of total organic carbon. Shading shows the time period when the headwaters were undisturbed; the horizontal dotted line indicates the appearance of a deforestation-related signal in 1927; and vertical dashed lines denote mean values for the periods before and after deforestation impacted the depositional record.

for landsliding (Kelliher et al., 1995; Gomez et al., 1999). The subdued topography of the bend obviates hysteresis in the relation between channel discharge and volume of inundation and transport, rather than diffusion, is the dominant process involved in the transfer of sediment from the channel to the flood plain.

3. Sampling and methods

Sediment samples from McPhail's bend were taken from the bottom or center of 0.1–0.5 m thick, compositionally similar (fine – medium silt) units in a single 9.7 m long sediment core with an established stratigraphy (Gomez et al., 1999), which had been kept in a cold room maintained at ~4°C. None of the sampled units showed evidence of post-depositional transformation (cf. Gomez et al., 2003), and the selective sampling strategy was adopted to obviate any inverse correlation that might arise between organic content and particle size. The sediment core was obtained near the center point of the bend as it appeared in 1868 (Fig. 7.1B), and all but one sample was associated with post-1850 flood events (Fig. 7.2A). A ^{14}C date (1820 ± 88 cal yr) obtained from wood fragments recovered at 9.3 m (Fig. 7.2A), coupled with the appearance of grass pollen at ~7 m, confirm that the sedimentary record preserved in the core extends back to the beginning of intensive European settlement (ca. 1820).

Twenty-one samples were split and analyzed. The samples contained little clay (13.9 ± 2.8%), and the average D_{50} of all but two samples was 11.2 ± 1.9 μm (sand content <2%). Sampled sediments associated with the 1950 and 1988 floods were somewhat coarser, $D_{50} = 22.5$ and 28.7 μm, respectively, and had a higher sand content (14%). Total organic carbon contents (percent by weight) were determined using a Europa Geo 20/20 isotope ratio mass spectrometer, interfaced to an ANCA-SL elemental analyzer in continuous flow mode (EA-IRMS). Each sub-sample was demineralized with 1 M HCl to remove inorganic carbon, rinsed with deionized water until neutral, then dried in an oven at 30°C, prior to being placed in a tin capsule for automated combustion. The carbon dioxide gas was resolved using gas chromatographic separation on a column at 85°C, and analyzed for total organic carbon. The analytical precision of the measurements is ±5%. We also employed organic phosphorus, the volcanic glass content of the sand fraction, and unsupported ^{210}Pb as indicators of soil loss, to determine when the effect of deforestation was translated to the flood plain sediments. Neither the carbon nor phosphorus analyses were focused on a particular size fraction within the sediment samples. Organic phosphorus was determined by sequential extraction (Ruttenberg, 1992). The analytical precision of these measurements is ±5% (Filippelli and Delany, 1996; Latimer and Filippelli, 2001). Particle size was measured with a CILAS 1064 laser granulometer (the error of mean values is ±0.6 μm), and the volcanic glass content of the sand fraction was ascertained using optical microscopy by point counting. The activity profile of unsupported ^{210}Pb in the sediment core was determined through its descendent ^{210}Po using alpha spectrometry, from a complementary suite of 18 samples. Errors associated with the determinations are of the order of 0.09 ± 0.045 Bq kg^{-1}.

We evaluated the influence of lowland flood plain storage on the export of carbon from the Waipaoa, Uawa, and Waiapu river basins in the period from 1932 to 1950 (Table 7.1), which is bracketed by the easily identifiable 1932 flood deposits and the construction

Table 7.1. Organic carbon associated with alluvial sediments sequestered on the lowland flood plains bordering major rivers draining the East Coast Continental Margin, New Zealand (1932–1950).

River	Active flood plain area[a] (km^2)	Sediment stored on flood plain (Mt)	Carbon content of sediment (%)	Carbon stored on flood plain (kt yr^{-1})	Suspended sediment yield[b] (Mt yr^{-1})	Carbon Yield[c] (kt yr^{-1})	Carbon stored on flood plain (%)
Waipaoa	24.9 (~ *7*)[e]	26.8	0.61[d]	9.1 (*3.6*)[e]	15	105 (*86.7*)[e]	8–9 (*4*)[e]
Uawa	4.7	3.1	1.00[f]	1.7	5	35	3–5
Waipau	36.2	23.5	1.79[g]	23.4	35	245	8–11

Italicized figures in parentheses, for the period May 1979 to September 1990 when the area of the active flood plains was restricted by flood control levées (after Gomez et al., 2003), are provided for comparison.

[a] Mantled by recent soils (Pullar and Penhale, 1970; Rijkse and Pullar, 1978; Rijkse, 1980), or (bracketed) within flood control levées.

[b] Hicks et al. (2000), and Hicks and Shankar (2003).

[c] Load weighted value for $C \approx 0.7\%$ (Gomez et al., 2003).

[d] Average for overbank sediments on McPhail's Bend deposited between 1932 and 1950 (Gomez et al., 2003). For comparison, the value for *B*- and *C*-horizons of Waipaoa series soils on the Poverty Bay Flats is 0.6% (Pullar, 1962).

[e] May 1979 to September 1990 (Gomez et al., 2003).

[f] Average for *B*- and *C*-horizons of Waipaoa series soils on the Tolaga Bay Flats (Rijkse and Pullar, 1978).

[g] Average for *B*- and *C*-horizons of Waiapu, Waitaia, Oweka, Waihoata series soils in the Waiapu River valley (Rijkse, 1980).

of flood-control levées (cf. Pullar, 1962; Pullar and Penhale, 1970). There are several uncertainties associated with our estimates of the amount of carbon sequestered in the flood plains of these three rivers as alluvial sediments accumulated following deforestation. First, we note that the average organic carbon content of the overbank sediments deposited between 1932 and 1950 at McPhail's bend is comparable to values for the *B*- and *C*-horizons of Waipaoa series (alluvial) soils at other locations on the Poverty Bay Flats (Pullar, 1962; Table 7.1), and to contemporary suspended sediment transported by the Waipaoa River at discharges greater than about five times the mean flow (Gomez et al., 2003). The higher amounts of organic carbon in the alluvial soils on the Uawa and Waiapu river flood plains likely are due to more pronounced soil development related to plant growth. These soils have had longer to develop because the frequency of flooding and rate of overbank deposition are less (Pullar and Rijkse, 1977; Hicks et al., 2004). They may also be related to the positive northerly precipitation gradient along the ECCM (mean annual rainfall increases from 1078 mm on the Poverty Bay Flats to 1962 mm in the Waiapu River valley) (cf. Post et al., 1982), or simply reflect the fact that spatial variability in the organic carbon content of the alluvial soils on the Uawa and Waiapu river flood plains has not been adequately accounted for. Secondly, the volume of sediment deposited on the Poverty Bay Flats during the 1932, 1938, 1944, 1948, and 1950 floods (on which Waipaoa series soils are formed; Pullar, 1962) is reliably known from numerous boreholes and other exposures (Pullar and Penhale, 1970). However, the depth of Waipaoa series soils on the Tolaga Bay Flats, and of the corresponding Waiapu, Waitaia, Oweka, Waihoata series soils in the Waiapu River valley is known at relatively few locations

(Rijkse and Pullar, 1978; Rijkse, 1980). Thirdly, because water and sediment gaugings did not begin until 1960 (Hicks et al., 2000), we rely on contemporary estimates of the long-term suspended sediment yield. Additionally, only the load-weighted carbon concentration of the Waipaoa River is known with any certainty (Gomez et al., 2003). To accommodate these uncertainties, we used a variety of plausible combinations of the fundamental values to generate a range of estimates for the annual percentage loss of organic carbon, transferred by geomorphological processes and associated with fluvial sediments, to flood plain storage (Table 7.1).

4. Response to deforestation

In the headwaters of the Waipaoa River basin deforestation increased the susceptibility of hillslopes to landsliding (Trustrum et al., 1999). Three indicators provide evidence that the effect of deforestation was translated to the flood plain sediments and point to the time the fundamental change in landscape dynamics was recorded in the stratigraphic record. First, two discontinuities occur in the ^{210}Pb profile at ~5 and ~3.8 m, respectively (Fig. 7.2B). It is clear, however, that in contrast to the published literature, the profile is indistinct, and there is no indication of a steady decline in unsupported ^{210}Pb content with depth (cf. Walling et al., 1996). Secondly, beginning with the 1927 flood, the proportion of organic phosphorus increases. Thirdly, there is a concurrent increase in the volcanic glass content of the sand fraction (Fig. 7.2C and D).

From the mid 1890s, deforestation had an increasingly pronounced impact on sediment production in the headwaters of the Waipaoa River basin (Hamilton and Kelman, 1952). The two discontinuities in the ^{210}Pb profile likely represent inputs of sediment that had reduced atmospheric ^{210}Pb. We associate this sediment with contributions made by material from lower down in the regolith profile as the incidence of landsliding increased and gullies were initiated following deforestation. Gully erosion involves incision into weathered bedrock. Shallow landsliding characteristically involves small, <1 m deep, planar failures that originate at the surface of the weathered bedrock or within the soil profile (Reid and Page, 2002). There is little inter-event variability (and no steady decline with depth) in the unsupported ^{210}Pb content of the alluvium because weathered bedrock is continually being released by gully erosion, and there is a low recurrence interval for shallow landsliding in any given location because regolith production is weathering-limited. Thus, in effect, the sediment source (from which the flood plain sediments originate) is refreshed continuously.

Organic phosphorus represents plant material and soil derived from the upper portions of the soil profile (Walker and Syers, 1976), and the amount of phosphorus in this geochemical pool is similar to that observed in other maritime steeplands (Filippelli and Souch, 1999). Volcanic ash from the 1850 BP Taupo and earlier eruptions is a significant component of soils throughout the headwaters of the Waipaoa River basin (Jessen et al., 1999). The increase in the organic phosphorus fraction and amount of volcanic glass in the flood plain sediments during the 1927 flood (Fig. 7.2C and D), thus, point to an increase in the intensity of shallow landsliding on hillslopes in the headwaters.

Prior to deforestation, landsliding would have been focused on the riparian zone, and there is no record of active gully erosion. Thus, the initial appearance of a deforestation-related signal in the ^{210}Pb activity profile likely is related to the failure of hillslopes that were strongly coupled to river channels, but possessed only a thin skeletal soil cover and retained little tephra (Henderson and Ongley, 1920; Hamilton and Kelman, 1952). Failures initiated on hillslopes with better-developed soils elsewhere in the headwaters, as well as the initiation of gullies, probably were conditional upon the loss of root strength, which lags deforestation. Our data indicate that the essential change began with the flood of 1927, when the amount of organic carbon associated with the flood plain sediments increased (Fig. 7.2E). Pullar (1962) probably equated the base of the 1932 flood deposits with the onset of "recent" (Waipaoa series) soil development on the Poverty Bay Flats, because these alluvial deposits were thicker and more extensive (cf. Fig. 7.2A), and they are easily recognizable in the field.

Less than a decade after the headwaters had been deforested (1880–1920), the amount of total organic carbon in overbank sediments deposited on McPhail's bend had increased by ~50% (Fig. 7.2E). Our interpretation of trends is based on small differences; however, the analytical precision of the measurements is high. Consequently, the differences are meaningful, and for total organic carbon, organic phosphorus, and the volcanic glass content of the sand fraction, a *t*-test shows that the difference in the mean values for the periods before and after deforestation is significant at the 1% level. Patterns of sedimentation elsewhere on the Waipaoa River flood plain also are similar to that on McPhail's bend (Gomez et al., 1999). We do not distinguish between different types of organic matter in the sediments (e.g. allochtonous and autochtonous plant and woody debris, or organic matter that is inseparable from mineral particles). However, previous research indicates that, commensurate with the small amount of time available for soil development between flood events, the organic matter content of the recent alluvium on the Waipaoa River flood plain is low (Pullar, 1962). The low total organic carbon content of the flood plain alluvium appears to reflect its derivation from regolith and soils developed on weathered sedimentary rocks of marine origin (Table 7.1; Gomez et al., 2003). Indeed, $\delta^{13}C$ values for the fine-grained rocks that are the primary source of the alluvium are indistinguishable from those for sediments on the flood plain (Gomez et al., 2003). Landsliding releases organic-rich soil, but the contribution made by material with a low carbon content derived from the lower portions of the soil profile dominates because the amount of topsoil displaced is a small proportion of the total volume of sediment mobilized. Some topsoil also is retained in the debris that remains on hillslopes. However, this material is released incrementally and may be dispersed by a different range of flows (Hicks et al., 2000).

5. Carbon sequestration

Steepland rivers are important sources for transporting organic carbon to Earth's oceans (Masiello and Druffel, 2001; Lyons et al., 2002; Gomez et al., 2003). However, our results indicate that from the perspective of organic carbon transferred by geomorphological processes and associated with fluvial sediments, the storage potential of the lowland flood plains, bordering steepland rivers draining the ECCM, is low. We estimate that in the period from 1932 to 1950, by which time "recent" alluvial soils had begun to accumulate (Pullar, 1962; Rijkse and Pullar, 1978; Rijkse, 1980), $\leq 10\%$ of the annual organic carbon flux delivered to the Pacific Ocean by rivers draining the ECCM, typically, was lost to lowland flood plain storage (Table 7.1). The contemporary (1979–1990) estimate for the Waipaoa River is even less (~4%) (Gomez et al., 2003) because the active flood plain area was reduced from the 1950s by the construction of artificial levées. Our data do not permit us to assess the influence of carbon sequestration along low-order channels, where a larger area (~5% as opposed to <2%, of total basin area) potentially is available for storage. On the ECCM, many of these headwater channels are choked with gravel and are aggrading (Jessen et al., 1999), but like low-order streams in other steepland basins, they appear to be capable of transporting the bulk of the fine sediment supplied to them (cf. Hovius et al., 2000; Fuller et al., 2003).

The low storage potential of the active flood plains bordering high-order channels on the ECCM accrues both from their size and from the magnitude–frequency characteristics of event sediment yields. In the Waipaoa and Waiapu river basins, most sediment is generated by processes involving scour by surface and channelized runoff, particularly gully erosion. There are few restrictions on sediment availability during all scales of runoff event, but frequent runoff events are relatively more important than large storms (which promote landsliding) to the long-term suspended sediment and particulate organic carbon yields (Gomez et al., 1999; Hicks et al., 2000; Gomez et al., 2003). In the Uawa River basin, frequent runoff events are characterized by relatively low suspended sediment yields (Hicks et al., 2004). Large magnitude events are more important to event sediment yields, as landslides occur only when a rainfall threshold is exceeded. However, even though sediment and organic matter transport is more closely tied to flood events, because storm hydrographs are flashy, there is little opportunity for sediment to be transferred to the flood plain. Consequently, the loss to storage represents a small proportion of the mean annual suspended sediment and sediment-related organic carbon yields. This may also be the case in other steepland river basins where landsliding is the dominant erosional process (cf. Hovius et al., 2000; Leithold and Blair, 2001; Masiello and Druffel, 2001; Fuller et al., 2003).

Whether or not flood plains are significant long-term sinks for organic carbon also depends upon the retention time of the reservoir, and in many fluvial systems, flood plain sediments are reworked over relatively short-time scales by channel migration (cf. Everitt, 1968). The active flood plains bordering the lowland reaches of many turbid steepland rivers, including those on the ECCM, are products of vertical accretion (Nolan et al., 1987; Gomez et al., 1999). Rivers with flood plains formed by vertical accretion may migrate laterally, but sediment derived from overbank flow can remain in storage for long periods of time (Nanson and Young, 1981). In the case of the Waipaoa River survey, data indicates that between 1920 and 1988, the planform morphology was stable with essentially no lateral migration (cf. Fig. 7.1B). The implication is that sediment and soil-derived biogenic elements may be sequestered in the coastal alluvial plains of the ECCM for lengthy periods of time. Moreover, their release may be contingent upon an allogenic process, such as sea level change, that promotes a change in fluvial style (cf. Butcher, 1990; Woolfe et al., 2000).

Finally, the potential of flood plains bordering the lower reaches of small, turbid, steepland rivers (such as the Waipaoa, Waiapu, and Uawa) to act as sinks for carbon mobilized by deforestation and, hence, their importance to the dynamics of carbon cycling, is contingent on the character of the carbon delivered to them. If the fluvial sediments are derived primarily from weathered sedimentary rock, the associated carbon likely will be old (cf. Kao and Liu, 1996; Masiello and Druffel, 2001). However, although steepland rivers may be effective conduits for transferring refractory carbon to the ocean, carbon sequestered in the terrestrial environment for long periods of time ($\sim 10^3$ yr) may not be entirely irrelevant in exchanges with the atmosphere when it, eventually, is released from storage (Cole and Caraco, 2001). This raises the intriguing possibility that the effects of anthropogenically accelerated upland erosion may continue to impact the carbon cycle long after the antecedent effects have been absorbed by the fluvial system downstream.

6. Conclusions

Using the samples taken from a 9.7 m long sediment core with an established stratigraphy, extracted from an untilled meander bend, we have documented the impact anthropogenic activity has had on the amount of organic carbon transferred by geomorphological processes and associated with fluvial sediments deposited on the Waipaoa River flood plain during the last 150 years (Fig. 7.2). During this time period the indigenous forest in the basin headwaters was almost eradicated by European colonists. When hillslope erosion intensified in the late 1920s following deforestation the organic carbon content of the overbank sediments increased by $\sim 50\%$. However, the amount of carbon in the alluvium is low ($< 1\%$ by weight), which likely reflects its derivation from weathered sedimentary rocks of marine origin.

We also evaluated the effect of flood plain storage on the export of sediment-related organic carbon from the Waipaoa and the two other major rivers draining the ECCM in the period (1932–1950) prior to the construction of flood control levée systems, when the effects of accelerated erosion in the headwaters had a discernable impact on the lowland flood plains (Table 7.1). Our results suggest that transmission losses to these flood plains are small and typically amount to $\leq 10\%$ of the annual organic carbon flux to the Pacific Ocean from the Waipaoa (8–9%), Waiapu (8–11%), and Uawa (3–5%) rivers. The low storage potential of flood plains bordering the lower reaches of these three turbid, steepland rivers accrues both from their relatively small sizes (amounting to $< 2\%$ of total basin area) and to the magnitude–frequency characteristics of event sediment yields, which retain the signature of the dominant erosion process operating in the basins.

Acknowledgements

This research was supported by the National Science Foundation (grants BCS-0136375 and BCS-0137570); the New Zealand Foundation for Research, Science and Technology (contract C09X0013); and Indiana State University. The study would not have been possible without the assistance of Hannah Brackley (Landcare Research), Michel D'Ath (Landcare Research), Dennis Eden (Landcare Research), Gabe Filippelli (IUPUI), Mike Glascock (University of Missouri Research Reactor), Dave Peacock (GDC), Ted Pinkney (Landcare Research), Karyne Rogers (IGNS), Brenda Rosser (Landcare Research), Yuko Siguta (ISU), Neil Whitehead (IGNS), and Joe Whitton (Landcare Research).

References

Alto, R., Maurice-Bourgoin, L., Dunne, T., Montgomery, D.R., Nittrouer, C.A., Guyot, J.L., 2003. Episodic sediment accumulation on Amazonian floodplains influenced by El Niño/Southern Oscillation. Nature 425, 493–497.

Butcher, S.W., 1990. The nickpoint concept and its implications regarding onlap to the stratigraphic record. In: Cross, T.A. (Ed.), Quantitative Dynamic Stratigraphy. Prentice-Hall, Englewood Cliffs, NJ, pp. 375–385.

Cole, J.J., Caraco, N.F., 2001. Carbon in catchments: connecting terrestrial carbon losses with aquatic metabolism. Mar. Freshwater Res. 52, 101–110.

Collins, A.L., Walling, D.E., Leeks, G.J.L., 1997. Use of the geochemical record preserved in floodplain deposits to reconstruct recent changes in river basin sediment sources. Geomorphology 19, 151–167.

Depetris, P.J., Kempe, S., 1993. Carbon dynamics and sources in the Paraná river. Limnol. Oceanogr. 38, 382–395.

Everitt, B.L., 1968. Use of the cottonwood in an investigation of the recent history of a floodplain. Am. J. Sci. 266, 417–439.

Filipelli, G.M., Delany, M.L., 1996. Phosphorus geochemistry of equatorial Pacific sediments. Geochem. Cosmoschim. Acta 60, 1479–1496.

Filippelli, G.M., Souch, C., 1999. Effects of climate and landscape development on the terrestrial phosphorus cycle. Geology 27, 171–174.

Fuller, C.W., Willett, S.D., Hovius, N., Slingerland, R.L., 2003. Erosion rates for Taiwan mountain basins: new determinations from suspended sediment records and a stochastic model of their temporal variation. J. Geol. 111, 71–87.

Gomez, B., Eden, D.N., Peacock, D.H., Pinkney, E.J., 1998. Floodplain construction by recent, rapid vertical accretion: Waipaoa river, New Zealand. Earth Surf. Proc. Land. 23, 405–413.

Gomez, B., Eden, D.N., Hicks, D.M., Trustrum, N.A., Peacock, D.H., Wilmshurst, J.M., 1999. Contribution of floodplain sequestration to the sediment budget of the Waipaoa river, New Zealand. In: Mariott, S.B., Alexander, J. (Eds), Floodplains: Interdisciplinary Approaches, Geological Society of London Special Publication 163, pp. 69–88.

Gomez, B., Trustrum, N.A., Hicks, D.M., Page, M.J., Rogers, K.M., Tate, K.R., 2003. Production, storage and output of particulate organic carbon: Waipaoa river basin, New Zealand. Water Resour. Res. 39 (6), 1161, doi: 10.1029/2002WR001619.

Hamilton, D., Kelman, E.H.H., 1952. Soil Conservation Survey of the Waipaoa River Catchment, Poverty Bay, New Zealand. Soil Conservation and Rivers Control Council, Ministry of Works and Development, Palmerston North, p. 103.

Henderson, J., Ongley, M., 1920. The geology of the Gisborne and Whatatutu subdivisions, Raukumara division, New Zealand. Geol. Surv. Bull. 21, 88.

Hicks, D.M., Shankar, U., 2003. Sediment from New Zealand Rivers, NIWA Miscellaneous Chart 236. National Institute for Water and Atmospheric Research, Wellington.

Hicks, D.M., Gomez, B., Trustrum, N.A., 2000. Erosion thresholds and suspended sediment yields: Waipaoa river basin, New Zealand. Water Resour. Res. 36, 1129–1142.

Hicks, D.M., Gomez, B., Trustrum, N.A., 2004. Event sediment characteristics and the generation of hyperpycnal plumes at river mouths: East Coast Continental Margin, North Island, New Zealand. J. Geol. 112, 471–485.

Houghton, R.A., 1995. Changes in the storage of terrestrial carbon since 1850. In: Lal, R., Kimble, J., Levine, E., Stewart, B.A. (Eds), Soils and Global Change. CRC Press Inc., Boca Raton, FL, pp. 45–65.

Hovius, N., Stark, C.P., Hao-Tsu, C., Jiun-Chuan, L., 2000. Supply and removal of sediment in a landslide-dominated mountain belt: central range, Taiwan. J. Geol. 108, 73–89.

Ittekkot, V., Safiullah, S., Arain, R., 1986. Nature of organic matter in rivers with deep sea connections: the Ganges–Brahmaputra and Indus. Sci. Total Environ. 58, 93–107.

Jessen, M.R., Crippen, T.F., Page, M.J., Rijkse, W.C., Harmsworth, G.R., McLeod, M., 1999. Land use capability survey of the Gisborne-East Coast region, Landcare Research Science Series, Vol. 21, p. 213.

Junk, W.J., 1985. The Amazon floodplain a sink or source for organic carbon? Mitteilungen aus dem. Geologische-Paläontologischen Institut der Universitat Hamburg, 58, pp. 267–283.

Kao, S.J., Liu, K.K., 1996. Particulate organic carbon export from a subtropical mountainous river (Lanyang Hsi) in Taiwan. Limnol. Oceanogr. 41, 1749–1757.

Kelliher, F.M., Marden, M., Watson, A.J., Arulchelvam, I.M., 1995. Estimating the risk of landsliding using historical extreme river flood data. NZ J. Hydrol. 33, 123–129.
Lal, R., 1995. Global soil erosion by water and carbon dynamics. In: Lal, R., Kimble, J., Levine, E., Stewart, B.A. (Eds), Soil and Global Change. Lewis Publishers, Boca Raton, FL, pp. 131–143.
Latimer, J.C., Filippelli, G.M., 2001. Terrigenous input and paleoproduction in the Southern Ocean. Paleoceanography 16, 627–643.
Leithold, E.L., Blair, N.E., 2001. Watershed control on the carbon loading of marine sedimentary particles. Geochim. Cosmochim. Acta 65, 2231–2240.
Lewis, K.B., Pettinga, J.R., 1993. The emerging, imbricate frontal wedge of the Hikurangi margin. In: Ballance, P.F. (Ed.), South Pacific Sedimentary Basins. Elsevier Science Publishers B.V., Amsterdam, pp. 225–250.
Lyons, W.B., Nezat, C.A., Carey, A.E., Hicks, D.M., 2002. Organic carbon fluxes to the ocean from high-standing islands. Geology 30, 443–446.
Mackay, J.A., 1982. Historic Poverty Bay 4th Impression. Poverty Bay–East Coast Centenniel Council, Gisborne, 471 pp.
Masiello, C.A., Druffel, E.R.M., 2001. Carbon isotope geochemistry of the Santa Clara river. Global Biogeochem. Cycles 15, 407–416.
Mazengarb, C., Speden, I.G., 2000. Geology of the Raukumara Area. Institute of Geological and Nuclear Sciences, Wellington, New Zealand, 1 sheet +60 pp.
Mazengarb, C., Francis, D.A., Moore, P.R., 1991. Tauwhareparae. Department of Scientific and Industrial Research, Wellington, New Zealand, 1 sheet +52 pp.
Milliman, J.D., Syvitski, J.P.M., 1992. Geomorphic/tectonic control of sediment discharge to the ocean: the importance of small mountainous rivers. J. Geol. 100, 525–544.
Morozova, G.S., Smith, N.D., 2003. Organic matter deposition in the Saskatchewan river floodplain (Cumberland marshes, Canada): effects of progradational avulsions. Sediment. Geol. 157, 15–29.
Murton, B.J., 1968. Mapping the immediate pre-European vegetation on the east coast of the North Island of New Zealand. Prof. Geogr. 20, 262–264.
Nanson, G.C., Young, R.W., 1981. Overbank deposition and floodplain formation on small coastal streams of New South Wales. Z. Geomorphol. 25, 332–347.
Nolan, K.M., Lisle, T.E., Kelsey, H.M., 1987. Bankfull discharge and sediment transport in northwestern California. IAHS Publication 165, 439–449.
Page, M.J., Reid, L.M., Lynn, I.H., 1999. Sediment production from Cyclone Bola landslides, Waipaoa catchment. NZ J. Hydrol. 38, 289–308.
Page, M.J., Trustrum, N.A., Brackley, H.L., Baisden, W.T., 2004. Erosion-related soil carbon fluxes in a pastoral steepland catchment, New Zealand. Agr. Ecosyst. Environ. 103, 561–579.
Post, W.M., Emanuel, W.R., Zinke, P.J., Stangenberger, A.G., 1982. Soil carbon pools and world life zones. Nature 298, 156–159.
Pullar, W.A., 1962. Soils and agriculture of the Gisborne plains. NZ Soil Bur. Bull. 20, 92.
Pullar, W.A., Penhale, H.R., 1970. Periods of recent infilling of the Gisborne plains basin. NZ J. Sci. 13, 410–434.
Pullar, W.A., Rijkse, W.C., 1977. Estimation of recent alluvial infilling of Tolaga Bay Flats, basin, using Waimihia formation and Taupo pumice as tephra marker beds. NZ J. Sci. 20, 49–53.
Reid, L.M., Page, M.J., 2002. Magnitude and frequency of landsliding in a large New Zealand catchment. Geomorphology 49, 71–88.
Richey, J., Hedges, J.I., Devol, A.H., Quay, P.D., Victoria, R., Martinelli, L., Forsberg, B.R., 1990. Biogeochemistry of carbon in the Amazon river. Limnol. Oceanogr. 35, 352–371.
Rijkse, W.C., 1980. Soils and agriculture of Waiapu Valley, East Coast, North Island, New Zealand. New Zealand Soil Survey Report 60, p. 86.
Rijkse, W.C., Pullar, W.A., 1978. Soils of Tolaga Bay Flats, East Coast, North Island, New Zealand. New Zealand Soil Survey Report 40, p. 82.
Ruttenberg, K.C., 1992. Development of a sequential extraction method for different forms of phosphorus in marine sediments. Limnol. Oceanogr. 37, 1460–1482.
Smith, L.C., Alsdorf, D.E., 1998. Control on sediment and organic carbon delivery to the Arctic Ocean revealed with space-borne synthetic aperture radar: Ob' river, Siberia. Geology 26, 395–398.
Stallard, R.F., 1998. Terrestrial weathering and the carbon cycle: coupling weathering and erosion to carbon burial. Global Biogeochem. Cycles 12, 231–257.

Trustrum, N.A., Gomez, B., Reid, L.M., Page, M.J., Hicks, D.M., 1999. Sediment production, storage, and output: the relative role of large magnitude events in steepland catchments. Z. Geomorphol. Suppl. Band 115, 71–86.

Ver, L.M.B., Mackenzie, F.T., Lerman, A., 1999. Biogeochemical responses of the carbon cycle to natural and human perturbations: past, present, and future. Am. J. Sci. 299, 762–780.

Walling, D.E., 1996. Erosion and sediment yield in a changing environment. In: Branson, J., Brown, A.G., Gregory, K.J. (Eds), Global Continental Changes: the Context of Palaeohydrology, Geol. Soc. London Special Publication 115, pp. 43–56.

Walling, D.E., He, Q., Nicholas, A.P., 1996. Floodplains as suspended sediment sinks. In: Anderson, M.G., Walling, D.E., Bates, P.D. (Eds), Floodplain Processes. Wiley, Chichester, pp. 399–440.

Walker, T.W., Syers, J.K., 1976. The fate of phosphorus during pedogenesis. Geoderma 15, 1–19.

Wilmshurst, J.M., Eden, D.N., Froggatt, P.C., 1999. Late Holocene forest disturbance in Gisborne, New Zealand: a comparison of terrestrial and marine pollen records. NZ J. Bot. 37, 523–540.

Woolfe, K.J., Larcombe, P., Orpin, A.R., Purdon, R.G., 2000. Spatial variability in fluvial style and likely responses to sea-level change, Herbert River, Queensland. Aust. J. Earth Sci. 47, 689–694.

Catchment Dynamics and River Processes: Mediterranean and Other Climate Regions
Celso Garcia, Ramon J. Batalla, Editors

Chapter 8

A decade of sediment transport measurements in a large Mediterranean river (the Tordera, Catalan Ranges, NE Spain)

Ramon J. Batalla[1,*], Celso Garcia[2] and Albert Rovira[1]

[1]*Department of Environment and Soil Sciences, University of Lleida, 25198 Lleida, Spain*
[2]*Department of Earth Sciences, University of the Balearic Islands, 07122 Palma de Mallorca, Spain*

Abstract

The sediment yield of the Mediterranean Tordera River (894 km^2) has been computed from field measurements of runoff, suspended sediment and bedload transport. Measurements were carried out during 1990s at three different points within the Tordera basin: the Upper Tordera (35 km^2), the Arbucies River (106 km^2), and the Lower Tordera (785 km^2). Almost one thousand samples of suspended load and bedload were used to estimate the total yield of two catchments representative of the main catchment lithologies, the Upper Tordera (metamorphic) and Arbucies (plutonic), and the sediment load at the entrance of the Lower Tordera reach. The annual load in suspension passing the downstream Lower Tordera section was estimated at 32,500 t, giving a specific sediment yield of 41 t km^{-2} yr^{-1}. Most of the sediment is supplied by the granitic catchments at a rate of 60 t km^{-2} yr^{-1}, three times higher than that of the metamorphic areas (20 t km^{-2} yr^{-1}). Annual bedload transport through the Lower Tordera section was estimated at 29,000 t yr^{-1}, giving a specific sediment yield of 37 t km^{-2} yr^{-1}. The catchments underlain by plutonic rocks supply more bedload (27 t km^{-2} yr^{-1}) than the metamorphic areas (22 t km^{-2} yr^{-1}). The total load at the Lower Tordera section was estimated at 61,600 t yr^{-1}, which gives a specific sediment yield of 78 t km^{-2} yr^{-1}. The total sediment load from the granitic catchments is 88 t km^{-2} yr^{-1} and doubles the sediment contribution from the metamorphic areas (41 t km^{-2} yr^{-1}). Most sediment is transported during short periods of time. In the Lower Tordera, 65% of the total load is carried in 10% of the time, in the Arbucies River the proportion raises to 82% and in the Upper Tordera it reaches almost 90%, for the same time duration. The plutonic areas appear to be the main source of sediment, both in suspension and as bedload, accounting for 90% (ca. 56,000 t) of the total annual load of the entire basin. The data indicate that the Lower Tordera shows a similar pattern of sediment yield to that described for Mediterranean semi-arid regions, whereas values estimated in the Arbucies River (mainly for bedload) and, especially, in the upper part of the catchment agree with data reported for more humid, Mediterranean mountainous environments.

Keywords: sediment yield, suspended load, bedload, Mediterranean river, lithology

*Corresponding author. *E-mail address:* rbatalla@macs.udl.es (R.J. Batalla)

1. Introduction

The total sediment load of a river is defined as the total amount of solid material passing through a given cross section, as bedload and suspended load. Annual sediment loads vary widely both between and within physiographic regions (Langbein and Schumm, 1958; Inbar, 1992). The magnitude of the sediment load depends on the supply of sediment from the catchment, which in turn is controlled by primary factors such as climate, basin area, geology and land-use, and the transport capacity of the river, resulting in transport rates and concentrations that vary over space and time. Soil texture, basin morphology, channel efficiency and channel or bank erosion have also been identified as primary controls on river sediment loads (Trimble, 1977). Simultaneous and systematic measurements of both components of the total load (bedload and suspended sediment load) are not commonly available, even under steady low flow conditions.

The sediment load of a river is a reflection of natural catchments dynamics but, in addition, it is of central interest, together with the dissolved load, for water quality and upstream land use management. Information on the sediment load of a river can be used for estimating erosion rates, to assess sediment dynamics during floods and to evaluate downstream geomorphic effects of instream human activities, such as gravel quarrying and reservoir operation. However, this type of information is generally not available, especially in many rivers of Mediterranean regions under different land use and lithology.

The geomorphological processes operating in the Mediterranean basin of the Tordera River have been studied by Maria Sala (University of Barcelona) since the end of 1970s (i.e. Sala, 1979, 1983). From 1990 to 1999, further research was conducted within the same framework (i.e. Batalla, 1993; Garcia, 1997; Rovira, 2001). The main aim of that research was to study the sediment transport dynamics in a series of tributaries and in the mainstem of the Tordera River, under different sedimentological and hydraulic conditions. For the first time, the three data sets representing water discharge, suspended sediment and bedload transport were analysed together and the results presented in this chapter, with the objective of estimating the overall sediment yield of the basin, its annual distribution and frequency and its relation with the main lithologies underlying the area.

2. Study area

2.1. *The Tordera basin*

The Tordera basin (894 km^2) is located in the northern part of the Catalan Coastal Ranges, 60 km north-east of Barcelona (Fig. 8.1). It collects water from the slopes of the Montseny and Guilleries massifs (northern part of the basin), and the Montnegre massif (southern part). The maximum height is reached at the Turó de l'Home (1712 m a.s.l., Montseny massif). The drainage basin is mainly composed of plutonic rocks (granodiorite, leucogranites), which occupy 57% of the basin area. An important metamorphic outcrop comprising slates, schist, gneiss and metamorphic calcareous rocks is located in the southern part of the Montseny massif and in the upper parts of the Montnegre massif. Four Holocene terrace levels where sand and fine gravel are predominant represent Quaternary deposits. Periglacial processes on hillslopes during the Pleistocene caused

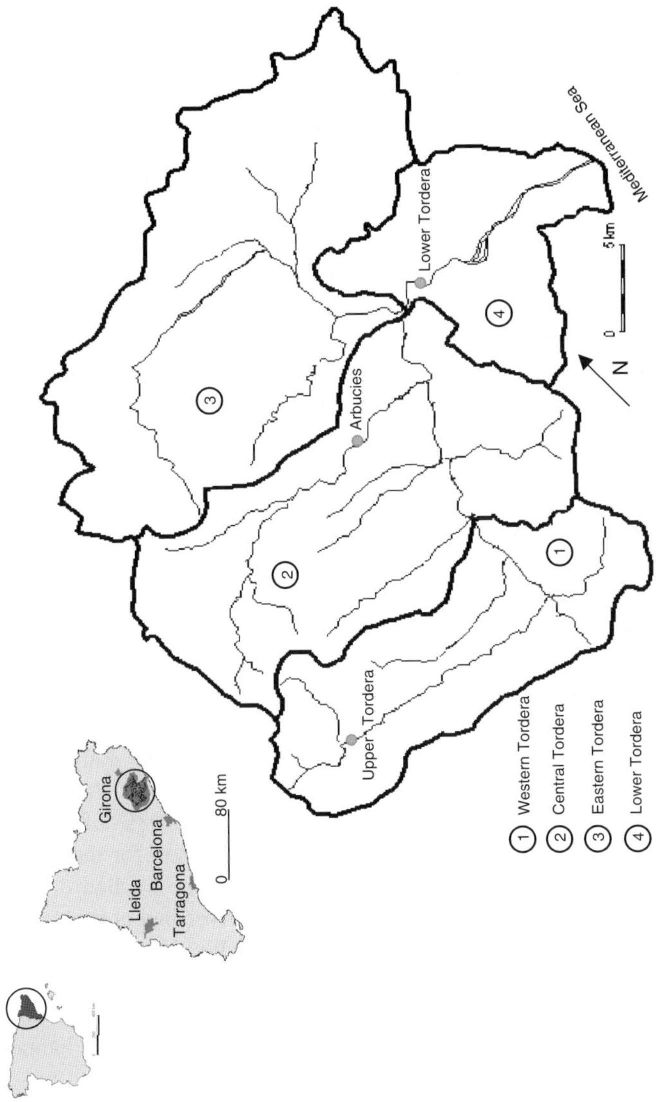

Figure 8.1. Location of the study area, including the monitoring sections and the sediment contribution areas in which the Tordera River basin has been divided for the purpose of this work. The drainage network scheme has been redrawn from Sala (1979).

intensive rock disintegration. Since then, considerable amounts of sandy material have been released from the bedrock and easily moved into the streams. The climate is classified as sub-humid Mediterranean with maritime trends (Sala, 1979). The mean annual rainfall ranges from 1000 mm at the summit areas to 600 mm on the coast. The main tributaries are the Santa Coloma (275 km^2) and Arbucies (114 km^2) rivers. The population is mainly located on the lower parts of the Montseny massif and on the alluvial floodplain.

2.2. Study sites

Sediment transport in the Tordera basin was investigated using field measurements at three points in its drainage network: the Upper Tordera (Garcia, 1997), mostly on metamorphic rocks, the Arbucies River (Batalla, 1993), one of the main tributaries mainly on plutonic rocks, and the Lower Tordera (Rovira, 2001), a section which integrates the water and sediment dynamics of the whole basin (Fig. 8.1).

The Upper Tordera River (at the La Llavina gauging station, 48 km^2) has a perennial flow. The mean annual runoff for 1952–1999 period is 21.3 hm^3, giving a mean specific runoff of 19 $l\,s^{-1}\,km^{-2}$. The median daily discharge (Q_{50}) is 0.27 $m^3\,s^{-1}$ with major floods reaching 45 $m^3\,s^{-1}$. The channel bed has a mean width of 5.5 m with gently sloping banks. The river-bed has a coarse surface layer, with some areas composed of fine-textured sediment accumulations, termed patches, with a characteristic grain size between 1 and 20 mm. These mobile bedforms depend on the fine sediment supplied by the stream network, mainly derived from an area of granodiorite (Garcia et al., 1999). The bed material is armoured: it has a median diameter (D_{50_s}) of 54 mm for the surface and a $D_{50_{sub}}$ of 16 mm for the subsurface. The dominant particle shape is a blade, due to the abundance of slate particles in the bed material. This has an important effect on bed armouring as a consequence of blade shape stability on the bed surface showing particle imbrication (Johansson, 1963; Laronne and Carson, 1976). The slope of the river reach is 0.02.

The Arbucies River drains an area of 114 km^2 (106 km^2 at the gauging station, for which hydrological records are available since 1967). Analysis of the series of daily discharges (1967–1992) shows that the river has discharge for 98% of the year, which represents 6–7 days of drought per year. The mean annual runoff for the same period is 33.7 hm^3, giving a mean specific runoff of 11 $l\,s^{-1}\,km^{-2}$. The median daily discharge is 0.54 $m^3\,s^{-1}$, with major floods reaching 65 $m^3\,s^{-1}$. The river-bed of the Arbucies River is mainly poorly sorted, according to the index of Folk and Ward (1957), and is composed of grains of quartz sand and fine gravels, with a bulk median diameter (D_{50}) of 2.2 mm. Bedload transport is mostly dominated by the continuous passage of sandy bedforms, even under very low and steady flows (0.250 $m^3\,s^{-1}$) (Batalla, 1997). The channel bed has a mean width of 5 m. Bed slope for the study reach is 0.01.

The Lower Tordera River (at the Fogars de Tordera gauging station, 785 km^2) has a continuous water flow during 85% of the time. Mean annual runoff for 1967–2002 period is 149 hm^3, which gives an mean specific runoff of 6 $l\,s^{-1}\,km^{-2}$. The median daily discharge is 1.75 $m^3\,s^{-1}$. Major floods can reach 1200 $m^3\,s^{-1}$, mostly occurring during the winter and autumn seasons which contribute more than 60% of the total water yield. The Lower Tordera exhibits a well-defined single thread channel, with a mean width of 70 m. The bed is mainly composed of coarse sand and fine gravels with a bulk median particle size (D_{50}) of 4.5 mm. As in the case of the Arbucies River, bedload transport is dominated by the passage of bedforms, even under very low flows (0.160 $m^3 s^{-1}$) (Rovira, 2001). Bed material can be classified as poorly sorted (according to the index developed by Folk and Ward (1957)). The slope of the reach upstream of the sampling site is 0.005.

3. Methodology

The sediment yield of the Tordera basin was computed from field measurements of sediment transport undertaken at three different points in the catchment during different periods in 1990s: the Upper Tordera (1995–1996), the Arbucies River, one of the most important tributaries of the Tordera River (1991–1992) and the Lower Tordera (1997–1998) (Table 8.1, Fig. 8.1). The Water Authorities of Catalonia supplied water level data and flood hydrographs recorded at the gauging stations of La Llavina (Upper Tordera), Arbucies and Fogars de Tordera (Lower Tordera). In addition, in the Upper Tordera, two pressure transducers housed in a stilling well at the sampling station and 70 m upstream, allowed the depth of flow to be measured and defined the water surface slope.

On the Upper Tordera, bedload transport was monitored by means of two automatic modified Birkbeck-type recording bed slot samplers (Reid et al., 1980), installed across the 4-m width of the stream (Fig. 8.2a). One slot was located in the thalweg at the centre of the channel; the other was placed midway between the left bank and the thalweg sampler. The bedload falls through a slot into an underlying container that continuously weighs the material. The slot width of each sampler was set at 130 mm, which represents between 65 and 70% of the surface GSD. The longitudinal slot length of the samplers is 450 mm, large enough to catch particles moving in saltation. Each collecting box has a depth of 80 cm and a capacity of 0.450 m^3. In total, 243 bedload samples were obtained during the study period (Garcia et al., 2000).

In the Arbucies River and the Lower Tordera, bedload transport was measured across the channel section by means of hand held Helley–Smith sampler with a 76 mm intake,

Table 8.1. Monitoring of sediment transport in the Tordera River basin between 1990 and 1999.

Sampling section and period		Number of samples	Range of discharges[a] ($m^3 s^{-1}$)	Time interval (%)
Upper Tordera (1995–1996)	Suspended load	161	0.1–34	81.0–0.03
	Bedload	243	1.9[b]–6.6	8.0–0.7
Arbucies (1991–1992)	Suspended load	242	0.1–9.8	88.0–0.2
	Bedload	71	0.3[b]–3.7	76.0–1.2
Lower Tordera (1997–1998)	Suspended load	170	0.1–25.2	90.0–1.2
	Bedload	75	0.2[b]–4.3	87.5–22.0

[a] The range of discharges under which bedload and suspended load were sampled and the time interval covered by samples within the overall period 1990–1999 are indicated.
[b] Threshold discharge for bed-material entrainment.

Figure 8.2. Sediment transport sampling equipment: (a) Bedload trap at the Upper Tordera and (b) Hand-held Helley–Smith sampler for bedload and US DH-48 Depth Integrating sample for suspended sediment, used at the Arbucies River and the Lower Tordera monitoring sections.

15 kg bag capacity and 0.45 mm mesh bag (Helley and Smith, 1971) (Fig. 8.2b). For particle sizes larger than 0.50 mm and smaller than 16 mm, the trap efficiency of the bedload sampler reaches 100%, irrespective of the transport rate, whereas the efficiency for larger size drops to less than 70% (Emmett, 1979). The sampling time ranged between 3 and 10 min, according to the observed bedload transport rate. In total, 146 bedload samples were taken (71 samples in the Arbucies river and 75 in the Lower Tordera) covering a range of discharges equalled or exceeded 87.5% of the time (1990–1999 period) (Batalla, 1997; Rovira, 2001) (Table 8.1).

In the Arbucies and the Lower Tordera sections, suspended sediment was sampled during low flows and small floods by means of a US DH48 depth-integrating sampler (Fig. 8.2b). During high flows, automatic water and sediment ISCO 3200 samplers located at the three measuring sections were used. In total, 160 samples of suspended sediment were taken on the upper Tordera, 242 samples on the Arbucies River and 170 samples on the Lower Tordera, covering a range of discharges equalled or exceeded between 80 and 90% of the time (1990–1999 period) (Table 8.1).

Sediment yield at the monitoring sections was calculated from load-rating relations between discharge and suspended sediment, and discharge and bedload using the flow duration curve method (Walling, 1984) (Table 8.2). Statistically significant relations were used to estimate suspended sediment concentrations and bedload transport rates, and thus loads, during periods or discharges for which measurements were unavailable. Since sediment-rating curves are based on instantaneous measurements of discharge, they will possibly underestimate the sediment loads (Walling, 1984), especially for the very high flows (i.e. in the Tordera those equalled or exceeded 0.5% of the time). On the Lower Tordera sampling section, the general relation between suspended concentration and discharge was not statistically significant. In order to overcome this situation, the data set was subdivided according to the type of flood within which samples were taken: (1) type A (so-called supply-rich floods), which included both first hydrograph peaks and single-hydrograph floods after a long period without floods, and (2) type B (so-called

Table 8.2. Example of computation of sediment yield (Arbucies River, suspended load, 1991–1992), following the rating curve method of Walling (1984).

Time interval (%)	Duration (%)	Mid-point (%)	Discharge ($m^3\ s^{-1}$)	Concentration ($mg\ l^{-1}$)	Load (t)
0–0.5	0.5	0.25	13.8	2216.3	4836.7
0.5–1	0.5	0.75	8.0	917.0	1156.8
1–2	1	1.5	6.0	577.1	1091.9
2–5	3	3.5	2.4	135.6	312.9
5–10	5	7.5	1.9	87.6	256.8
10–20	10	15	1.2	45.0	174.5
20–30	10	25	0.9	27.2	77.2
30–40	10	35	0.8	21.2	51.4
40–50	10	45	0.7	16.9	35.7
50–60	10	55	0.6	13.4	24.5
60–70	10	65	0.5	10.9	17.5
70–80	10	75	0.5	9.9	15.0
80–90	10	85	0.4	8.3	11.2
90–95	5	92.5	0.4	8.3	5.6
95–98	3	96.5	0.4	6.2	2.1
98–99	1	98.5	0.3	5.4	0.6
99–99.5	0.5	99.25	0.3	5.4	0.3
99.5–100	0.5	99.75	0.3	5.4	0.3
				Total	8071

exhaustion floods), which included both second and third hydrograph peaks and successions of floods. The existence of an exhaustion effect over flood periods and after the first peak within single storm hydrographs was documented (Rovira, 2001). A rating curve between suspended sediment concentration and discharge was, therefore, derived for each flood type and they were subsequently used for computing the sediment yield. The statistical bias of all the rating curves was corrected by the factor $\exp(2.65s^2)$ to improve the accuracy of the estimates of river load (Ferguson, 1986).

In addition, and in order to establish the overall sediment yield for the whole Tordera basin, the catchment was divided into three areas, whose sediment yields were estimated by extrapolating the data collected from the Upper Tordera measuring section to the metamorphic areas (including areas with clay sediments), and the data collected from the Arbucies measuring section to the plutonic areas (including areas with sandy sediments). The lithology was extracted from the work of Sala (1979). The alluvial and basaltic zones occupy 10% of the catchment and they have not been included in the computation. The three areas are (Fig. 8.1):

- The *Western Tordera*, which includes the catchments of the Upper Tordera and Vallgorguina, as the most significant tributaries. It covers the western part of the basin and it is underlain by metamorphic (50%) and plutonic (50%) rocks.
- The *Central Tordera*, which includes the catchments of the central part of the basin (e.g. Arbucies, Gualba, Breda and Fuirosos), and is mostly on granitic rocks (91%).
- The *Eastern Tordera*, which includes the catchments of the eastern part of the catchment (e.g. the Santa Coloma and Sils) and it is mostly on granitic rocks and sandy deposits (91%).

4. Results

4.1. Water discharge

The total water discharge from the Upper Tordera for 1990–1999 period was 201 hm^3, yielding a mean annual runoff of 22.3 hm^3 (Table 8.3) and a mean specific discharge of 0.015 $m^3\,s^{-1}\,km^{-2}$ (Table 8.3, Fig. 8.3). Maximum runoff occurred in 1996–1997, one of the sampling years, and the minimum in 1998–1999. The inter-annual coefficient of variation was 54%. The median specific discharge (Q_{50}) was 14.7 $l\,s^{-1}\,km^{-2}$ (Fig. 8.4), which is equivalent to a median daily discharge of 0.7 $m^3\,s^{-1}$. The winter season contributed 36% of the total annual runoff, while summer only supplied 8%. The flow duration curve (1990–1999) indicates that a discharge of 0.08 $m^3\,s^{-1}$ (1.6 $l\,s^{-1}\,km^{-2}$) was equalled or exceeded 90% of the time, while a discharge of 5.7 $m^3\,s^{-1}$ (118 $l\,s^{-1}\,km^{-2}$) was equalled or exceeded 1% of the time. A discharge of 0.26 $m^3\,s^{-1}$ (5.4 $l\,s^{-1}\,km^{-2}$) was equalled or exceeded 50% of the time (Fig. 8.4).

The total water discharge from the Arbucies catchment for 1990–1999 period was 177 hm^3, giving a mean annual runoff of 22 hm^3 (Table 8.3). The wettest year was 1992–1993, one of the sampling years, whereas the driest year was 1998–1999. The inter-annual coefficient of variation was 37%. The median specific discharge was 6.6 $l\,s^{-1}\,km^{-2}$ (Fig. 8.4), which is equivalent to a median daily discharge of 0.7 $m^3\,s^{-1}$. The winter season contributed 35% of the total annual runoff, while summer only supplied 10%.

Table 8.3. Water yield at the three monitoring sections in the Tordera River basin during the study period 1990–1999.

	Upper Tordera	Arbucies	Lower Tordera	Upper Tordera	Arbucies	Lower Tordera
	Runoff ($hm^3 yr^{-1}$)			Specific runoff ($l s^{-1} km^{-2}$)		
1990–1991	28	23	n.d.	19	7	n.d.
1991–1992[a]	28	30	179	18	9	7
1992–1993[a]	28	34	124	18	10	5
1993–1994[a]	10	16	45	7	5	2
1994–1995	17	20	n.d.	11	6	n.d.
1995–1996[a]	25	29	141	17	9	6
1996–1997	46	n.d.	197	30	n.d.	8
1997–1998[a]	10	17	57	6	5	2
1998–1999[a]	9	9	41	6	3	2
Mean	22	22	112	15	7	5

n.d. hydrological data are not available.
[a] Years with complete hydrological data at the three locations used to compute the mean solid load.

The flow duration curve (1990–1999) indicates that a discharge of $0.1 m^3 s^{-1}$ ($0.9 l s^{-1} km^{-2}$) was equalled or exceeded 90% of the time while a discharge of $3.8 m^3 s^{-1}$ ($35.6 l s^{-1} km^{-2}$) was equalled or exceeded 1% of the time. A discharge of $0.45 m^3 s^{-1}$ ($4.2 l s^{-1} km^{-2}$) is equalled or exceeded 50% of the time (Fig. 8.4).

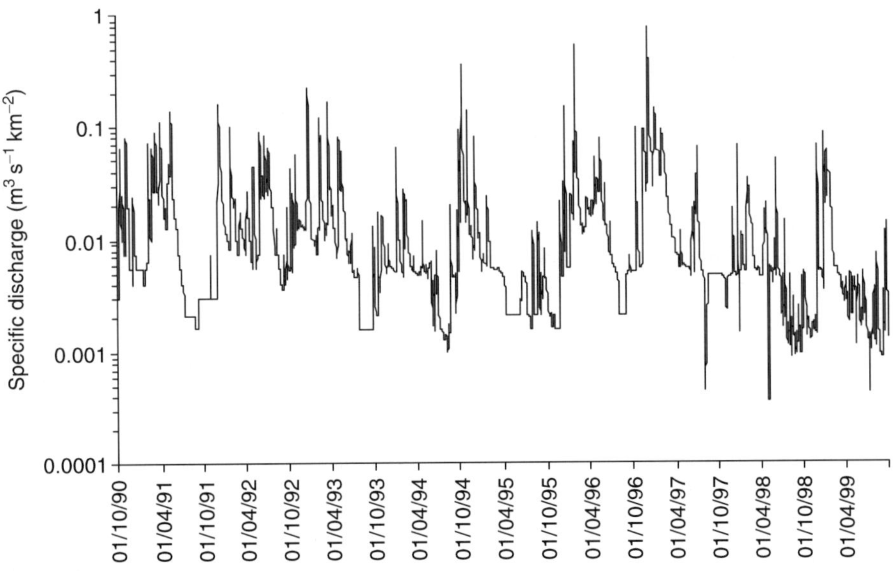

Figure 8.3. Continuous water flow at the La Llavina gauging station (Upper Tordera) for the study period 1990–1999.

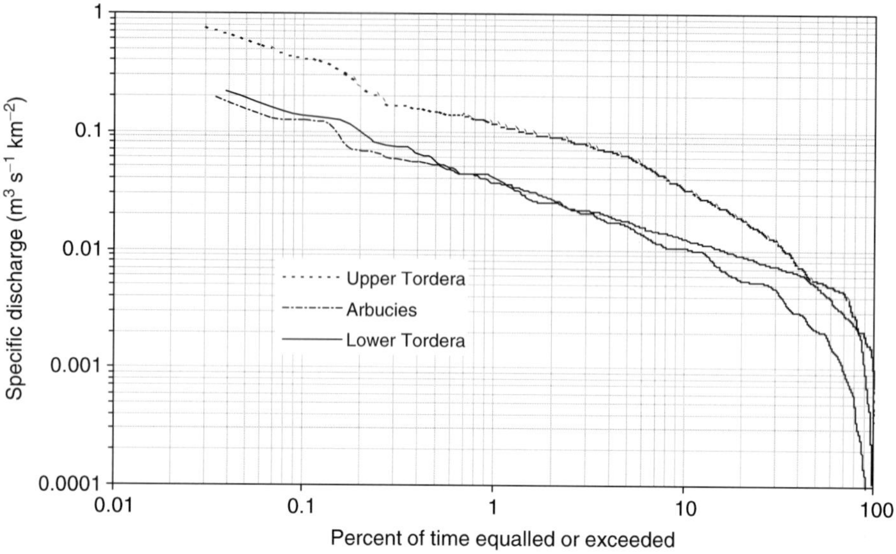

Figure 8.4. Flow frequency curves derived from daily discharges at the monitoring sections in the Tordera River for the study period 1990–1999.

The total water discharge at the Fogars de Tordera gauging station for 1990–1999 period was 784 hm^3 (Table 8.3). The mean annual runoff was 112 hm^3, with a maximum in 1996–1997 and minimum in 1998–1999. The inter-annual coefficient of variation was 54%. The median specific discharge was 4.5 $l\ s^{-1}\ km^2$ (Fig. 8.4), which is equivalent to a median daily discharge of 3.5 $m^3\ s^{-1}$. The winter season contributed 39% of the total annual runoff, while summers only contributed 10%. The flow duration curve (1991–1999) indicates that a discharge of 0.1 $m^3\ s^{-1}$ (0.12 $l\ s^{-1}\ km^{-2}$) was equalled or exceeded 90% of the time while a discharge of 31.4 $m^3\ s^{-1}$ (40 $l\ s^{-1}\ km^{-2}$) was equalled or exceeded 1% of the time. A discharge of 1.6 $m^3\ s^{-1}$ (2 $l\ s^{-1}\ km^{-2}$) is equalled or exceeded 50% of the time (Fig. 8.4).

Annual runoff, and thus specific discharges, were much higher in the metamorphic areas of the Tordera basin, represented by the Upper Tordera (464 mm yr^{-1}), in comparison with the areas mostly underlain by plutonic rocks, such as the Arbucies River (207 mm yr^{-1}) and the Lower Tordera (142 mm yr^{-1}). The inter-annual runoff variability is remarkable in the three areas, reflecting the sub-humid Mediterranean climate of the region, with moderate floods mostly in autumn and winter, and continuous streamflow during most of the summer. Annual evapotranspiration is estimated to be around 50% in the upper catchment and 75% in the lower areas of the basin.

4.2. Sediment yield

The sediment yield of the Tordera River was estimated from suspended load and bedload measurements obtained at three measuring sections: the Upper Tordera in the metamorphic headwaters, the Arbucies River representing the plutonic bedrock catchments, and the Lower Tordera at the entrance to the downstream reach of the basin, which integrates the sediment production from the whole catchment. Sediment yield was calculated for 1990–1991 to 1998–1999 period, although only the 6 years with a full set of hydrological data available for the sections were used in the analysis (Table 8.4). The comparisons of sediment concentrations and loads are presented in relation to specific discharge (q in $m^3\ s^{-1}\ km^{-2}$).

4.2.1. Suspended load

The suspended sediment load includes fine particles that travel in suspension in the water column. Suspended sediment concentrations (C_{ss}) showed a positive and statistically significant relation with discharge (q) ($p < 0.01$) in the three monitoring sections of the Tordera River (Fig. 8.5). A significant degree of scatter exists, however, which may be related either to seasonal causes or to hysteretic effects during flood events (Batalla and Sala, 1994; Rovira, 2001). The statistical bias of the relation between q and C_{SS} due to the scatter of points was calculated to be 7% in the Lower Tordera, 73% in the Arbucies River and 76% in the Upper Tordera, and this was subsequently corrected using the factor developed by Ferguson (1986). The scatter of the concentration data was especially marked in the Upper Tordera section. There, the low exponent of the concentration/discharge relation (0.70) indicates that the increment of the suspended load was not strongly related to the increase of discharge during floods (i.e. hydraulically dependent), but to the sediment supplied from the catchment. In contrast, the exponents of the concentration/discharge relation in the granitic catchments (Arbucies 1.62, and Lower Tordera 1.28 for the primary floods and 1.44 for the secondary ones) indicates a higher increment of suspended load after a given increment of discharge, a fact that can be related to a larger availability of fine sediment in the channel.

The mean suspended concentration (estimated as the mean of the measured concentrations) was 165 mg l^{-1} for Arbucies, 148 mg l^{-1} for the Lower Tordera, and 68 mg l^{-1} for the Upper Tordera, although sampled discharges were considerable higher for the Upper Tordera (specific discharge q averaged 0.078 $m^3\ s^{-1}\ km^{-2}$) than for its plutonic counterpart (q in the Arbucies River averaged 0.016 $m^3\ s^{-1}\ km^{-2}$), and for the whole catchment (q averaged 0.012 $m^3\ s^{-1}\ km^{-2}$ in the Lower Tordera). For a given discharge, the concentrations were, on mean, double at Arbucies and the Lower Tordera, where more sediment was probably available to be transported during floods. As indicated above, this situation may reflect the intensive rock disintegration as a result of periglacial processes that occurred during the Pleistocene, which especially affected the plutonic rocks of the basin (Sala, 1979). Since then, considerable amounts of sandy material have been released from the bedrock and readily moved to the streams, thus still supplying surplus amounts of sediment and currently yielding higher concentrations. The maximum concentrations were recorded in the Lower Tordera (2.8 g l^{-1}) and Arbucies (2.2 g l^{-1}) under very similar specific discharges. These values were considerably higher than the maximum concentrations obtained in the Upper Tordera (0.63 g l^{-1}) even under a higher specific discharge.

The mean annual suspended sediment load passing the downstream Lower Tordera section was calculated at around 32,500 t yr^{-1}, which represents a specific sediment

Table 8.4. Sediment load and yield at the three monitoring sections in the Tordera River basin during the study period 1990–1999.

	Upper Tordera	Arbucies	Lower Tordera	Upper Tordera	Arbucies	Lower Tordera
	Load (t yr^{-1})			Specific load (t km^{-2} yr^{-1})		
Suspended load						
1990–1991	857	1425	n.d.	18	13	n.d.
1991–1992[a]	770	8071	34,061	16	76	43
1992–1993[a]	826	3329	27,335	17	31	35
1993–1994[a]	135	318	2263	3	3	3
1994–1995	655	5808	n.d.	14	55	n.d.
1995–1996[a]	1196	9772	109,625	25	92	140
1996–1997	3045	n.d.	102,083	63	n.d.	130
1997–1998[a]	111	835	5497	2	8	7
1998–1999[a]	164	174	3608	3	2	5
Mean	533	3750	30,398	11	35	39
Bias corrected mean[b]	939	6487	32,526	20	61	41
Bedload						
1990–1991	698	1420	n.d.	15	13	n.d.
1991–1992[a]	608	2806	42,876	13	26	55
1992–1993[a]	752	2432	34,782	16	23	44
1993–1994[a]	67	624	15,002	1	6	19
1994–1995	455	1911	n.d.	9	18	n.d.
1995–1996[a]	2146	3517	38,549	45	33	49
1996–1997	6212	n.d.	42,018	129	n.d.	54
1997–1998[a]	51	971	16,750	1	9	21
1998–1999[a]	115	329	10,630	2	3	14
Mean	623	1780	26,432	13	17	34
Bias corrected mean[b]	1041	2901	29,075	22	27	37
Total load						
1990–1991	1554	2844	n.d.	32	27	n.d.
1991–1992[a]	1378	10,877	76,937	29	103	98
1992–1993[a]	1578	5761	62,117	33	54	79
1993–1994[a]	202	941	17,264	4	9	22
1994–1995	1110	7719	n.d.	23	73	n.d.
1995–1996[a]	3342	13,289	148,175	70	125	189
1996–1997	9258	n.d.	144,101	193	n.d.	184
1997–1998[a]	162	1806	22,247	3	17	28
1998–1999[a]	279	503	14,239	6	5	18
Mean	1157	5530	56,830	24	52	72
Bias corrected mean[b]	1980	9389	61,601	41	89	78

n.d. hydrological data are not available.

[a] Years with complete hydrological data at the three locations used to compute the mean solid load.

[b] Statistical bias of the load rating curve has been corrected using the factor developed by Ferguson (1986).

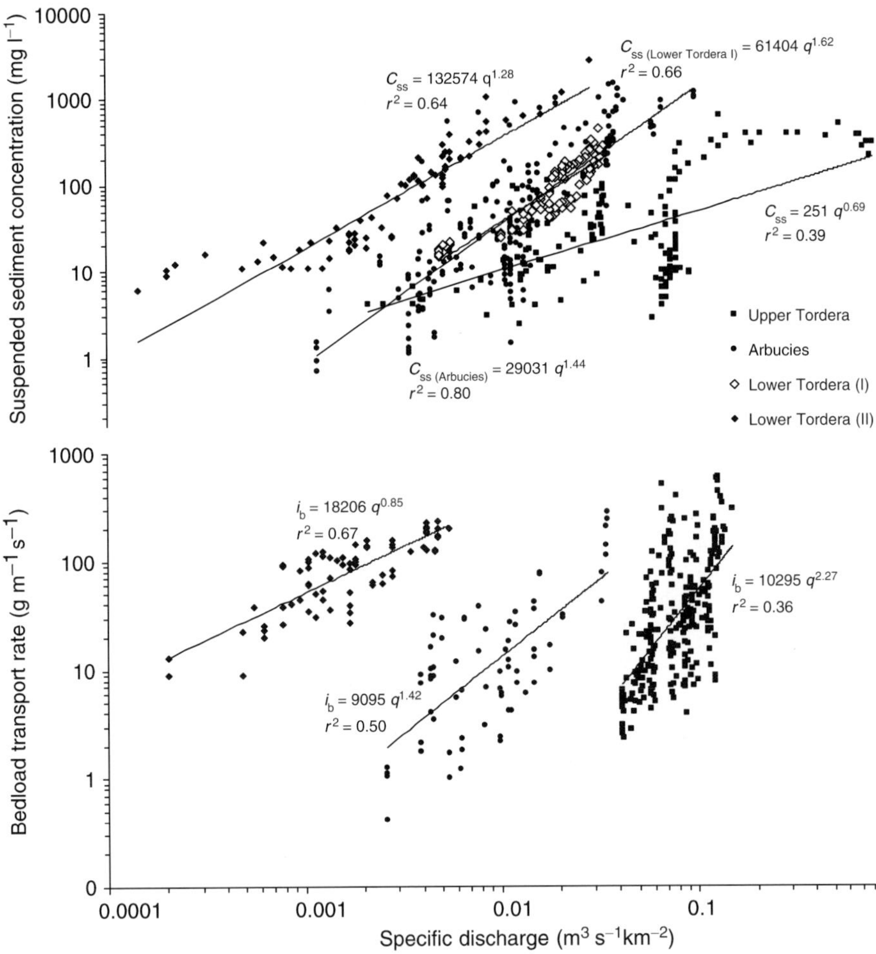

Figure 8.5. Suspended (C_{ss} vs. q) and bedload (i_b vs. q) rating curves for the three monitoring sections in the Tordera River for the study period 1990–1999.

yield of 41 t km^{-2} yr^{-1} (Table 8.4, Fig. 8.6). Inbar (1992) described similar values in the Mediterranean basin of Alexander in Israel. The suspended load varied greatly between years (the coefficient of variation reached 135%), with a minimum of 3600 t in the year 1998–1999 (4.6 t km^{-2} yr^{-1}) and a maximum of 110,000 t in 1995–1996 (140 t km^{-2} yr^{-1}). Most of this sediment was supplied by the granitic catchments of the region (*ca.* 60% of the basin area), of which the Arbucies River is a representative example. This basin has a mean annual suspended sediment yield on the order of 60 t km^{-2} yr^{-1}, which conforms the 100 t km^{-2} yr^{-1} reported by Walling and Webb (1983, 1996) and Lvovich et al. (1991) for Mediterranean areas on the Iberian Peninsula and is three times higher than the suspended sediment yield of the Upper Tordera in the metamorphic area of the catchment (20 t km^{-2} yr^{-1}). The suspended yield of the Lower Tordera appears to be thus more controlled by the granitic-bedrock sub-catchments. The inter-annual

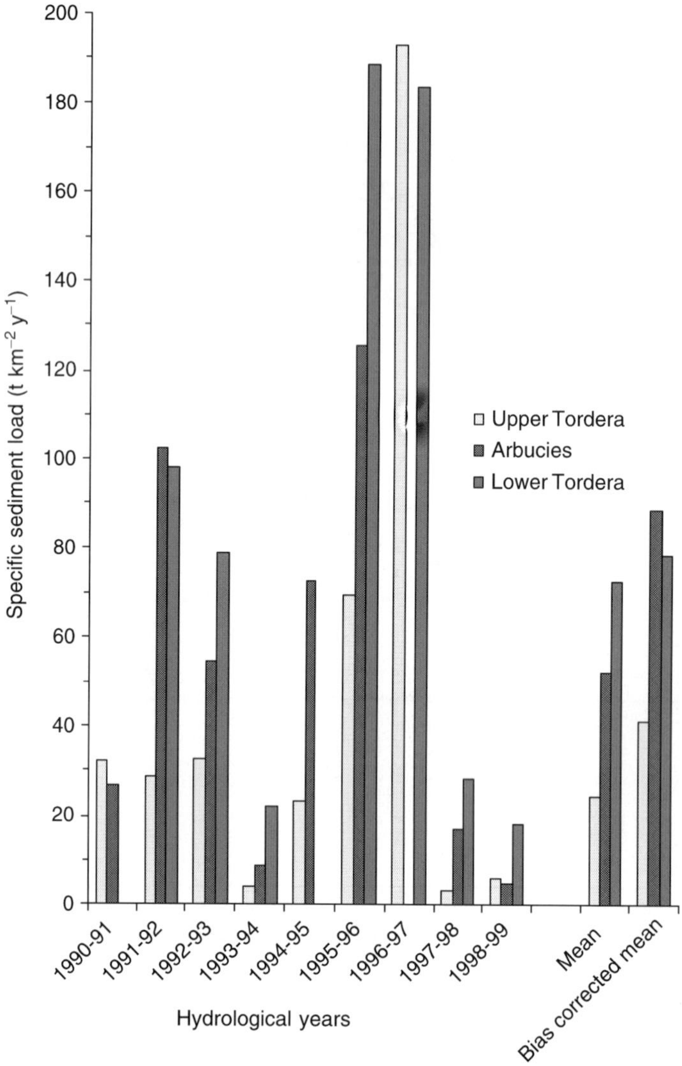

Figure 8.6. Annual specific solid load at the three monitoring sections in the Tordera River basin.

variability was high in all cases. The coefficient of variation was 112% in Arbucies and 86% in the Upper Tordera (La Llavina).

Most suspended sediment in the Tordera basin is transported during short periods of time, corroborating the high temporal variability of this class of sediment load (Fig. 8.7). In the Lower Tordera and Arbucies, 90% of the suspended load is carried in 10% of the time, while in the Upper Tordera, 80% of the fine load is transported during the same period of time. In all three cases, almost all sediment (>95%) is transported by discharges greater than those equalled or exceed 50% of time. These results emphasise the role of floods in the transport of suspended sediment in these Mediterranean rivers where, for

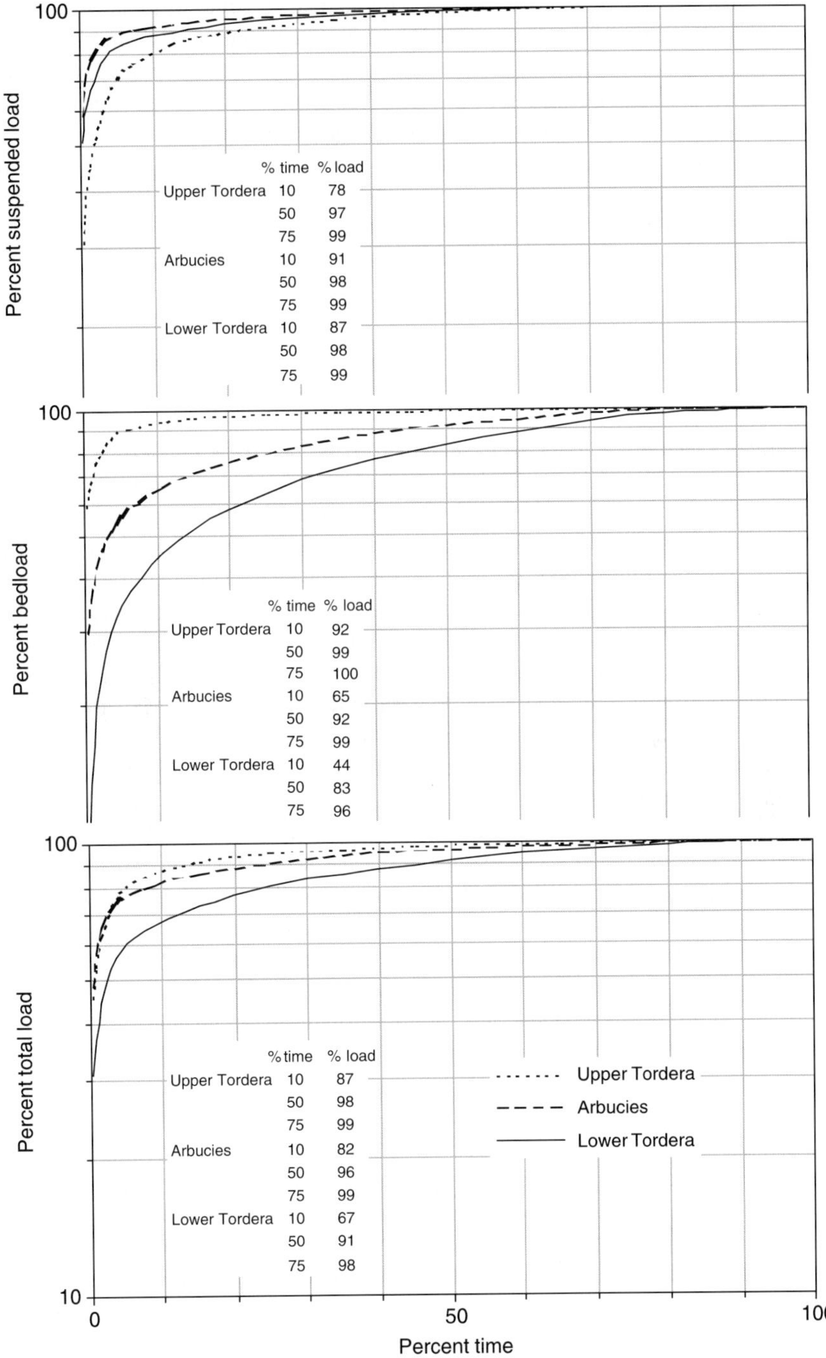

Figure 8.7. Sediment load frequency curves at the three monitoring sections in the Tordera River basin. Only 6 years with available hydrological data at all three sections have been used in the computation (1991–1992, 1992–1993, 1993–1994, 1995–1996, 1997–1998, and 1998–1999).

instance, discharges equalled or exceeded only 0.5% of time were able to transport between 40% in the case of the Upper Tordera and 50% in the cases of the Arbucies and the Lower Tordera, of the fine sediment load. In contrast, rivers in more humid environments need a higher duration of time (2%) to transport similar proportion (50%) of the annual suspended load (i.e. Creedy River, UK; Walling, 1984).

The degree of correlation between the annual suspended loads of the three monitoring stations for 6 years of comparison is high ($0.80 > r^2 < 0.84$) (i.e. a high coincidence exists between the three locations for the years with minimum, mean and maximum loads), reflecting the efficiency of the drainage network in transferring fine particles once they are entrained by the water in the channel of this free-flowing water system (there are no dams in the area).

4.2.2. Bedload

Bedload includes coarse particles (from sand to gravels and coarser materials) that travel in permanent contact with the streambed. As with the suspended sediment concentrations, the bedload transport rates (i_b) show a positive and statistically significant relation with specific discharge (q) ($p < 0.01$) in the three monitoring sections of the Tordera River (Fig. 8.5). A considerable degree of scatter exists, however. This may be related to the entrainment threshold of the material and sediment availability, both factors of great significance for the gravel-bed of the Upper Tordera (Garcia et al., 1999), and the migration of bedforms, which is an important phenomenon in the sandy-gravel beds of the Arbucies and the Lower Tordera rivers (Rovira, 2001; Batalla and Martin-Vide, 2003). The statistical bias of the relation between q and i_b due to the scatter of points, was calculated to be 10% in the Lower Tordera, 63% in the Arbucies River and 65% in the Upper Tordera. The underestimation was subsequently corrected by means of the factor developed by Ferguson (1986). The scatter of transport rates is especially important in the Upper Tordera section, where the exponent of the Eq. (2.27) indicates that bedload transport is much more dependent of the increment of discharge than in the Arbucies (1.42) and the Lower Tordera (0.85). On one hand, the bedload in the Upper Tordera is controlled by patch dynamics but, especially, by the breaking of the armour layer, which releases substantial amounts of subsurface material to be transported as bedload (Garcia et al., 1999). On the other, and in contrast, the bedload yield in the Arbucies River and in the Lower Tordera is strongly influenced by the continuous movement of the sandy particles on the channel bed. This enables low flows to carry bedload, thus expanding the range of flows during which this type of sediment transport occurs. Field observations indicated that a water depth of only around 10 cm is needed to initiate the movement of sand particles in both sections (slopes of 0.5 to 0.9%). This water depth is equalled or exceeded 80% of the time in the case of the Arbucies and 90% of the time in the case of the Lower Tordera (Batalla et al., 1995; Rovira et al., 2004).

The mean bedload transport rates range from 30 to 100 g m^{-1} s^{-1} (submerged weight) for the sandy-gravel beds of the Arbucies and the Lower Tordera rivers, respectively, although in the Lower Tordera, measurements were made under lower discharges. In the Upper Tordera, the mean bedload transport rate reaches 65 g m^{-1} s^{-1}, a rate associated with discharges averaging 0.08 m^3 s^{-1} km^{-2}, which are much higher than those sampled in its sandy bedded counterparts. A given bedload rate requires much less stream power (e.g. discharge) in the loose riverbed of the Lower Tordera (Rovira, 2001) than in the sandy-gravel bed of the Arbucies River (Batalla, 1997) and, especially, than in the armoured gravel-bed river of the Upper Tordera (Garcia et al., 1999) (Fig. 8.5). As in the case of the suspended load, this situation may be related to the intensive rock disintegration that occurred during the Pleistocene, still supplying surplus amounts of sandy sediment and yielding higher contemporary rates of bedload transport. As a result, and in terms of the capability of the water column to carry bedload, the flows are one order of magnitude more efficient in the Lower Tordera than in Arbucies and two orders of magnitude more efficient in the Lower Tordera than in the Upper Tordera. The maximum bedload rates were obtained in the Upper Tordera (600 g m^{-1} s^{-1} at a flow of 0.120 m^3 s^{-1} km^{-2}), followed by the Arbucies (280 g m^{-1} s^{-1} at a flow of 0.03 m^3 s^{-1} km^{-2}) and the Lower Tordera (230 g m^{-1} s^{-1} at a flow of 0.005 m^3 s^{-1} km^{-2}), the last two results being obtained under smaller specific discharges.

The mean annual total bedload flux passing the downstream Lower Tordera section was estimated to be around 29,000 t yr^{-1} (in submerged weight), giving a specific sediment yield of 37 t km^{-2} yr^{-1} (Table 8.4, Fig. 8.6). These results are, for instance, in agreement, with those obtained by Schick and Lekach (1993) in the Mediterranean basin of Nahal Yael (Israel). Annual bedload transport varies greatly between years (the coefficient of variation reaches 52%), with a minimum of 10,600 t in the year 1998–1999 (13 t km^{-2} yr^{-1}) and a maximum of 43,000 t in 1991–1992 (55 t km^{-2} yr^{-1}). Although high, the inter-annual variability is smaller than that described for the suspended sediment. The granitic catchments of the region (e.g. Arbucies) supply more bedload to the lower reaches (27 t km^{-2} yr^{-1}) than the metamorphic areas (22 t km^{-2} yr^{-1}). Inter-annual variability is high in all cases. The coefficient of variation reaches 75% in the Arbucies and 130% in the Upper Tordera (La Llavina).

Most bedload in the Tordera basin is transported during short periods of time, corroborating the high temporal variability of this type of sediment transport in Mediterranean catchments. However, the importance of high flow is less than for the fine sediment load (Fig. 8.7). In the Lower Tordera, 45% of the bedload is carried in 10% of the time, in the Arbucies River 65% of the bedload is transported in 10% of the time, and in the Upper Tordera 95% of the bedload is transported in 10% of the time. For comparison, in the Santa Clara River (California, USA), 97% of the sediment is transported during 1.2% of the time (Envicom, 1979). Bedload transport is thus more continuous in the Lower Tordera, where sand particles move even under very low discharges (the discharge representing the threshold of entrainment of D_{50} is equalled or exceeded 90% of the time), than in the Upper Tordera, where only discharges equalled or exceeded 15% of the time are able to move the D_{50} of the bed-material, and thus bedload transport is more hydraulically dependent. The Arbucies River occupies an intermediate position, reflecting the sand and gravel sediment mixture of its river-bed. The results obtained for the amount of bedload carried during 50% of the time points in the same direction: i.e. 85% in the Lower Tordera, 95% in the Arbucies and almost 100% in the Upper Tordera.

The results also illustrate the distinct role of floods in the transport of bedload in these rivers. For instance, flows equalled or exceeded only 0.5% of the time are able to transport 60% of the total bedload of the Upper Tordera, 30% in the Arbucies and 10% in the Lower Tordera.

The degree of correlation between the annual bedload fluxes of the Arbucies and the Lower Tordera is high ($r^2 = 0.90$) (e.g. there is a high coincidence between the two locations in years with minimum, mean and maximum load show), reflecting the rapid transfer of bedload of similar size between the two sampling sections separated by only 12 km. No significant correlation exists between the annual bedload yields of the Upper Tordera and the Lower Tordera sections ($r^2 = 0.45$). This latter result may reflect the different nature of the bedload in the two reaches, in relation to grain-size and the sedimentological structure of the river-bed and it may especially reflect the distance between them (45 km). Coarse sediment is deposited in the upper river sections and only sporadically reaches the valley bottom and the basin outlet. The bedload yield of the Lower Tordera would, therefore, appear to be more influenced by inputs from the granitic-bedrock catchments (e.g. the Arbucies and the neighbour basin of the Santa Coloma, the main tributary of the Lower Tordera), than the contribution of coarse material coming from the metamorphic areas.

4.2.3. Total load

The total load includes all sediment that travels through a given river section and accounts for particles moving in suspension, by saltation and as bedload. It can be estimated by adding the suspended sediment load and the bedload rates, and it is expressed as a solid flux rate (kg s^{-1}). In the case of the Tordera, the total load was only calculated for discharges under which both transport components were sampled and are presented for the Upper Tordera and the Arbucies River, due to their similar catchment areas but different dominant lithology (Table 8.5). The samples have been grouped by discharge increments of 0.5 $m^3\ s^{-1}$, therefore, only a small number of data points are used in the analysis (Fig. 8.8). The mean total sediment load is higher in the Upper Tordera (1.51 kg s^{-1}) than in the Arbúcies River (0.92 kg s^{-1}), although the overall composite samples were collected under higher discharges in the Upper Tordera (ca. 4.5 $m^3\ s^{-1}$) than in the Arbucies (ca. 2 $m^3\ s^{-1}$). As was the case with the two loads separately, the total sediment

Table 8.5. Characteristical values of total sediment load (TSL) at the monitoring sections of the Upper Tordera and the Arbucies River during the study period 1990–1999.

	Load (kg s^{-1})	Upper Tordera	Arbucies
Bedload	Maximum	1.37 (7.25)[a]	0.98 (3.75)
	Minimum	0.02 (1.75)	0.04 (0.25)
	Mean	0.49 (4.50)	0.35 (2.00)
Suspended load	Maximum	2.78 (7.25)	1.52 (3.75)
	Minimum	0.05 (2.75)	0.01 (0.25)
	Mean	1.00 (4.50)	0.56 (2.00)
Total load	Maximum	4.20 (6.25)	2.50 (3.75)
	Minimum	0.12 (2.25)	0.04 (0.25)
	Mean	1.51 (4.50)	0.92 (2.00)

[a] In parenthesis the discharge in $m^3\ s^{-1}$ under which those concentrations were measured.

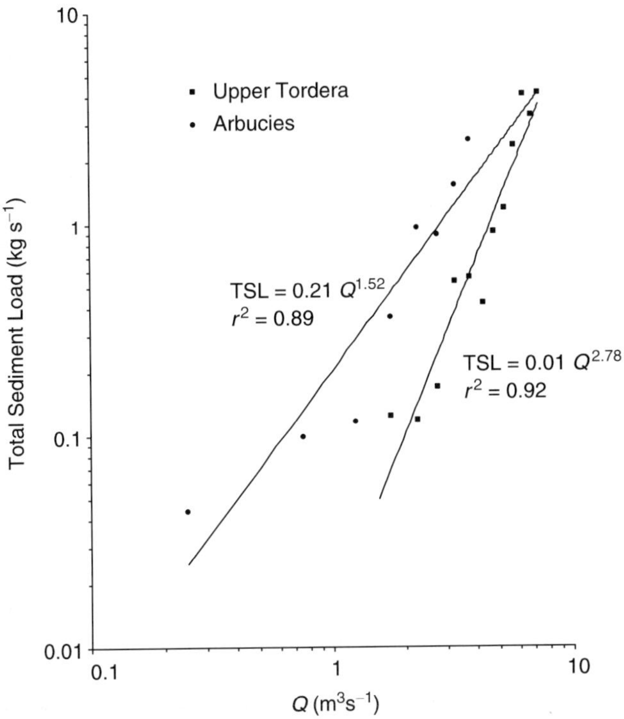

Figure 8.8. Sediment total load rating curves for two monitoring sections in the Tordera River for the study period 1990–1999.

load (TSL) showed a positive and statistically significant relation with the discharge ($p < 0.01$) for the two monitoring sections (Fig. 8.8). The statistical relation between discharge and total sediment load reflects the general behaviour of the two separate components. There is: (1) a more constant sediment transport in the case of the Arbucies River, especially due to the low entrainment threshold of particles in the river-bed; and (2) more hydraulically dependent dynamics in the case of the Upper Tordera, a consequence of the more stable riverbed structure, which is rarely entrained under discharges lower than 2 $m^3 s^{-1}$. Despite this different behaviour, however, the total concentration in the two rivers tended to converge at high flow discharge, at around 7 $m^3 s^{-1}$, probably representing a situation with complete riverbed mobilisation and high sediment supply from their respective catchments.

The mean annual total sediment load passing the downstream Lower Tordera section was estimated at around 61,600 t yr^{-1}, with a specific sediment yield of almost 80 t $km^{-2} yr^{-1}$ (Table 8.4, Fig. 8.6). The total bedload transport varied widely between years (the coefficient of variation reaches 97%), with a minimum of 14,200 t in 1998–1999 (18 t $km^{-2} yr^{-1}$) and a maximum of 148,000 in 1995–1996 (190 t $km^{-2} yr^{-1}$). The total sediment load from the granitic catchments was 88 t $km^{-2} yr^{-1}$, twice the sediment contribution from the metamorphic areas (41 t $km^{-2} yr^{-1}$) (Table 8.4). Overall, the total load is almost equally distributed between fine sediment in suspension and coarse sediment from the river-bed. However, the contribution of suspended sediment (70%) is higher than that for bedload (30%) in the Arbucies River. This result is partly contradictory to the data reported by Batalla et al. (1995), in which bedload accounted for a larger part of the total load. The rating curves between discharge and solid load used in this chapter are the same as published previously elsewhere (e.g. Batalla et al., 1995), but the study period is different (1967–1992). For comparison, however, total load is exactly the same (*ca.* 9500 t yr^{-1}).

Most of the total sediment load in the Tordera River was transported during short periods of time (Fig. 8.7), in accordance with the results for the suspended and bedload reported above. In the Lower Tordera, 65% of the total load was carried in 10% of the time, in the Arbucies 80% of the total load was carried in 10% of the time, and in the Upper Tordera 90% of the total load was carried in 10% of the time. Overall, sediment transport is thus more constant in the Lower Tordera, where almost all classes of sediment move with even low discharges. In contrast, the sediment load of the Upper Tordera is more dependent on increased flow (e.g. floods), capable of mobilising particles from the river-bed. Again, the Arbucies basin shows an intermediate position. The values for the amount of total load carried during 50% of the time evidence a similar trend with 90% in the Lower Tordera, 95% in the Arbucies and almost 100% in the Upper Tordera. In addition, the results illustrate the role of floods in the transport of bedload in Mediterranean rivers. For instance, in the Tordera events equalled or exceeded only 0.5% of the time were able to transport 50% of the total bedload in the Upper Tordera, 45% in the Arbucies River and 30% in the Lower Tordera.

From these results, it can be concluded that the total sediment yield from the Lower Tordera catchment reflected the overall basin sediment dynamics, but was more closely related to the sediment input from the granite-bedrock parts of the catchment (e.g. the Arbucies and Santa Coloma rivers, etc.), which act as the main sediment source for the basin, and is not so dependent on the contribution from the metamorphic areas (Table 8.4). The total sediment load passing the Lower Tordera section (TSL_{LT} in t km^2 yr^{-1}) can be related to both the sediment load of the Upper Tordera section (TSL_{UT}) and the Arbucies section (TSL_{AR}), by means of the following equation, which is statistically significant at $p < 0.01$:

$$TSL_{LT} = 8.53 + (1.6\ TSL_{UT}) + (0.51\ TSL_{AR})$$
$$(r^2 = 0.98, N = 6, \text{statistical errors: } 4.7, 0.31, 0.16, \text{ respectively})$$

As indicated in the methodology section, and with the objective of establishing the overall sediment yield for the whole Tordera basin in relation to the main lithologies of the catchment, the catchment has been divided into three areas (the Western Tordera, Central Tordera, and Eastern Tordera), for which their sediment loads are well represented by concentrations and rates associated with the metamorphic and plutonic areas (Fig. 8.1, Table 8.6).

Using this approach, the mean annual total sediment load at the basin outlet (Lower Tordera at Fogars) was estimated to be around 62,200 t yr^{-1} (Table 8.6), a value very close to the value estimated from direct measurements (61,600 t yr^{-1}, Table 8.4). Consistency of these results confirms the accuracy of the overall calculations. The Eastern Tordera supplied almost 40% of the annual total load of the Lower Tordera, followed by

Table 8.6. Sediment yield in the Tordera River in relation to the main sediment contributing zones and lithology.

			Area	Suspended load		Bedload		Total yield	
			(km^2)	($t\ yr^{-1}$)	(%)	($t\ yr^{-1}$)	(%)	($t\ yr^{-1}$)	(%)
Western Tordera	Metamorphics		98	1921	25	2127	45	4047	33
	Plutonics		93	5753	75	2576	55	8328	67
		Total	191	7674	100	4702	100	12,376	100
Central Tordera	Metamorphics		25	490	3	543	7	1033	4
	Plutonics		256	15,667	97	7014	93	22,682	96
		Total	281	16,157	100	7557	100	23,714	100
Eastern Tordera	Metamorphics		25	490	3	543	6	1033	4
	Plutonics		288	17,626	97	7891	94	25,517	96
		Total	313	18,116	100	8434	100	26,549	100
	Total		785	41,946		20,693		62,639	

the Central Tordera (36%) and the Western Tordera (25%). The plutonic zones in the three areas (occupying 49% in the Western Tordera and 91% of the Central and Eastern parts of the basin) appear to be the main source of sediment, both in suspension and as bedload, accounting for 90% of the annual total load (ca. 56,000 $t\ yr^{-1}$).

At the Lower Tordera measuring section, the suspended sediment load accounted for 2/3 of the total load (41,000 $t\ yr^{-1}$), whereas bedload was responsible for the other 1/3 (20,600 $t\ yr^{-1}$). Both components were of the same order of magnitude as loads calculated from direct measurements. However, in comparing the two sets of results, the Tordera River mainstem appears to be a sink for fine sediment (ca. 9000 $t\ yr^{-1}$) and a source of bedload (ca. 8000 $t\ yr^{-1}$). On one hand, the river floodplain could act as a sedimentation zone for suspended load, a phenomenon that may have occurred during high magnitude floods (e.g. January 1996, December 1997) (Fig. 8.3). On the other, the reasons for the increased bedload are not entirely clear, although they could be related to the almost constant bedload detected at the Fogars measuring sections, which, in turn, may have overestimated the bedload transport, and thus, the final value of sediment load.

The annual contribution of bedload transport to the total sediment yield of the Lower Tordera drainage basin is very high. In sand bed channels, bedload commonly constitutes a small proportion (between 1 and 20%) of the total load (Lane and Borland, 1951; Simons and Senturk, 1977). The continuous movement of the sand fraction on the channel bed influenced the bedload yield in the Lower Tordera. Field observations indicated that movement of fine sediment on the channel bed occurred during 90% of the time. The relatively high bedload yield was mainly due to the high frequency of discharges, which were capable of transporting high amounts of sandy sediment. This natural process may accelerate the progressive incision of the drainage network caused by widespread afforestation of the basin headwaters and by the extensive gravel mining that took place in the river during 1960s and 1970s. Clear evidences of incision can be seen in the area of Fogars de Tordera, with examples of undercut infrastructures (e.g. the A7 motorway bridge collapse during a flood in 1971, the bridge to the village of

Hostalric undercut by 3 m, and the new gauging station at Fogars de Tordera, which is seriously damaged since it commenced operation in 1992). In addition, the incision process might be reinforced by the tectonic activity in the study area. This is a tectonically active area, with frequent but low seismic activity, with recorded earthquakes usually below VII degrees on the Richter scale (Udias, 1982). Insufficient field data are currently available to verify this hypothesis.

5. Summary and conclusions

This chapter provides information on the sediment yield of the Tordera River, a large non-regulated Mediterranean catchment, in which research on sediment transport was undertaken between 1990 and 1999. One thousand measurements of suspended load and bedload were used to estimate the total yield of two catchments, which are representative of the main lithologies in the region (the Upper Tordera and the Arbucies), and the total sediment yield of the Lower Tordera. The measurements covered a wide range of discharges in the three monitoring sections, ranging from baseflow up to small and medium floods. Overall, the data indicated that the Lower Tordera shows a similar pattern of sediment yield to that documented for other rivers in Mediterranean semi-arid regions, whereas the values obtained in the Arbucies River (mainly for bedload) and, especially, in the upper part of the catchment are more similar to data reported for more humid Mediterranean mountainous environments. The main conclusions of the work are as follows:

(1) The mean annual suspended sediment load passing the downstream Lower Tordera section was estimated at 32,500 t, giving a specific suspended sediment yield of 41 t km^{-2} yr^{-1}. Most of this sediment is supplied by the granitic catchments of the basin at a rate of 60 t km^{-2} yr^{-1}, three times higher than that of the metamorphic areas (20 t km^{-2} yr^{-1}).

(2) The mean annual bedload flux crossing the Lower Tordera section was estimated at 29,000 t yr^{-1}, giving a specific sediment yield of 37 t km^{-2} yr^{-1}. More bedload is supplied by catchments underlain by plutonic rocks, e.g. the Arbucies, (27 t km^{-2} yr^{-1}), than by those in the metamorphic areas (22 t km^{-2} yr^{-1}).

(3) Altogether, the mean annual total sediment load in the Lower Tordera section was estimated at around 61,600 t yr^{-1}, with a specific sediment yield of almost 80 t km^{-2} yr^{-1}. Overall, the total load is almost equally distributed between fine sediment in suspension and coarse sediment transported as bedload. The mean annual total sediment load from granitic catchments is 88 t km^{-2} yr^{-1}, which is double than the sediment contribution from the metamorphic areas (41 t km^{-2} yr^{-1}).

(4) Most sediment in the Tordera River is transported during short periods of time. In the Lower Tordera, 65% of the total load is carried in 10% of the time, in the Arbucies River the proportion raises to 82% and in the Upper Tordera it reaches almost 90%, for the same time duration. As a whole, sediment transport is more constant in the Lower Tordera and the Arbucies, where sediment moves under almost all discharges, in comparison with the Upper Tordera where sediment transport, especially coarse bedload, is more sporadic and exhibits a stronger hydraulic (e.g. flood) dependency.

(5) The Eastern Tordera supplies almost 40% of the annual total load to the Lower Tordera, followed by the Central Tordera (36%) and the Western Tordera (24%). The plutonic

areas appear to be the main source of sediment, both in suspension and as bedload, accounting for the 90% (*ca.* 56,000 t) of the annual total load for the entire basin.

Acknowledgements

Funds for the development of the Tordera research programme were mainly provided through two projects funded by the Spanish Ministry of Education and Science (AMB93-0418 and HID96-0971). The Water Authorities of Catalonia supplied hydrological data for the Tordera River gauging stations. We are indebted to the GRAM team at the University of Barcelona for their assistance during the fieldwork. Revisions undertaken by Olav Slaymaker, University of British Columbia, and Des Walling, University of Exeter, greatly improved the first version of the manuscript.

This chapter is for us an opportunity to express our own debt of gratitude to Maria Sala, as a teacher, a colleague both in science and in life, and a friend. Thank you, Maria.

References

Batalla R.J., 1993. Contribució del transport de sorres al balanç de sediment d'una conca granítica mediterrània. Ph.D. Thesis, Universitat de Barcelona, Spain.

Batalla, R.J., 1997. Evaluating bed-material transport equations from field measurements in a sandy gravel-bed river. Earth Surf. Proc. Land. 21, 121–130.

Batalla, R.J., Sala, M., 1994. Temporal variability of suspended sediment transport in a Mediterranean sandy gravel-bed river. In: Olive, L.J., Loughran, R.J., Kesby, J.A. (Eds), Variability in Stream Erosion and Sediment Transport. IAHS Publication 224, p. 299.

Batalla, R.J., Sala, M., Werritty, A., 1995. Sediment budget focused on solid material transport in a subhumid Mediterranean drainage basin. Z. Geomorphol. 29 (2), 249–264.

Batalla, R.J., Martin-Vide, J.P., 2003. Thresholds of particle entrainment in a sandy gravel-bed river. Catena 44 (3), 223–243.

Emmett, W.W., 1979. A field calibration of the sediment trapping characteristics of the Helley Smith bedload sampler. US Geol. Surv. Prof. Paper, 1139.

Envicom, 1979. Santa Clara River Sand and Gravel Extraction Master EIR Study, Draft Environmental Impact Report: Conditional Use Permits 1670 and 3390.

Ferguson, R.J., 1986. River loads underestimated by rating curves. Water Resour. Res. 21, 74–76.

Folk, R.L., Ward, W.C., 1957. Brazos river bar: a study in the significance of grain size parameters. J. Sediment. Petrol. 27, 3–26.

Garcia, C., 1997. Transporte de fondo en un río de gravas y su relación con la dinámica del lecho fluvial. Ph.D. Thesis, Universitat de Barcelona, Spain.

Garcia, C., Laronne, J.B., Sala, M., 1999. Variable source areas of bedload in a gravel-bed stream. J. Sediment. Res. 69 (1), 27–31.

Garcia, C., Laronne, J.B., Sala, M., 2000. Continuous monitoring of bedload flux in a mountain gravel-bed river. Geomorphology 34, 23–31.

Helley, E.J., Smith, W., 1971. Development and Calibration of a Pressure-Difference Bedload Sampler. United States Geological Survey Open File Report.

Inbar, M., 1992. Rates of fluvial erosion in basins with a Mediterranean type climate. Catena 19, 393–409.

Johansson, C.E., 1963. Orientation of pebbles in running water. A laboratory study. Geogr. Ann. 45, 85–112.

Lane, E.W., Borland, W.M., 1951. Estimating bedload. Trans. Am. Geophys. Union 32 (1), 121–123.

Langbein, B.S., Schumm, S.A., 1958. Yield of sediment in relation to mean annual precipitation. Trans. Am. Geophys. Union 39, 1076–1084.

Laronne, J.B., Carson, M.A., 1976. Interrelationships between bed morphology and bed material transport for a small gravel bed channel. Sedimentology 23, 67–85.

Lvovich, M.I., Karasik, G.Y., Bratseva, N.L., Medvedeva, G.P., Maleshko, A.V., 1991. Contemporary Intensity of the World Land Intracontinental Erosion. USSR Academy of Sciencies, Moscow.
Reid, I., Layman, J.T., Frostick, L.E., 1980. The continuous measurement of bedload discharge. J. Hydraul. Res. 18, 243–249.
Rovira, A., 2001. Balanç de sediment i dinàmica fluvial en un riu de règim hidrològic transitori (Tram final de la Tordera). Ph.D. Thesis, Universitat de Barcelona, Barcelona.
Rovira, A., Batalla, R.J., Sala, M., 2004. Fluvial sediment budget of a Mediterranean River: the Lower Tordera (Catalan Coastal Ranges, NE Spain). Catena 60, 19–42.
Sala, M., 1979. La cuenca del Tordera. Estudio geomorfológico. Ph.D. Thesis, Universitat de Barcelona, Spain.
Sala, M., 1983. Fluvial and slope processes in the Fuirosos basin, Catalan Ranges, North East Iberian Coast. Z. Geomorphol. 27 (4), 393–411.
Schick, A.P., Lekach, J., 1993. An evaluation of two ten-year sediment budgets, Nahal Yael, Israel. Phys. Geogr. 14 (3), 225–238.
Simons, D.B., Senturk, F., 1977. Sediment transport technology. Water Resources Publications, Fort Collins, CO.
Trimble, S.W., 1977. The fallacy of stream equilibrium in contemporary denudation studies. Am. J. Sci. 277, 876–887.
Udias, A., 1982. Rasgos sismotectónicos y estructura de la corteza de la región de Catalunya-Pirineos. In: Càtedra de Geofísica, Universidad Complutense de Madrid (Ed.), La sismicidad en la zona comprendida entre 40° N - 44° N y 3° W - 5°. E. Publ. 190, pp. 149–155.
Walling, D.E., 1984. Dissolved loads and their measurements. In: Hadley, R.F., Walling, D.E. (Eds), Erosion and Sediment Yield: Some Methods of Measurements and Modelling. Geo Books, London, pp. 111–177.
Walling, D.E., Webb, B.W., 1983. Patterns of sediment yield. In: Gregory, K.J. (Ed.), Background to Paleohydrology. Wiley, New York, pp. 69–100.
Walling, D.E., Webb, B.W., 1996. Erosion and sediment yield: a global overview, Symposium on Erosion and Sediment Yield: Global and Regional Perspectives. Exeter. IAHS Publication 236, pp. 3–19.

Catchment Dynamics and River Processes: Mediterranean and Other Climate Regions
Celso Garcia, Ramon J. Batalla, Editors

Chapter 9

Upland gravel-bed rivers with low sediment transport

Michael Church* and Marwan A. Hassan

Department of Geography, The University of British Columbia, Vancouver, British Columbia, Canada V6T 1Z2

Abstract

Harris Creek is a cobble–gravel stream occupying an upland valley with low sediment supply and a strongly seasonal, nival flow regime. Bed material transport is generally restricted to phase 1 (sand transport) and phase 2 (partial transport of gravels). Potentially mobile sediment patches are sparse, the armour ratio is greater than three, and the larger grains on the bed have formed reticulate networks that increase hydraulic resistance to flow. The grain size that can withstand the usual range of shear forces imposed by the flow is accordingly reduced. The consequence of these conditions is low bed material transport in a channel that has generally been stable over a period of more than 40 years. These characteristics, hence the overall channel morphology, are a direct consequence of the low and episodic bed material supply from a limited number of sources along the channel. Hence, catchment-scale control of sediment supply creates distinctive reach-scale channel morphology and conditions important site scale modifications of the channel bed that restrict sediment movement to bring it into balance with the supply. We propose that this is a characteristic condition of upland cobble–gravel channels in landscapes that are not significantly disturbed. Harris Creek represents the normal fluvial regime for such settings.

Keywords: bed material transport, channel stability, cobble–gravel channel, partial sediment transport, sediment supply, selective sediment transport, upland channels

1. Introduction

Sediments delivered to the fluvial system from headwater slopes commonly accumulate in upland valleys where channel gradients suddenly decline. The headmost extended alluvial reaches are found here. They receive sediments from steep headwater channels and slopes having a wide range of grain sizes. Because of competence limitations on the reduced valley-floor gradient, the coarser fractions remain there for long periods so that the channels typically have gravel or cobble–gravel beds. Such channels have been studied in glaciated (Carling, 1989; Andrews, 1994; Habersack and Laronne, 2001) and non-glaciated (Jackson and Beschta, 1982; Lisle et al., 2000; Emmett and Wolman, 2001; Ryan and Emmett, 2002) hill and montane country, and in Mediterranean mountains

*Corresponding author. *E-mail address:* mchurch@geog.ubc.ca (M. Church).

(Tacconi and Billi, 1987; Garcia et al., 1999, 2000). These regions have in common strongly seasonal flow regimes with occasional extreme flows. An important defining condition of such channels, then, is that delivery of bed material sediments to them is highly episodic. Sand and finer materials may be more or less abundant, according to the geology, hence weathering of the source terrane and colluvial deposits, but these materials pass downstream relatively quickly and do not define the character of the channels. Their morphology is adapted to the condition of low bed material transport. Adaptations to low transport include strong downstream size gradation, sparse sediment patchiness, and the development of surface structures (Church et al., 1998) that substantially increase the strength of the bed. Consequently, the channels remain more or less stable for extended periods of time. It may then surprise observers when they are suddenly destabilized by the arrival in the channel of a substantial charge of material following some major headwater sediment delivery event or streambank disturbance, or by the occurrence of an extreme flood.

Upland cobble–gravel channels have prompted a substantial reexamination of sediment transport theory, which classically was developed for relatively high transport rates. Bed material transport in these channels typically falls very near or below the conventionally adopted threshold condition. In such circumstances, the rate of change of transport with flow is much more sensitive than it is under high-transport conditions since the phenomenon involves a rapidly increasing proportion of the bed area as flow increases. This, in turn, implies that bed sediment is only partially entrained (Nizery and Braudeau, 1953; Wilcock and McArdell, 1993, 1997; Haschenburger and Wilcock, 2003). Furthermore, the transport usually is size selective, so that the gradation of transported sediment is not the same as the gradation of the bed material. These phenomena have recently been incorporated into sediment transport theory (Parker et al., 1982; Diplas, 1987; Parker, 1990; Wilcock and McArdell, 1997; Wilcock and Crowe, 2003), but the role of bed structures still has not been properly parameterized.

We have, for some years, studied one such channel – Harris Creek – in the Okanagan Plateau of southern British Columbia, giving special attention to the conditions of sediment transport and to the development of sediment structures. We have supported the field studies by experimental and computational investigations. The purpose of this chapter is to relate transport and morphology to the sediment source conditions in the contributing drainage basin. We first present our summary description of the sediment transport regime and consequent morphology of Harris Creek and, by extension, of similar upland streams. We intend to define the "normal" regime of these upland gravel-bed channels and to identify the conditions that create it.

2. Study context

2.1. Study catchment

Harris Creek is a tributary of Bessette Creek and Shuswap River near Lumby, British Columbia (Fig. 9.1). The drainage area is 253 km^2, with 220 km^2 above the principal research site. The elevation ranges from 530 to 2005 m, but about half the area lies on a southward rising plateau surface above 1200–1650 m.

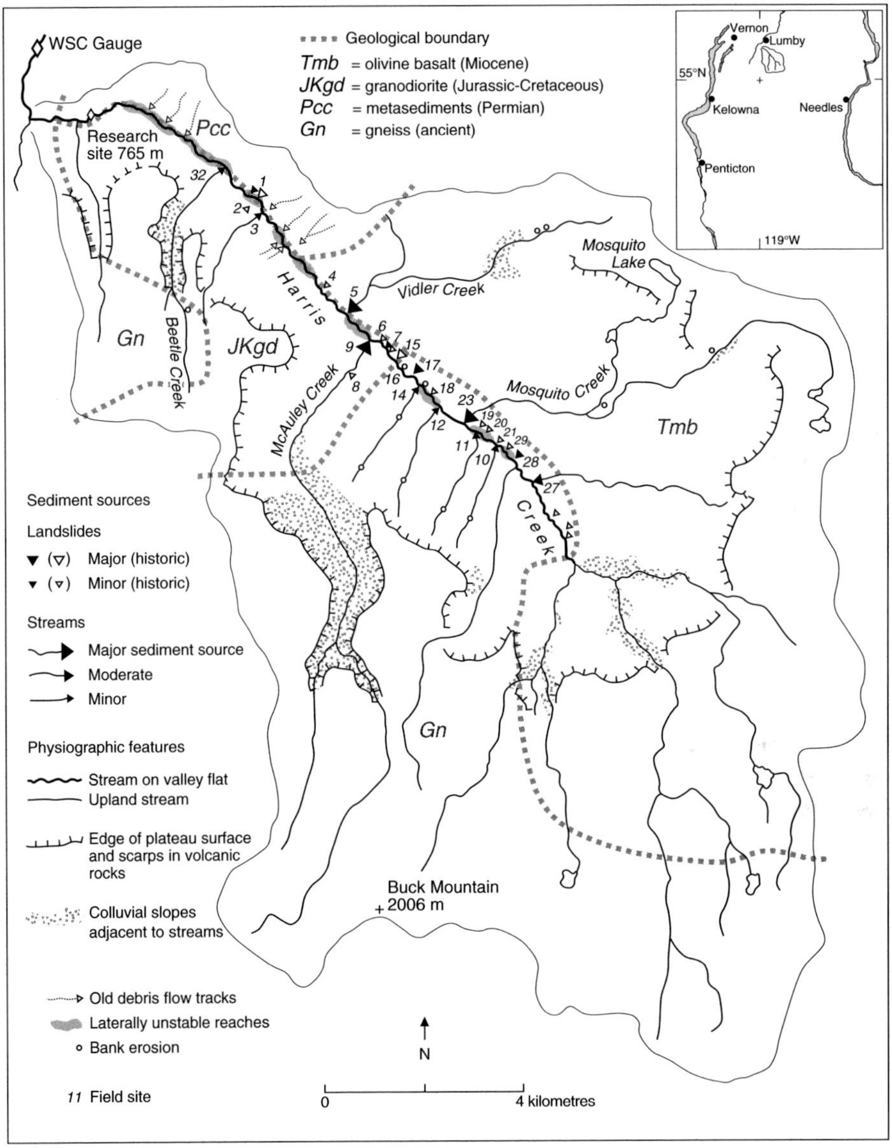

Figure 9.1. Map of the drainage basin of Harris Creek, showing the geology (after Jones, 1959), location of the plateau edge, principal sediment sources (numbers follow Ryder and Fletcher, 1991), the location of the principal research site for sediment transport, and of the WSC gauge: inset, location map.

The plateau accumulates a substantial snowpack over winter. Snowmelt creates a spring freshet in May and early June (Fig. 9.2). In some years, the highest flows are created by rain on snow or by rain alone, particularly by strong convective cells within late spring frontal storms. A 20-year-record of flows is available from a Water Survey of Canada

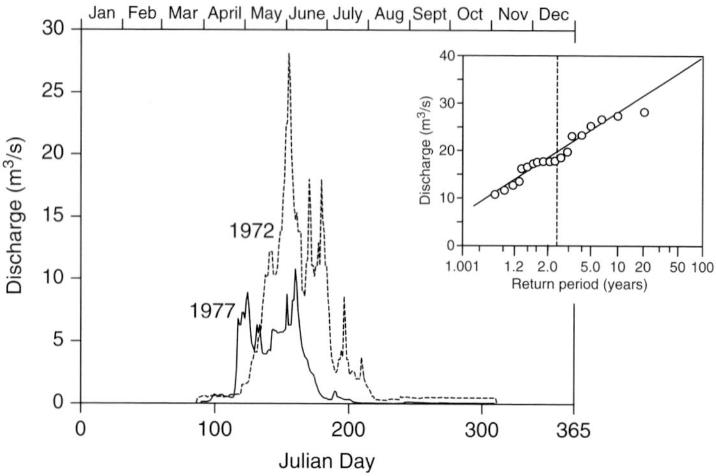

Figure 9.2. Hydrological regime of Harris Creek: typical annual hydrographs for low and high flow years (WSC data from Stn. 08LC005; Bessette Creek near Lumby): inset, flood frequency plot, adjusted by area-ratio to the principal research site on lower Harris Creek.

gauge (Stn. 08LC005) located near the basin outlet.[1] The maximum flow was recorded in May or early June in all but 1 year of the record (when the peak was a rainstorm flood on 2nd July). Mean annual flood at the outlet of Harris Creek is 22 $m^3\ s^{-1}$ (19.5 $m^3\ s^{-1}$ at the sediment transport research site (Fig. 9.2), by area-ratio adjustment).

Bedrock geology (Jones, 1959; Figure 1) consists of felsic plutonic rocks and gneiss of the Shuswap terrane and Tertiary volcanics of the Kamloops Group, including tuffs, breccias, basalt and other lavas. The contrasting lithologies are responsible for the topographic asymmetry of the upper valley. Tertiary gravels are also present in restricted outcrop. These varied lithologies yield a wide range of detritus, from silt to boulders. Lava flows form prominent scarps along the plateau edges (Fig. 9.1) which, in several places, have been subject to massive failures in the distant past. The most important determinant of the surface geology and of sediment supply to Harris Creek, however, is late Pleistocene glaciation. All but the steepest slopes are drift covered, till being widespread on the valleysides and plateau. The texture of the till varies according to the underlying bedrock lithology. Thicker drift, consisting of glaciofluvial gravels, glaciolacustrine silts and till, underlies the south-facing lower slopes along the main valley (Fig. 9.3A). Because of strong insolation onto the south slope and the consequent pattern of glacier melt in the valley, glacial ablation deposits are almost entirely restricted to that slope, the north-facing slopes having only shallow till over bedrock. Small alluvial fans have formed at the mouths of the principal tributaries and a Holocene floodplain has developed along lower Harris Creek, but there generally have been only minor landscape changes in Holocene time.

[1] The station is actually on Bessette Creek downstream from the confluence with Harris Creek, but Harris Creek constitutes more than 86% of the total contributing watershed.

Figure 9.3. Sediment sources in Harris Creek basin. (A) Prospect of Harris Creek valley, looking north from the plateau rim near McAuley Creek. Extensive glacial sediments on the south-facing slope are evident in the landslide scars; active slide 17 is indicated by the arrow; (B) slide 1; till over glaciolacustrine sediment actively undercut by Harris Creek; (C) stream bank of Harris Creek that has been under attack. Here, strong root reinforcement in the upper bank resists major erosion; (D) gravel deposit upstream of a logjam. The jam has failed and the river (out of view to the left) has degraded through the deposit. The stream bank here continues to be an active source of sediment. (Photographs by J.M. Ryder).

Stream courses in Harris Creek can be classified into three groups (Ryder and Fletcher, 1991):

(1) On the plateau surface, small streams with low gradients pass through numerous ponds and bogs. Consequently, sediment supply from the plateau is negligible.
(2) Steep channels (6°–17°) descend from the plateau through V-shaped valleys that are directly coupled to adjacent colluvial slopes from which they recruit sediment via debris slides and flows, and dry ravel.
(3) On the main valley floor, the gradient of Harris Creek is between 1° and 2°, but it is intermittently coupled to the valleyside where it flows against Quaternary sediments.

2.2. Sediment sources and sediment budget

Sediments arriving in Harris Creek are recruited almost entirely from discrete sources, including episodic debris flow from the steeper tributaries, slumps and slides from Quaternary sediments along the tributaries and lower valleyside (Fig. 9.3B), and episodic bank erosion (Fig. 9.3C). Ryder and Fletcher (1991) have mapped the principal sources (Fig. 9.1). Three active landslides contributed about 2100 tonnes of sediment in 1 year (more than 1900 tonnes from one slide), but only 6 tonnes altogether in two preceding years (Hou, 1997). Fluvial sediment transport from the tributaries was on the order of 400 tonnes per annum in the same year as the notable landslide input (Hou, 1997), the bulk of it deriving from Mosquito Creek, a steep right-bank tributary. Sediments delivered to the upland valley floor may be moved relatively quickly downstream, or may become stored in the alluvial fans, or in bars or floodplain sediments along Harris Creek, where they may remain for decades to many centuries (on the evidence of mature forest cover).

Recent land use has effected some modification of sedimentation and of stream morphology. Early in the 20th century, substantial areas of forest were burned to facilitate mineral prospecting, an act that appears to have mobilized a considerable volume of sediment. In the mid-twentieth century, the valley floor was logged and machinery was driven through the channel of Harris Creek. This broke down stream banks and destabilized some reaches (see Fig. 9.1). It also led to the creation of logjams along the creek, which subsequently trapped much sediment. Today, the logjams have been breached, but substantial sediment from that era remains stored on the valley floor adjacent to the contemporary channel (Fig. 9.3D). One degrading logjam contributed 120 tonnes of sediment to Harris Creek in a high yield year, but only 13 tonnes in the preceding 2 years (Hou, 1997).

Recent logging in the drainage basin has been far from the stream channels and has contributed only minor fine sediment from road surface drainage. Channel reaches that exhibit persistent low-level instability (Fig. 9.1) lie adjacent to or directly downstream from continuing sources of sediment. In these reaches, stream bank erosion along Harris Creek yielded 1120 tonnes in the high yield year and 290 tonnes in the preceding 2 years. Available data of sediment delivery to Harris Creek are summarized in Table 9.1. The data

Table 9.1. Data of sediment yield to Harris Creek and fluvial bed material transport (tonnes per annum).

Year	1991	1992	1993
Tributary streams (fluvial)			
McAuley Creek			40 (17–94)
Mosquito Creek			330 (300–370)
Landslides directly to Harris Creek			
Landslide 1	1.7 ± 0.4	1.1 ± 0.2	67 ± 15
Landslide 7	3.4 ± 0.6	3.0 ± 0.4	2000 ± 60
Landslide 17	Stable throughout the period		
Landslide 28	1.1 ± 0.1	0	39 ± 7
Failed logjam	11 ± 0.4	2.1 ± 0.3	120 ± 24
Streambank erosion	290 ± 50	0	1100 ± 170
Total mass wasting	310 ± 50	6.2 ± 0.5	3300 ± 180
Total, including tributary inputs			3700 (3500–3900)
Fluvial bed material transport	280–660	30–60	3200–7600

After Hou (1997), results have been rounded to two significant digits. Landslides are identified by number in Figure 9.1.

are based on only 3 years' observations and cannot be considered to represent reliable averages or extremes, but they do emphasize the episodic nature of significant sediment delivery to Harris Creek.

3. Sediment transport regime in Harris Creek

3.1. Sediment flux relations

We have studied bed material transport over a channel-spanning bar in lower Harris Creek (the "research site": see Fig. 9.1) using pit traps. We sampled suspended sediment by taking large water samples. At this site, median size (D_{50}) of the subsurface material is 16–19 mm, and of the surface material, 45–64 mm (see Fig. 9.10 for size distributions), yielding armour ratios in the range 2.5–4.0. The traps in fact sampled bedload, which we assume to be equivalent to bed material transport in the cobble gravel channel. Details of the trap construction and operation are given in Church et al. (1991) and in Hassan and Church (2001), while a critical evaluation of their performance is given in Sterling and Church (2002). The latter authors concluded that trap efficiency is uniformly high for all materials larger than sand. Trap locations are shown in Figure 9.4. It is important to recognize that the traps sampled only a fraction of the channel width, so that our measurements represent specific flux at various places on the channel bed rather than total transport in the stream. Measurements were continuous throughout the spring freshet period in 1989 and 1991.

Total load and size-specific rating curves were determined for each trap in each season, examples of which are shown in Figure 9.5. Salient features of these flux relations are as follows:

(a) All of the observed transport rates are extremely low: in fact, about half of our values fall below the small reference transport rate adopted by Parker et al. (1982) to signify the practical threshold of motion. That threshold is not surpassed until shear stress is about 15 Pa, when we detect material up to $D = 8$ mm in motion.
(b) The rating relations are extremely sensitive. We find that the exponent of the rating relation, b, typically takes values $4 < b < 12$.
(c) The threshold for sand movement is near 3 $m^3\ s^{-1}$, corresponding to shear stress of about 4 Pa, while the onset of gravel transport occurs near 4.5 $m^3\ s^{-1}$, corresponding to shear stresses near 15 Pa: consequently, two phases of transport are clearly evident in the data; a sand transport phase is followed by the onset of gravel transport at higher flows.
(d) Within-season anti-clockwise hysteresis is evident in the gravel transport regime, implying that the bed material is mobilized near the peak flows and continues to move on the declining stage (Fig. 9.6).
(e) In detail, the trap rating curves are not simple power relations. At trap 3B, for example, gravel transport appears no longer to vary systematically at flows above 10 $m^3\ s^{-1}$ (Fig. 9.5). From data of other traps we infer that this behaviour reflects a shift in the locus of transport within the channel rather than stabilization of the total transport or loss of trap efficiency.
(f) The relations are not stable from year to year (Fig. 9.5).

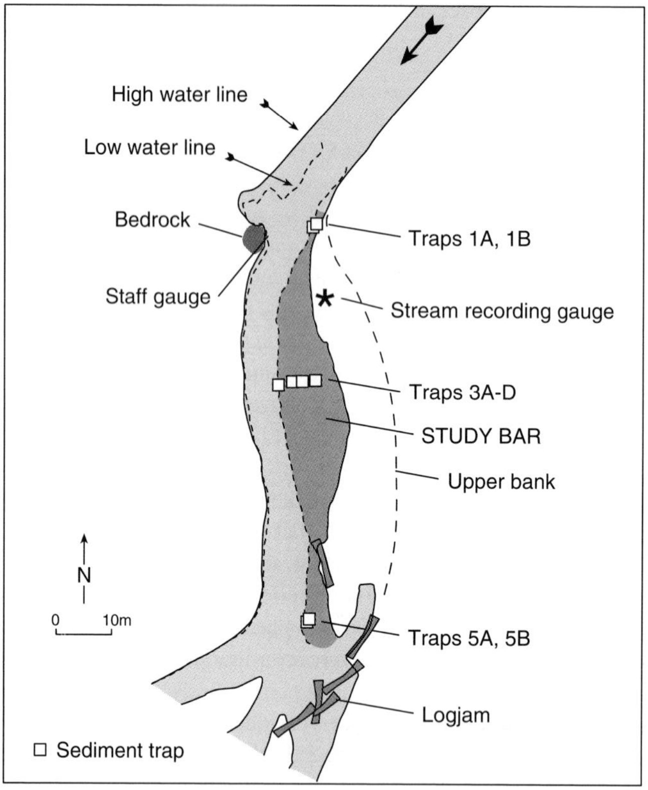

Figure 9.4. Map of the sediment transport study bar ("research site" of Fig. 9.1) in lower Harris Creek, showing bedload trap locations.

The two-phase transport observed by us corresponds with transport stages previously identified in gravel-bed channels with low transport (Emmett, 1976; Jackson and Beschta, 1982; Andrews, 1983). Ashworth and Ferguson (1989) and Warburton (1992) identified three phases of transport, dividing the gravel regime into size-selective and fully mobile phases. The sensitivity of transport to changing flow conditions has been noted in other cobble–gravel streams (Nanson, 1974; Adenlof and Wohl, 1994; Whiting et al., 1999; Emmett and Wolman, 2001).

It is interesting to ask how our observations compare with predictions from classical bedload transport formulae since the marginal conditions for transport depart dramatically

---

Figure 9.5. Bedload rating curves derived from trap observations in Harris Creek. Panels A–C indicate transport of selected sand, granule and pebble fractions, while panel D indicates total transport. Data points represent observations at trap 3B (Fig. 9.4) in 1989. Data are presented in non-standard units (kg/h) in order to preserve comprehensible quantities. Rating relations are fitted to data below 10 $m^3 s^{-1}$, except the 16–22 mm relation which is, in any case, not significant. Additional relations (data suppressed for clarity) were defined by observations at traps 3B and 1A in 1991. The 1991 statistics are for trap 3B. CI is the $\alpha = 0.10$ confidence interval about the estimate of b, the slope of the rating relation.

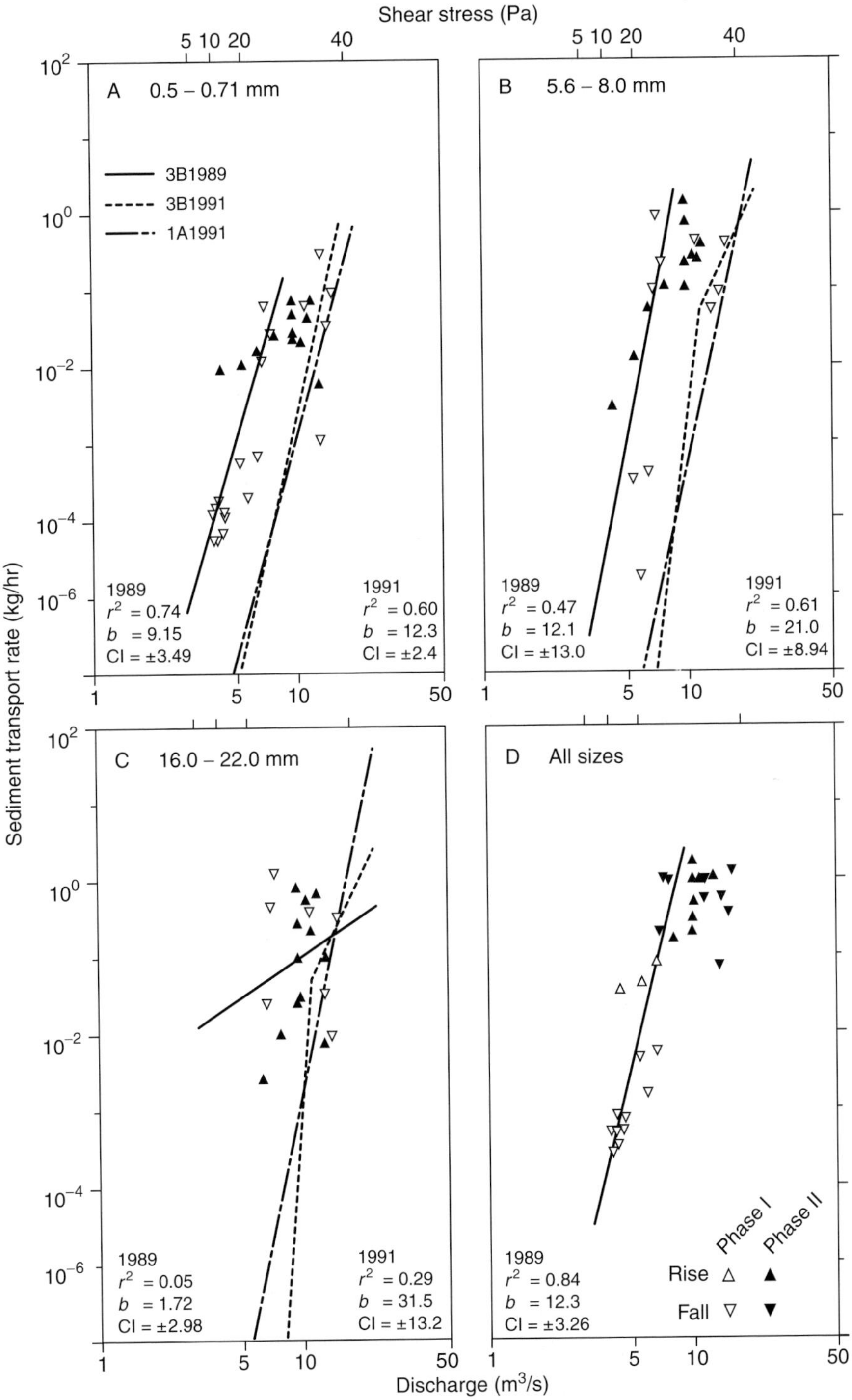
Shear stress (Pa)
5 10 20 40
Sediment transport rate (kg/hr)
Discharge (m³/s)
A 0.5 – 0.71 mm
3B1989
3B1991
1A1991
1989
r² = 0.74
b = 9.15
CI = ±3.49
1991
r² = 0.60
b = 12.3
CI = ±2.4
B 5.6 – 8.0 mm
1989
r² = 0.47
b = 12.1
CI = ±13.0
1991
r² = 0.61
b = 21.0
CI = ±8.94
C 16.0 – 22.0 mm
1989
r² = 0.05
b = 1.72
CI = ±2.98
1991
r² = 0.29
b = 31.5
CI = ±13.2
D All sizes
1989
r² = 0.84
b = 12.3
CI = ±3.26
Phase I
Phase II
Rise
Fall

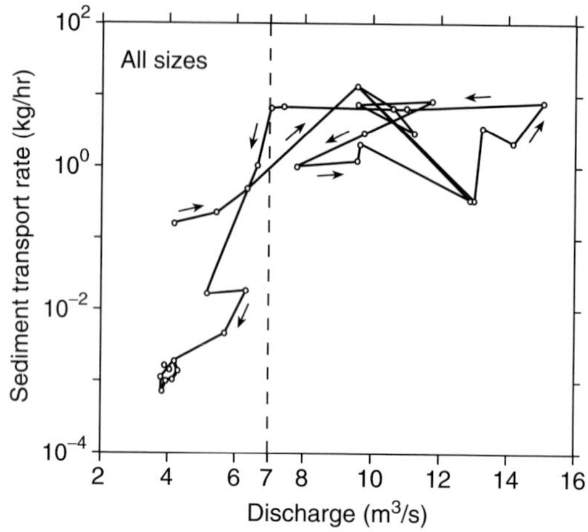

Figure 9.6. Trajectory of total transport at trap 3B (1989) plotted on arithmetic coordinates in order to emphasize the hysteretic nature of the relation.

from those to which formulae are calibrated. In Figure 9.7, we compare observations from trap 3B in 1989 with estimates from the Meyer-Peter and Müller formula (1948), a formula specifically developed to estimate gravel transport and possibly the most widely tested formula in existence, and with Parker's (1990) surface-based formula. Evidently, the

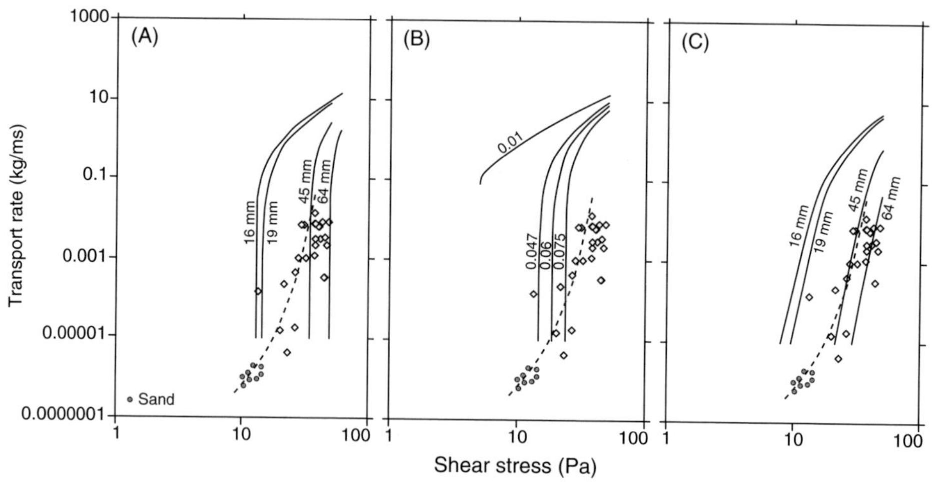

Figure 9.7. Comparison between bedload transport observations (data of trap 3B, 1989) and the predictions from the Meyer-Peter and Müller (1948); Parker (1990) bulk formulae. The dashed line in the plots is the rating curve from Figure 9.5D. (A) Meyer-Peter and Müller formula: predicted transport for various reference grain sizes and the critical Shields number specified by Meyer-Peter and Müller. (B) Meyer-Peter and Müller formula: predicted transport for D_{50} at the trap site and various values of critical Shields number. (C) Parker formula: predicted transport for various reference grain sizes.

formulae overpredict transport dramatically for the bulk bed material – which is actually coarser than the material transported – although they appear to estimate the threshold for sand movement fairly well. A tolerable prediction is achieved for the gravel transport (phase 2) regime when the surface grain size is adopted, but then phase 1 transport is underestimated or is not predicted at all. In comparison with classical experimental results (upon which the Meyer-Peter formula is based), bedload transport in Harris Creek is somehow severely constrained. Parker's formula is based on field measurements, but grain sizes in transport were representative of those on the bed, unlike the present case.

The total bed material transport in Harris Creek was estimated by Hou (1997) on the basis of a program of Helley-Smith sample measurements taken at a road bridge about 100 m upstream from the research site. He estimated the following rating equation:

$$G_b = 7.1 \times 10^{-7} Q^{4.1} \quad (1)$$

$$n = 26;\ R^2 = 0.86;\ F = 138;\ p < 0.001;\ \text{standard error range is } 0.72\text{–}1.38x$$

where G_b is bedload transport in kg s^{-1}, Q is stream discharge in m^3 s^{-1}, n is number of samples, R^2 is the coefficient of determination, F is Snedecor's significance statistic, and p is the probability of non-significance. The relation is excellent, probably in reflection of the fact that it is a gross scale correlation for the entire channel.

Hou estimated total transport in each of the three seasons by convolving his rating curve with the flow duration derived from the gauge at the research site. His results are reported in Table 9.1 as ranges which take into account the statistical error of the rating equation. The results show two orders of magnitude in the range of transport over just 3 years, and they are consistent with the yield data derived from the various sediment sources.

3.2. *Threshold of motion*

Classical Shields theory predicts that the shear stress required to mobilize coarse grains (>8 mm) from the streambed is linearly proportional to the grain size. This condition represents dynamical similitude with respect to the forces necessary to entrain grains of varying size. In contrast, most flume studies have indicated that all grain sizes move off at essentially the same critical shear stress (Wilcock, 1992). In this circumstance, it is supposed that the largest grains present on the bed control the entrainment of all the grains. The result implies that sediment entrainment is a self-similar process; that is, there is no regime of size-selective mobilization.

Our results from Harris Creek (Fig. 9.8) fall between these extremes, implying that some range of grain sizes controls the entrainment process. A similar result was obtained experimentally by Wilcock and McArdell (1993; their Fig. 8). Considering sizes finer than the median of the surface – which represents virtually everything that moved – we obtain $\tau_{ri} = 6.2 D_i^{0.48 \pm 0.023}$, in which τ_{ri} is the critical shear stress for entrainment of grains of size D_i, and the range is the $\alpha = 0.10$ confidence interval for the exponent.[2] It appears that

[2] This result is different than that quoted in Church and Hassan (2002). In that paper we reported the best fit single relation for all our data, including sizes greater than the surface median.

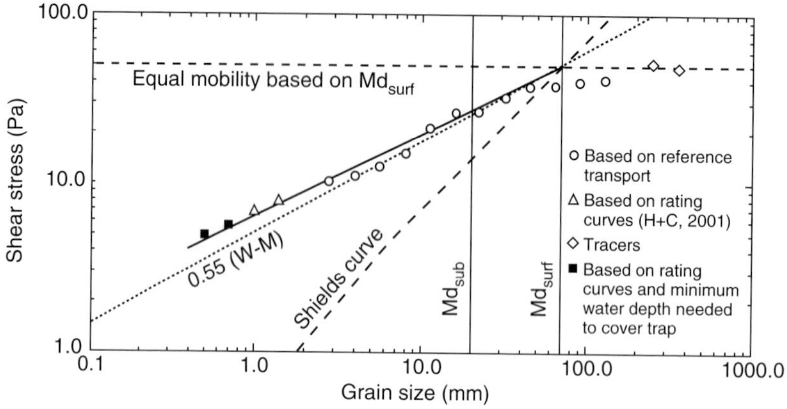

Figure 9.8. The relation between shear stress required to initiate the movement of sediments and grain size. See Church and Hassan (2002) for details of the construction of the Harris Creek plot. Shields' function is given as modified by Miller et al. (1977). The experimental result of Wilcock and McArdell (1993) is shown for comparison. Reproduced with permission of AGU.

sediment mixture properties or bed structures, or both, control the entrainment process. The force necessary to move the largest material in Harris Creek diverges from the power function fitted to the majority of the range. The break occurs near the surface median grain size and is the point at which our general relation, Shields' relation, and equal mobility based on the surface D_{50} coincide. Large material, as befits its high exposure, is entrained at forces smaller than those predicted by our general relation or by Shields, and the largest sizes appear to be essentially equally mobile. Presumably, this reflects the role of classical effects of hiding/exposure (Fenton and Abbott, 1977) superimposed on the local constraints to grain mobility. The result saves, to some extent, the practice of using the surface D_{50} as the reference grain size for single estimates of the entrainment force based on measurement of the bed material, but it is valid only for high transport rate computations.

3.3. *Partial transport regime*

We have plotted the fractional transport rate $p_i q_b$, in which p_i is the fraction of transported sediment in the ith size range, scaled by f_i, the fraction in the ith size range present in the *subsurface* bed, following the convention of Wilcock and Southard (1988). Figure 9.9A displays results selected to illustrate the changing character of the transported sediment as flow changes.

We observe three regions in the plot. At the left hand side is a region in which p_i/f_i declines below 1.0 for progressively finer sizes, indicating that these sizes are present as bedload in *smaller* proportion than in the bed. We interpret these results to indicate that sands are escaping into saltatory or suspended motion and are bypassing the bedload trap. The size range exhibiting this behaviour is consistent with grain size distributions (GSDs) of sediment sampled in suspension (Fig. 9.10). The limit size for bypassing increases with

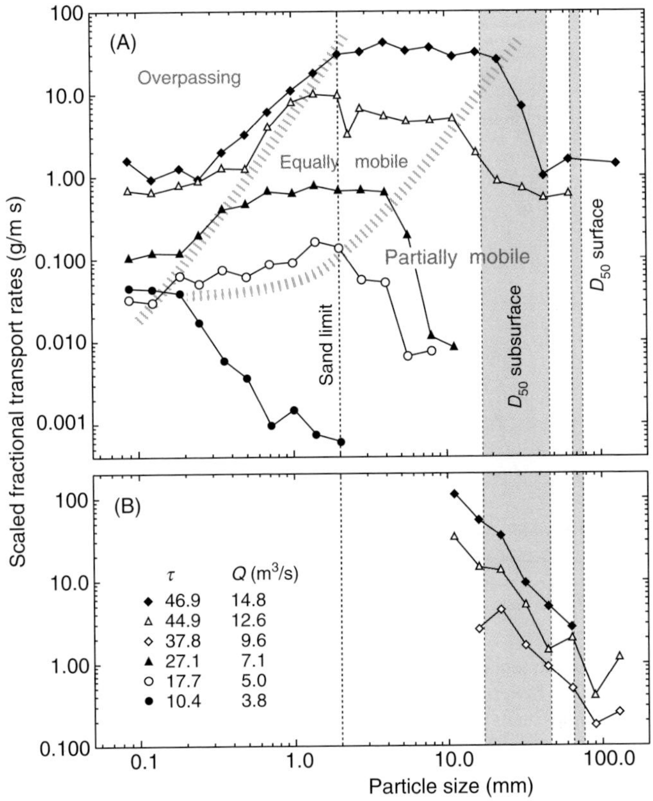

Figure 9.9. (A) Fractional transport ratio scaled by the proportion of each grain size present in the bed for observations over a range of flows at trap 3B. The figure actually shows plots of $(p_i/f_i)g_b$, but the total transport over all sizes, g_b, is a constant for each data set, so the plots demonstrate the variation in p_i/f_i. The inclusion of g_b is simply a device to separate the individual plots. The theoretical value of p_i/f_i for equal mobility is 1.0, but values in this diagram vary because of the retention of q_b in the ratio. In any case, practical values of p_i/f_i at equal mobility are slightly larger than 1.0 because of the under-representation in p_i of the non-equally mobile sizes. See Church and Hassan (2002; p. 19-4) for further discussion of this ratio (after Church and Hassan, 2002; Fig. 2b: reproduced with permission of AGU). (B) Same plot, standardized by the *surface* grain size distribution. This plot is restricted to the pebble and cobble gravel sizes since smaller material is not exposed on the bed surface. Values of shear stress are Pa.

flow, reaching about 2 mm at the highest flows observed at the illustrated trap, consistent with the results of a study for trap efficiency (Sterling and Church, 2002).

At the right hand side of Figure 9.9A is another region of steadily declining transport ratios. These indicate declining participation of the indicated grain sizes in the transport and directly demonstrate "partial transport" – the condition in which only some of the grains present in the bed are mobile at any one time. The condition becomes more severe as grain size increases. Furthermore, at each flow $< 14\ m^3\ s^{-1}$ there is an upper limit of grain sizes smaller than the surface median size for which $p_i/f_i > 0$; that is, an upper limit of grain sizes that take part in the motion at all, indicating size selective transport at modest flows. This observation confirms the size dependency of the threshold of motion displayed

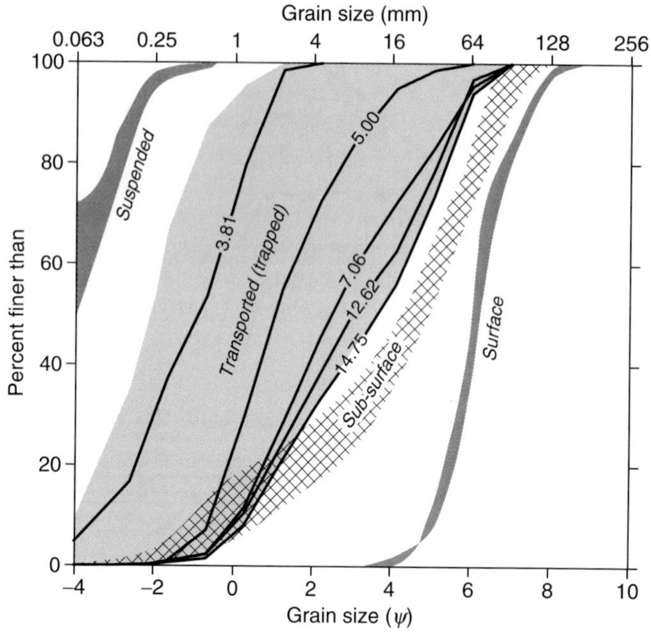

Figure 9.10. Range of grain size distributions of transported ("trapped") bedload in Harris Creek, and of bed surface and subsurface material. The sand fraction of suspended material is also shown. Individual GSDs represent the samples analysed in Figure 9.9A: the numbers represent the discharge at which each sample was taken.

in Figure 9.8 and is also consistent with the observed variation in the upper limit of tracer stone sizes observed to move in different years (see Church and Hassan, 2002).

Between these two extremes lies a range with approximately constant p_i/f_i, indicating equal mobility of grain sizes in this range; that is, these sizes take part in the motion in approximately the same proportion as their presence in the bed. This range decreases as flows become smaller, eventually pinching out at the lowest flows for which transport is detected, when the motion is restricted to sand. Gravels enter the range of equal mobility only when flows exceed about 7 $m^3 s^{-1}$.

At no stage in our observations did we observe equal mobility/bypassing of all size fractions on the bed (the onset of phase 3 transport). According to our threshold diagram, we should observe that condition when shear stress on the bed exceeds 50 Pa, a condition that was very nearly reached. In 1990, when we maintained only hydrological observations, the stream experienced an exceptionally high flow estimated to have been greater than 30 $m^3 s^{-1}$. In that flood, surface clast structures (Section 4.3) consisting of the largest stones on the bed were destroyed, indicating that phase 3 was reached.

In the foregoing discussion, the normalization by f_i is based on subsurface material. This is supposed to represent the long-term, average distribution of transported sediment, hence seems most appropriate given varying flows and surface conditions. But, in fact, material is directly entrained from the bed surface. Normalization by the surface fractions presents a more severe picture as illustrated in Figure 9.9B. In comparison with surface grain sizes, no gravel sizes experienced full mobilization at any measured flow. In effect, the surface remained substantially immobile for all flows up to, at least, 15 $m^3 s^{-1}$.

3.4. Grain size distribution of transported material

Most bedload samples are taken rapidly in small capacity samplers. Given the characteristically low-transport rates in Harris Creek and like streams, and the well-known non-steadiness of bedload transport (reviews in Gomez, 1991; Kuhnle, 1996), reliable GSD cannot usually be extracted from the samples. The large capacity of pit traps overcomes this constraint. Our samples were collected over periods that varied between 20 min and 26 h, according to the flow and transport rate, and so represented a substantial integration of the mobile sediment. Sample weight varied between 20 g and 50 kg. Considering maximum grain size in each sample, about one-third of the samples surpass the 0.1% criterion of Church et al. (1987) for the ratio between the weight of the maximum grain and the total sample weight, while another one-third surpasses the 1% criterion and the remainder surpass the 5% criterion. The range of recovered GSD is shown in Figure 9.10, along with individual GSDs corresponding to the sampling events displayed in Figure 9.9A.

Transported GSDs again confirm the picture implied in that figure. As flow increases, progressively larger grain sizes and a wider absolute range of sizes take part in the motion. At the lower end of the distribution, sand is lost to suspension. Only a small fraction of suspended material is coarser than 0.25 mm, however (never more than 5%), whereas almost no material finer than 0.25 mm travels on the bed at flows greater than 5 $m^3 s^{-1}$, so there is almost complete separation of the load components. In fact, there is relatively little suspended material in transport: the bed was almost always visible through the water during our work.

We obtained no observation in which the transported GSD matched that of the subsurface material, and the median size of the surface material was just entrained at the highest flows. Transport remained size selective in the stream at all flows up to 15 $m^3 s^{-1}$, a flow with return period of about 1.3 years.

4. Bed condition

4.1. Bed material

The bed material is the integrated result of the transport processes just described. The subsurface bed material in Harris Creek generally varies between medium fine sand (+0.125 mm) and about 256 mm ($\psi = 8$; Fig. 9.10), although larger clasts occur. In four samples of 600 kg each, the median grain size varied between 22 and 45 mm, and the Trask sorting coefficient varied between 2.4 and 3.9. Values near 2.4 represent recently transported material in finer sediment patches (Section 4.2), while values near 3.9 represent the apparently stable bar surface at the research site. The bar tail was much finer than the head and midsections of the bar. The main difference amongst the subsurface deposits is the presence of a larger sand fraction in the finer samples, lending them a slight bimodality. The sand fraction in all the subsurface samples remained at about 20% or less, so the subsurface in Harris Creek is framework supported.

Surface accumulations were bulk sampled in a pool and in a riffle immediately upstream of the study bar. The distributions are similar (Fig. 9.10), with median grain sizes of 64 and 76 mm, respectively, and a maximum grain size of about 512 mm ($\psi = 9$). Sorting is 2.4 and 2.7. Material finer than 11 mm ($\psi = 3.5$) is rare on the bed surface, except in lee-side deposits at the bar tail and in small patches including granules. The surface material is a censored version of the bulk deposits, the major truncation point in the surface environment being about 16 mm, which is approximately the D_{45} of the bulk material.

4.2. Sediment patchiness

It is well known that the bed material grain size varies significantly in stream channels. It has been noted that distinct "patches" of relatively homogeneous sediment occur, and that the finer patches represent a large part of the mobile bed material in the reach (Paola and Seal, 1995). Patches of various size and genesis have been identified (reviewed in Laronne et al., 2001).

In Harris Creek, patches on three scales are easily identified

(i) *Micropatches*. Minor accumulations (<1 m diameter) of fine material (sand and granules) behind obstacles and in depressions on bar surfaces, as illustrated near traps 3 in Figure 9.11B. They may be many grains deep but their aggregate thickness amounts only to centimetres. These are the usual "shadow deposits" that are widespread in stream channels.

(ii) *Mesopatches*. Accumulations of mobile material of order 3 m diameter, including granules and pebble to small cobble gravel in slack water sites. These deposits are found dominantly along the channel edge, on riffle tails, and on flow separation lines. They typically are tens of centimetres deep. The example illustrated in Figure 9.11A is in a separation zone generated by a pool on the upstream side of a bedrock outcrop, and was one of the "surface accumulations" sampled.

(iii) *Macropatches*. Bar scale (e.g. order 10 m) accumulations of sediment with a range of grain size, from sand to cobble gravel. These deposits are typically found in bar top and bar tail situations (Fig. 9.11B), and are often located behind protective barriers such as large woody debris. Logjam deposits (Fig. 9.3D) are included in this category. Thickness varies from centimetres to as much as metres.

Distinct patches of the kind described here are, overall, not common along the stream. Except on major bar tops (themselves not common), the bed presents a more uniform cobble surface (Fig. 9.12A). Figure 9.12 shows a long rapid reach with no great topographic differentiation of the bed. Where there is topographic differentiation, as in pools, bed material texture is more continuously varied than in patches. Figure 9.11C shows varying sediment texture in a pool and onto the adjacent bar top. The largest clasts are found in the pool, and maximum grain size then declines onto the bar top. In general grain size changes in correspondence with the change in water depth and, one supposes, the shearing forces imposed by the flow.

Figure 9.11. Sediment patchiness in Harris Creek. (A) Vertical air photograph showing a mesoscale patch of relatively fine gravel superimposed on a riffle near the right bank at the upstream end of the research site (the 1.2 m scale bars point toward the deposit). This deposit was recent and was sampled for bulk bed material. Arrow at the top of the photo indicates traps 1. (B) Lower part of the research site showing microscale patches of granule gravel near traps 3 (indicated by the linear arrows: 3A is at the bottom and 3D is at the top), and a macroscale patch of granule gravel and sand on the distal end of the bar, partly protected by the downed trees on the low flow channel edge. (C) Vertical view defining zones in the channel with distinctive grain size. The largest grains occupy the talweg. The photos were taken with a balloon-borne, gimbal-mounted camera with radio shutter release suspended about 30 m above the channel using helium-filled balloons.

We expect that finer sediment patches are associated with the mobile bed material of the creek, and their relative sparsity is a function of the characteristically low rate of transport. We will return to this topic in our final discussion.

4.3. Clast structures

A more remarkable feature of the channel bed in Harris Creek is the patterned distribution of the surface clasts (Church et al., 1998) (Fig. 9.12). A distinctive pattern is not highly developed on the bar surface at the research site, but it is prominent for several hundred metres upstream where the channel exhibits only very subdued pool and riffle features.

The larger clasts on the bed surface form irregular reticulate networks within which finer material subsists. The usually incomplete cell-like structures are formed mainly by

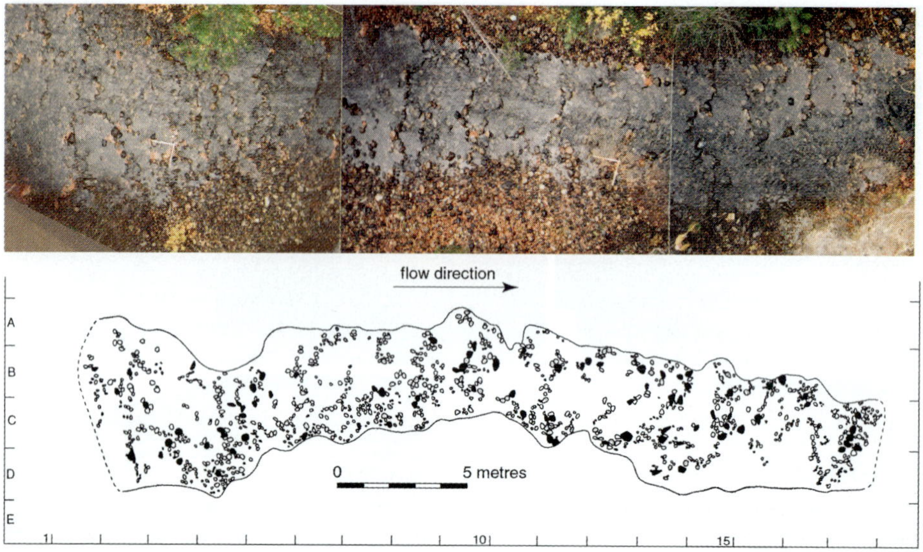

Figure 9.12. (A) Vertical air photograph (scale bars are 1.2 m) showing cellular sediment structures exposed by the low flow. Balloon control lines are visible. The illustrated reach is an extended rapid with little large-scale topographic variation, typical of significant stretches of the channel. (B) Map of stones that stand proud of the general bed level, defining the sediment structures. The map was constructed by tracing from stereo air photos of which a montage is shown in (A). The solid black clasts are those > 260 mm ($\psi \approx 8$) in projected major axis diameter, which correspond approximately with the D_{99} of the surface material in the channel. Stones larger than D_{75} of the surface material were mapped (map constructed by S. Tribe). (From Fig. 9.1 in Church et al., 1998: reproduced with permission of AGU.)

stones larger than surface D_{84} (128 mm; $\psi = 7$) and have characteristic spacing of order 1 m. Hence the ratio of structure diameter to constituent clast size is of order 10:1. Successively considering linear and circular geometries for the stone arrangements, that ratio implies that the constituent clasts occupy between 15 and 25% of the bed. This is just the range of fractional areas indicated by Rouse (1965); see also Laronne and Carson (1976) to contribute most of the boundary frictional resistance to flow.

The principal structural elements are upstream or downstream facing arcuate ridges, protruding one stone proud of the general bed level, and one or a few stones wide. The features tend to be obliquely to transversely oriented with major:minor axial ratio of about 2:1. Individual ridges might be up to 20 stones in length and typically span one-quarter to three-quarters of the channel. Elements of such structures have been described previously. Most fundamentally, they are imbricate features. Imbrication (Johnston, 1922; Johansson, 1976) is a pervasive feature of nearly all gravel-bed channels. Compact clusters (Dal Cin, 1968; Brayshaw, 1984) form the articulation points of the stone network, and from these stone lines (Laronne and Carson, 1976; Martini, 1977) extend and ramify into the cell-like network. The pattern appears to consist of arrangements of large stones around a random spatial distribution of the very largest ones. The areas of finer material within the net may constitute the main reservoir of surface-exposed, normally mobile material – the main active sediment micropatches. Casual observations of similar features have previously been reported by McDonald and Banerjee (1971), Gustavson (1974) and Bluck (1987). We were able to duplicate the features in scaled experiments (further described in the next section) in a laboratory flume.

The features can form only in partial (phase 2) transport, since it is necessary that a substantial number of clasts remain immobile on the bed surface for a more or less protracted time. We suppose that the development of the bed structures must take some time under these conditions. We were casually able to confirm this conjecture when a major flood (estimated $Q_{max} > 30\ m^3\ s^{-1}$) mobilized the bed and destroyed the structures in 1990. The structures reformed over the following two seasons under the influence of flows that approached but did not surpass the level of the mean annual flood, in which phase 3 transport was not reached.

4.4. The hydraulic role of the bed structures

In order to examine the conditions of formation further and to investigate the hydraulic role of the bed structures, we resorted to experiment using a sediment mixture modelled on that in Harris Creek. In the first series of experiments (Church et al., 1998), we fed no sediment into the flume. We ran each experiment for about 100 h, during which sediment transport declined by more than three orders of magnitude. Results were replicable. The fraction of the bed material charge removed from the flume in each run was less than 10% and degradation never appreciably exceeded $1.0D_{84}$ equivalent of the bed material. At the commencement of each run, the initial surface coarsened as fine material was transported away or sifted into voids between the large particles. The latter typically rolled into contact with static neighbours and stopped, leading to cluster development and later ramification into lines and nets. The main development occurred within the first 24 h of the run. Fully developed networks were very similar in all respects to those observed in the field except that they were somewhat more elongate than the geometry at Harris Creek, possibly because of the simple lateral geometry of the flume.

In these experiments, sediment flux remained within the partial transport regime for the range of flows tested. The median size of the transported material near the end of the experiment – that is, after full structural development – was much less than the median size of the surface bed material (Church et al., 1998), but the surface material itself was not so coarse as would be expected for the imposed shear force. We established this fact by examining the ratio

$$q_b^* = \left[\frac{(\tau_b - \tau_{cs})}{\tau_b - \tau_{ct}} \right]^{1.5} \tag{2}$$

which is the ratio of the transport rate for the developed surface compared with the transport rate estimated for a surface as fine as the load (Dietrich et al., 1989). τ_b is the shear stress imposed on the bed; τ_{cs} is the critical shear stress for the surface material, which is supposed to be proportional to the D_{50} size of that material, and τ_{ct} is the critical shear stress for the transported material. (We employ here a usual functional dependence for sediment transport whereby $q_b \infty \tau^{1.5}$, whereas we have earlier shown that much more sensitive relations are found in Harris Creek. However, the introduction of the critical shear stress amends this apparent discrepancy, as well as being convenient for our present purpose. The formula given above can be developed directly from the well-known Meyer-Peter and Müller bedload formula, for example.)

We form q_b^* as the ratio of transport near the end of the experiment to that after 1 h – the latter being assumed to represent transport over an undeveloped surface. $q_b^* \approx 0.001$. Eq. (2) then yields the approximate equality $D_{50s}/D_{50t} = \tau_b/\tau_{ct}$, the expected ratio of surface to transported grain sizes. Over 14 experiments at varying slopes and flows, we obtained an expected mean value $D_{50s}/D_{50t} = 5.60 \pm 0.29$, whereas the *observed* ratio (from bed and load sampling) was 3.11 ± 0.20. We assign the difference to the effect of the structures in increasing the reluctance of the coarser material to move, so that the ambient shear forces were withstood by smaller material than indicated by the classical theory.

Further consideration of Eq. (2) shows that the transport after full structure development was very near the threshold value, and that an effective increase in the critical shear stress created this condition. q_b^* approaches zero within a reasonable range of τ^* (the Shields number) only when the numerator in Eq. (2) approaches zero (i.e. an indeterminate condition is not reasonable). For the mean conditions in our experiments ($D_{50s} = 4.2$ mm; $D_{50t} = 1.35$ mm; $\tau_b = 5.1$ Pa), the upper limit of τ_{cs}^* is 0.075. The mean value of τ_b^* (using the observed D_{50s}) was 0.079. Conversely, if τ_{cs}^* is assigned the classical value of 0.06, $q_b^* = 0.12$, an increase of two orders of magnitude! Evidently, an increase in τ_{cs}^* by 30% above the classical Shields datum (67% above the nominal threshold of 0.045 for widely graded mixtures) is sufficient to reduce transport to near zero. The bed surface condition is a very sensitive mediator of bed material transport.

In Harris Creek, $\tau_c^* = 0.075$ may still be an underestimate, since that value still overestimates sediment transport (Fig. 9.7B), but the question of the relevant grain size for this comparison enters into consideration.

It seems reasonable that the decline in sediment transport should be associated with an increase in the resistance to flow, since the energy that would otherwise do sediment transporting work must be dissipated in some other way. Using measured velocity profiles taken during our runs, we estimated the roughness length, k_s, at several stages in some of our experiments. The results are shown in Figure 9.13. Roughness length is a conceptual length commonly supposed to be a small multiple of bed surface grain size, which indexes the grain resistance to flow. In our experiments, the final value of D_{84s} was about 14 mm (0.014 m), suggesting $0.028 < k_s < 0.056$ (i.e. $2–4D_{84}$). It seems that k_s based on individual clasts could not conceivably exceed 0.1. Observed values appreciably exceed that value and increase coherently with the development of the bed structure.

We initiated further experiments in the same fashion (zero sediment feed) but, after 16 h, when reasonable structural development had occurred, we commenced feeding sediment at some fraction of the observed transport rate during the immediately preceding period (see Hassan and Church, 2000). Feed rate varied between 50 and 150% of the observed transport. Transport remained near the 16-h rate and the bed structures persisted, but they were progressively buried by patches of finer sediment which grew larger as feed rate increased. Again, the actual transport rates remained much smaller than predicted by the ratio of bed surface grain size to transported grain size, the difference reflecting the increase in bed stability introduced by the structures. In these experiments, we determined that between 17 and 47% of the resistance to flow was carried by the bed structures (Hassan and Church, 2000).

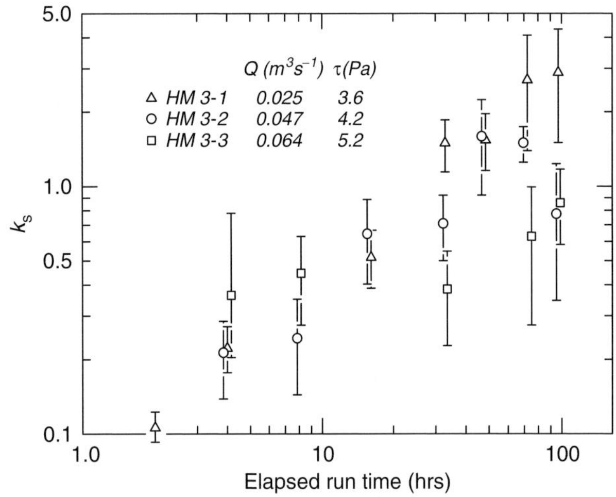

Figure 9.13. Evolution of k_s in each of three experiments with relatively low, intermediate and high flow. The error bars (2 s ranges) are wide due to the variability of the velocity profiles (Wilcock, 1996). The variation may be assumed to be due to structure development since larger scale form roughness was absent from these experiments. (From Fig. 9.4 in Church et al., 1998; reproduced with permission of AGU.)

5. Channel stability

Channel changes are the consequence of bed material displacement by stream currents. The corollary of low bed material transport must be a low rate of channel change. We have compared aerial photographs taken in 1947 and in 1987 to determine the character and extent of changes in the channel of Harris Creek in the reach between "landslide 1"and "landslide 28" (Fig. 9.1), a distance of 7 km encompassing much of the upland valley reach.

In 1987, 83% of the total stream length in our 7 km reach occupied the same channel as in 1947 (Fig. 9.14). The basis for this figure is that if the 1987 channel occupied all or part of the 1947 channel (i.e. if it had moved laterally less than one full channel width), the channel was classified as "unchanged". Significant changes during the 40-year period are restricted to 12 individual sites, each covering a distance equivalent to one or several bar lengths. These sites represent avulsions around sediment accumulations below landslides that chronically feed sediment into the river, and sites where logjams formed in the 1950s following floodplain logging. Remarkably, all four active landslide sites in the reach (Fig. 9.1) were active in 1947. They represent sites of chronic shallow failures in till underlain by glaciolacustrine silt at places where the channel has impinged on footslopes underlain by glacial material (Fig. 9.3B). All of the slides (active and inactive) occur on the right bank, which is at the base of the drift-mantled south-facing slope. Only at landslide 1 has the channel shifted away from the toe of an active slide within the study period. Elsewhere, particularly in front of inactive slides 18–21, there is an old channel at the slide base, while the active channel is across the valley. There is evidently

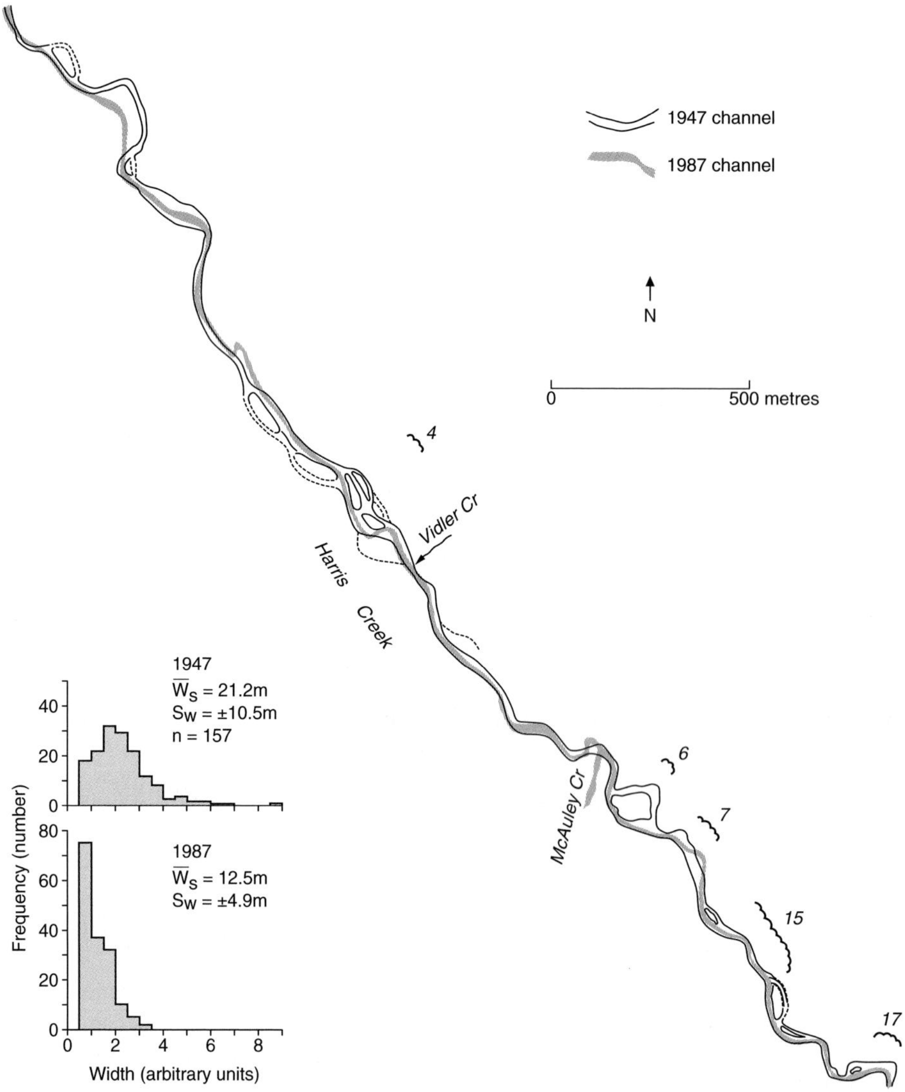

Figure 9.14. Map of Harris Creek channel near McAuley Creek confluence (see Fig. 9.1), showing the 1947 and 1987 channels superimposed. Significant landslides are marked and numbered as in Figure 9.1. This represents about one-third of the reach over which the comparison between the two dates was conducted. Inset: frequency histograms of channel width on the two dates. Measurements were made from the maps, but at much larger scale than illustrated. Vegetation overhang may have induced some bias in the results. However, the 1987 result compares favourably with field measurements. Width units are arbitrary units of measurement.

an intimate relation between slide activity and channel impingement on the failing slope such that slide activity declines or ends after it has shed sufficient material to drive the channel away.

Elsewhere along the reach there are distinct zones where the channel widens and bar development is more extensive. The channel widens from 10 m to about 30 m and bank erosion has occurred (Fig. 9.1 shows locations). From the head of the reach to Vidler Creek, these zones of modest lateral instability are associated with upstream sediment sources – either the landslides or tributaries or debris flow tracks. Farther downstream, they are principally associated with the old logjams. The jams themselves are now breached and the channel has degraded through the sediment wedge formed there. However, there is still a substantial amount of recently deposited sediment stored along the sides of the contemporary channel (Fig. 9.3D).

The 1987 channel is distinctly narrower than the 1947 channel (Fig. 9.14, inset), the difference being associated with a reduced volume of in-channel bed material stored in active bars. Furthermore, the 1947 photos reveal significant patches of recently established bush growing up on formerly active bar surfaces, whereas the forest along the 1987 channel is mainly young mature woodland. Evidently, bed material transport declined through much of the 20th century. It appears that substantial sediment input was associated with the early 20th century fires and that the landscape has been "relaxing" ever since that period.

Table 9.1 reveals landslides and streambank erosion to be about equally important as sources of sediment to the river, at least during the short period of record. However, the landslides are much more episodic in sediment delivery than streambanks. It is likely that the tributary streams are even more steady producers of sediment and, in the longer run, are about as important as each of the other two sources. In most seasons, it appears that most of the sediment yielded to Harris Creek derives from streambank erosion, the consequence of bar development and erosion as bed material is slowly staged downstream, and from tributary streams. The data of Table 9.1 also show that, within the limitation of the data, Harris Creek today appears to be neither accumulating nor losing significant volumes of sediment, though it would take far better measurements than we have available to confirm overall grade in this low transport upland channel.

6. Discussion

Harris Creek appears to be approximately in equilibrium with its limited sediment supply. The creek presents a range of characteristics that identify it as an essentially stable cobble–gravel channel with low bed material transport. Salient characteristics include the following:

- limited bar development and limited evidence for contemporary lateral instability – a high proportion of the stream channel remains in the position occupied 40 years earlier;
- sparse occurrence of distinctive sediment patches of all scales, implying limited occurrence on the surface of frequently mobilized sediment;
- extensive development and persistence of surface stone networks composed of the coarsest material in the streambed;
- high armour ratio.

Furthermore, measurements of bed material transport reveal that transport is nearly always restricted to phases 1 and 2, so the armoured surface remains intact. All sizes are not transported with equal facility, so that the GSD of transported sediment is characteristically finer than that of the bed material, and much finer than the surface armour.

We propose that the characteristics enumerated above are typical of gravel-bed channels with low bed material transport and define a distinctive regime of sediment transport that occurs in many gravel-bed channels. Only the first of the characteristics is direct evidence for stability, however, and it cannot be appraised unless historical information is available. Visually, the most striking features are the relatively low variability of surficial sediments – that is, the relative lack of obvious patches and of bed topography – and the highly structured surface. There is also a notable absence of deep pools and the usual pool-riffle morphology is substantially attenuated. Sediment patches serve to facilitate differential mobility of mixed sediments by maintaining the exposure of finer, relatively mobile sediments (Paola and Seal, 1995). If overall transport is low, texturally differentiated patches will be relatively rare and the bed will be stable. If overall transport is relatively high (but still restricted to phase 2), the incidence of such patches will be high, the bed will tend to take on the overall character of the patches, and will be comparatively mobile. Correspondingly, the channel will become less stable and more varied topographically as it is deformed to accommodate transient bed material deposits.

It has been argued that, as transport through a gravel-bed channel increases, the size distribution of surface material regresses toward that of the bulk material beneath (assumed to be representative of the mixture of sizes transported through the channel in the long term), so that the armour ratio declines (see Parker and Klingeman, 1982). The surface armour is regarded as a mechanism for adjusting the relative mobility of different sizes in the presence of a competence limitation so that sediment equal mobility is maintained. The picture proposed in the last paragraph is, rather one, in which, as transport increases, the armour becomes *spatially* mixed with an increasing incidence of patches of frequently re-entrained sediment. In the limit of high (i.e. phase 3) transport, the entire bed becomes covered with mobile sediment and the armour, if still present, is buried. There is no longer any differential mobility of different sizes. For this condition to be reached, it is a sufficient condition that the armour be destroyed, but it is not a necessary one.

For any transport rate below the phase 3 threshold, the latter picture implies that the sediment being transported through the channel is systematically different in GSD than the sediment resident in the bed. A further implication, then, is that material of different sizes moves through the system at different speeds, so that residence time systematically varies. In particular, sand and fine gravel are remobilized every year and move through Harris Creek relatively quickly, whereas cobbles remain resident in the bed for a long time. The sediment yielded to the stream must be systematically finer, then, than the sediment resident in the streambed. This circumstance is not possible under the assumption of equal sediment mobility.

The coarse material that is left in the channel – material evidently near the limit of competence of the stream to move – is slowly worked into imbricated arrangements including stone clusters, lines and nets, which further increase the stability of the material and of the channel bed by increasing hydraulic roughness and by increasing the reluctance of individual, imbricated grains to move. This phenomenon constitutes self-organized stabilization of the channel, and it is achieved for a surface considerably finer than would otherwise be necessary to bring transport into balance with supply (cf. Church et al., 1998; Buffington and Montgomery, 1999). It coincidently prolongs the residence of this material in the channel. But there must be some recruitment of coarse material to the channel over the long term; otherwise there would be none there at all. So an interesting new question arises: in channels that characteristically experience partial and selective transport, do these coarse materials continue to accumulate? Does the bed become coarser (and even more stable) over long periods of time? There are two ways in which this circumstance might be avoided:

(i) bed-breaking floods, such as that observed by us in 1990 in Harris Creek, might transport sufficient large material to establish a long term equilibrium between sediment supply and sediment output. Equal mobility might then be a long-term statistical phenomenon rather than a short-term dynamical one. Church et al. (1991) showed that this is the case for the sand fraction in Harris Creek. Alternately,
(ii) attrition of grains might balance the effect of coarse sediment influx by breaking clasts down into finer ones. Particle attrition has usually been thought of in terms of abrasion during transport. We suggest that attrition is much more likely to occur as the result of weathering processes during particle storage in the channel (cf. Jones and Humphrey, 1997). In Harris Creek, freeze-thaw weathering is apt to be an effective mechanism for particle attrition.

These possibilities represent interesting hypotheses for new research.

The ultimate reason for the low rate of sediment transport through Harris Creek is the low rate of supply from the upland landscape. We have shown that significant sediment sources are restricted to a countable number of point sources. Much of what is yielded from them is washload (derived from the glaciolacustrine silts). Nonetheless, a wide range of grain sizes is episodically delivered to Harris Creek, so that a competence-limited, structured channel results. Hence, the character of the channel is determined in large measure by the volume and character of the sediment supplied from the land surface. The conclusion is reinforced by the appearance that, earlier in the 20th century, sediment supply was more prolific and the channel was correspondingly more active. In recent years, the adjustment to reduced sediment supply apparently has been achieved principally by surface coarsening and structural modification, rather than by incision (compare Lisle et al., 1993).

We suggest that a condition of low-sediment delivery and competence-limited transport are the usual conditions for upland gravel-bed channels in landscapes that are relatively little disturbed. They lead to the development of semi-stable channels exhibiting structurally robust arrangements of the bed materials. Such features have in the past been represented as indicating "sediment starvation". But they constitute reach and section scale adjustments consistent with the sediment supply that establish and maintain a long-term equilibrium in sediment movement through the channel. We claim that this represents the normal regime and condition of such channels.

Acknowledgements

Many people, mostly students, have helped us in the field and laboratory over 15 years and we recognize that, without their careful work, this project and, in particular, the synthesis presented in this chapter would not have been possible. We are pleased that all of the students have subsequently become professional earth scientists. Our collaborators include Dr. Zhihui Hou, whose Ph.D. research included collection of the important data on sediment yield to Harris Creek, Dr. June Ryder, who conducted the sediment source study, Dr. John Wolcott, who established the basic observations at the research site and initiated the experimental studies, Shannon Sterling and Selina Tribe, both of whom contributed important parts of the research. We thank Eric Leinberger for the line drawings and Darren Ham for formatting the photo illustrations. Basil Gomez and Trevor Hoey furnished helpful reviews, which we appreciate. Church acknowledges the continuing financial support of the Natural Sciences and Engineering Research Council of Canada through its Discovery Grant program.

References

Adenlof, K.A., Wohl, E.E., 1994. Controls on bedload movement in a subalpine stream of the Colorado Rocky Mountains. Arct. Alp. Res. 26, 77–85.

Andrews, E.D., 1983. Entrainment of gravel from naturally sorted river bed material. Geol. Soc. Am. Bull. 94, 1225–1231.

Andrews, E.D., 1994. Marginal a bed load transport in a gravel bed stream, Sagehen creek, California. Water Resour. Res. 30, 2241–2250.

Ashworth, P.J., Ferguson, R.I., 1989. Size-selective entrainment of bedload in gravel-bed streams. Water Resour. Res. 25, 627–634.

Bluck, B.J., 1987. Bed forms and clast size changes in gravel-bed rivers. In: Richards, K. (Ed.), River Channels: Environment and Process, Institute of British Geographers Special Publication 17. Blackwell, Oxford, pp. 159–178.

Brayshaw, A.C., 1984. Characteristics and origin of cluster bedforms in coarse-grained alluvial channels. In: Koster, E.H., Steel, R.J. (Eds), Sedimentology of Gravels and Conglomerates, Canadian Society of Petroleum Geologists Memoir 10, pp. 77–85.

Buffington, J.M., Montgomery, D.R., 1999. Effects of hydraulic roughness on surface textures of gravel-bed rivers. Water Resour. Res. 35, 3507–3521.

Carling, P.A., 1989. Bedload transport in two gravel-bedded streams. Earth Surf. Processes Land. 14, 27–39.

Church, M., Hassan, M.A., 2002. Mobility of bed material in Harris Creek. Water Resour. Res. 38 (11), 19-1–19-12, doi: 10.1029/2001WR000753.

Church, M., McLean, D.G., Wolcott, J.F., 1987. River bed gravels: sampling and analysis. In: Thorne, C.R., Bathurst, J.C., Hey, R.D. (Eds), Sediment Transport in Gravel-Bed Rivers. Wiley, Chichester, pp. 43–88.

Church, M., Wolcott, J.F., Fletcher, W.K., 1991. A test of equal mobility in fluvial sediment transport: behaviour of the sand fraction. Water Resour. Res. 27, 2941–2951.

Church, M., Hassan, M.A., Wolcott, J.F., 1998. Stabilizing self-organized structures in gravel-bed stream channels: field and experimental observations. Water Resour. Res. 34, 3169–3179.

Dal Cin, R., 1968. Pebble clusters: their origin and utilization in the study of paleocurrents. Sediment. Geol. 2, 233–241.

Dietrich, W.E., Kirschner, J.W., Ikeda, H., Iseya, F., 1989. Sediment supply and the development of the coarse surface layer in gravel-bedded rivers. Nature 340, 215–217.

Diplas, P., 1987. Bedload transport in gravel-bed streams. J. Hydraul. Eng. 113, 277–292.

Emmett, W.W., 1976. Bedload transport in two large, gravel-bed rivers, Idaho and Washington, U.S. Water Resources Council, Sedimentation Committee, Third Interagency Sedimentation Conference, Denver CO, pp. 1100–1115.
Emmett, W.W., Wolman, M.G., 2001. Effective discharge and gravel-bed rivers. Earth Surf. Processes Land. 26, 1369–1380.
Fenton, J.D., Abbott, J.D., 1977. Initial movement of grains on a stream bed: the effect of relative protrusion. Proc. R. Soc. London, Ser. A 352, 523–537.
Garcia, C., Laronne, J.B., Sala, M., 1999. Variable source areas of bedload in a gravel-bed stream. J. Sediment Res. 69, 27–31.
Garcia, C., Laronne, J.B., Sala, M., 2000. Continuous monitoring of bedload flux in a mountain gravel-bed river. Geomorphology 34, 23–31.
Gomez, B., 1991. Bedload transport. Earth-Sci. Rev. 31, 89–132.
Gustavson, T.C., 1974. Sedimentation on gravel outwash fans, Malaspina Glacier foreland, Alaska. J. Sediment Petrol. 44, 374–389.
Habersack, H.M., Laronne, J.B., 2001. Bed load texture in an alpine gravel bed river. Water Resour. Res. 37, 3359–3370.
Haschenburger, J.K., Wilcock, P.R., 2003. Partial transport in a natural gravel bed channel. Water Resour. Res. 39 (1), 4-1–4-9, doi: 10.1029/2002WR001532.
Hassan, M.A., Church, M., 2000. Experiments on surface structure and partial sediment transport on a gravel bed. Water Resour. Res. 36, 1885–1895.
Hassan, M.A., Church, M., 2001. Sensitivity of bed load transport in Harris Creek: seasonal and spatial variation over a cobble-gravel bar. Water Resour. Res. 37, 813–825.
Hou, Z., 1997. Sediment budget of gold and magnetite and their distribution in stream sediments in lower Harris Creek, south-central British Columbia, Canada. Ph.D. Thesis, The University of British Columbia, Canada, p. 228.
Jackson, W.L., Beschta, R.L., 1982. A model of two-phase bedload transport in an Oregon Coast Range stream. Earth Surf. Proc. Land. 7, 517–527.
Johansson, C.E., 1976. Structural studies of frictional sediments. Geografiska Ann. 58A, 201–301.
Johnston, W.A., 1922. Imbricated structure in river-gravels. Am. J. Sci. Ser. 5 (2), 387–390.
Jones, A.G., 1959. Vernon map area. Geol. Surv. Canada Mem., 296.
Jones, L.S., Humphrey, N.F., 1997. Weathering-controlled abrasion in a coarse-grained meandering reach of the Rio Grande: implications for the rock record. Geol. Soc. Am. Bull. 109, 1080–1088.
Kuhnle, R.A., 1996. Unsteady transport of sand and gravel mixtures. In: Carling, P.A., Dawson, M.R. (Eds), Advances in Fluvial Dynamics and Stratigraphy. Wiley, Chichester, pp. 183–201.
Laronne, J.B., Carson, M.A., 1976. Interrelationships between bed morphology and bed-material transport for a small gravel-bed channel. Sedimentolology 23, 67–85.
Laronne, J.B., Garcia, C., Reid, I., 2001. Mobility of patch sediment in gravel bed streams: patch character and its implications for bedload. In: Mosley, M.P. (Ed.), Gravel-bed Rivers V. New Zealand Hydrological Society, Wellington, NZ, pp. 249–289.
Lisle, T.E., Iseya, F., Ikeda, H., 1993. Response of a channel with alternate bars to a decrease in supply of mixed-size bed load: a flume experiment. Water Resour. Res. 29, 3623–3629.
Lisle, T.E., Nelson, J.M., Pitlick, J., Madej, M.A., Barkett, B.L., 2000. Variability of bed mobility in natural gravel-bed channels and adjustments to sediment load at local and reach scales. Water Resour. Res. 36, 3743–3755.
Martini, I.P., 1977. Gravelly flood deposits of Irvine Creek, Ontario, Canada. Sedimentology 24, 603–622.
McDonald, B.C., Banerjee, I., 1971. Sediments and bedforms on a braided outwash plain. Can. J. Earth Sci. 8, 1282–1301.
Meyer-Peter, E., Müller, R., 1948. Formulas for Bedload Transport, Proceedings of the Third International Association of Hydraulic Research, Second Meeting, Stockholm, pp. 39–64.
Miller, M.C., McCave, I.N., Komar, P.D., 1977. Threshold of sediment motion under unidirectional currents. Sedimentology 24, 507–527.
Nanson, G.C., 1974. Bedload and suspended load transport in a small, steep mountain stream. Am. J. Sci. 274, 471–486.
Nizery, A., Braudeau, G., 1953. Variation de la granulométrie de charriage dans une section de rivière, IAHR, International Hydraulic Conference, Minneapolis, pp. 49–60.
Parker, G., 1990. Surface-based bedload transport relation for gravel rivers. J. Hydraul. Res. 28, 417–436.

Parker, G., Klingeman, P.C., 1982. On why gravel bed streams are paved. Water Resour. Res. 18, 1409–1423.
Parker, G., Klingeman, P.C., McLean, D.L., 1982. Bedload and size distribution in paved gravel-bed streams. J. Hydraul. Div., Am. Soc. Civ. Eng. 108, 544–571.
Paola, C., Seal, R., 1995. Grain size patchiness as a cause of selective deposition and downstream fining. Water Resour. Res. 31, 1395–1407.
Rouse, H., 1965. Critical analysis of open-channel resistance. ASCE. J. Hydraul. Div. 91 (HY4), 1–25.
Ryan, S.E., Emmett, W.W., 2002. The Nature of Flow and Sediment Movement in Little Granite Creek Near Bondurant, Wyoming, USDA Forest Service, Rocky Mountain Research Station, General Technical Report, RMRS-GTR-90, p. 48.
Ryder, J.M., Fletcher, W.K., 1990. Exploration geochemistry – sediment supply to Harris Creek (82L/2), BC Ministry of Energy, Mines and Petroleum Research, Geological Fieldwork 1990, Paper 1991-1, pp. 301–306.
Sterling, S.M., Church, M., 2002. Sediment trapping characteristics of a pit trap and the Helley-Smith sampler in a cobble gravel bed river. Water Resour. Res. 38 (8), 19-1–19-11, doi: 10.1029/2000WR000052.
Tacconi, P., Billi, P., 1987. Bed load transport measurements by the vortex-tube trap on Virginio Creek, Italy. In: Thorne, C.R., Bathurst, J.C., Hey, R.D. (Eds), Sediment Transport in Gravel-Bedded Rivers. Wiley, Chichester, UK, pp. 583–616.
Warburton, J., 1992. Observations of bedload transport and channel changes in a proglacial mountain stream. Arct. Alp. Res. 24, 195–203.
Whiting, P.J., Stamm, J.F., Moog, D.B., Orndorff, R.L., 1999. Sediment transporting flows in headwater streams. Geol. Soc. Amer. Bull. 111, 450–466.
Wilcock, P.R., 1992. Experimental investigation of the effect of mixture properties on transport dynamics. In: Billi, P., Tacconi, P., Thorne, C.R., Hey, R.D.O. (Eds), Dynamics of Gravel-bed Rivers. Wiley, Chichester, pp. 109–131.
Wilcock, P.R., 1996. Estimating local bed shear stress from velocity observations. Water Resour. Res. 32, 3361–3366.
Wilcock, P.R., Crowe, J.C., 2003. Surface-based transport model for mixed-size sediment. J. Hydraul. Eng. 129, 120–128.
Wilcock, P.R., McArdell, B.W., 1993. Surface-based fractional transport rates: mobilization thresholds and partial transport of a sand-gravel sediment. Water Resour. Res. 29, 1297–1312.
Wilcock, P.R., McArdell, B.W., 1997. Partial transport of a sand/gravel sediment. Water Resour. Res. 33, 233–245.
Wilcock, P.R., Southard, J.B., 1988. Experimental study of incipient motion in mixed size sediment. Water Resour. Res. 24, 1137–1151.

Catchment Dynamics and River Processes: Mediterranean and Other Climate Regions
Celso Garcia, Ramon J. Batalla, Editors

Chapter 10

Maintenance of an obstruction-forced pool in a gravel-bed channel: streamflow, channel morphology, and sediment transport

Richard D. Woodsmith[1,*] and Marwan A. Hassan[2]

[1]*USDA, Forest Service, Pacific Northwest Research Station, 1133 N, Western Ave., Wenatchee, WA 98801, USA*

[2]*Department of Geography, University of British Columbia, Vancouver, B.C. V6T 1Z2, Canada*

Abstract

Maintenance of pool morphology in a stream channel with a mobile bed requires hydraulic conditions at moderate to high flows that route bed load through the pool as it is delivered from upstream. Through field measurements of discharge, vertical velocity profiles, bed load transport, and streambed scour, fill, and grain-size distribution, we found that maintenance of a pool associated with a large, in-channel obstruction was more adequately explained by analogy to scour processes at bridge abutments than by the concept of velocity reversal. The cross-sectional area through the deepest part of the pool remained remarkably stable throughout a wide range in discharge magnitude. This occurred despite transport of significant quantities of bed load from upstream through the pool, well above and well below bankfull discharge and on rising as well as falling hydrograph limbs. There was no tendency for near-bed velocity or shear stress (averaged over the cross-sectional width of active bed load transport) at the pool centre to exceed that at the upstream riffle tail (pool head) or downstream riffle head (pool tail) as discharge increased up to at least 1.3 times bankfull. Fractional rates indicated that at the pool centre significant bed load transport was initiated and approached 100 g m^{-1} s^{-1} at notably lower mean bed shear stress than at the pool head or tail. Furthermore, incipient motion analyses suggested that mean bed shear stress entrainment thresholds were lower at the pool centre than at the pool head or tail. These findings indicated that total entrainment force in the pool was underestimated by average bed shear stress alone. Through inference, rather than direct measurement, we concluded that, as at bridge abutments, turbulent effects generated by interaction of streamflow with the obstruction added a component of total entrainment force. We further inferred that this combination of mean bed shear stress and instantaneous turbulent force was responsible for entrainment and transport of sediment, thereby maintaining pool morphology. This conceptual model of pool maintenance through a combination of mean bed shear stress and large-scale turbulent force suggests that flow obstructions in gravel-bed streams may be a dominant factor, perhaps as important as mean hydraulic variables or caliber and volume of sediment supply, in controlling local channel morphology and local bed load dynamics.

Keywords: pool-riffle, scour and fill, bed load, flow obstructions, mountain streams

---

*Corresponding author. *E-mail address:* rwoodsmith@fs.fed.us (R.D. Woodsmith).

1. Introduction

Pool-riffle sequences are a fundamental morphologic expression of interactive adjustments among hydraulics, bed scour, and sediment transport and deposition in lower- to moderate-gradient streams. They are also responsible for generating a diverse range of hydraulic and biological niches, which are critical in sustaining high-quality river habitats. Pools can be loosely defined as topographic depressions in the channel bed (Montgomery and Buffington, 1997). Although a minimum size is necessarily implied or stated explicitly (Woodsmith and Buffington, 1996), there is no general agreement on a precise definition.

Understanding formation and maintenance of pools is problematic, because at low discharge pools are relatively low energy environments, characterized by slower, convergent flow, lower water surface slopes, and finer bed surface material than are riffles. Nevertheless, during high discharge events when bed load is being transported, if stream competence is not at least as large in pools as it is immediately upstream, then sediment continuity will not be maintained and pools will aggrade. Pools commonly persist despite a mobile bed and annual bed load yields far larger than pool volume. In gravel-bed streams in valley bottoms, pool formation and maintenance has been attributed to scour along a series of alternate bars by converging flow opposite the slip face of each bar, whereas riffles form on the stoss side of these bars. Downstream of the pool, flow diverges, shear stress is reduced, and bed load is deposited at another bar (Richards, 1976; Church and Jones, 1982). Understanding the controls on streambed topography and pool maintenance in free-formed, alluvial channels has also been approached through detailed modeling of the interactive adjustments of velocity, boundary shear stress, and gravitational and centrifugal forces influencing sediment transport and water surface and bed topography (Dietrich et al., 1979; Nelson and Smith, 1989).

Perhaps the best known model of pool-riffle maintenance is provided by the "velocity (or shear stress) reversal hypothesis," which describes a reversal in the location of maximum near-bed velocity or shear stress from riffles to pools as discharge increases to about bankfull. According to this hypothesis, as discharge increases during high flow events, energy slope, near-bed velocity, and shear stress increase more rapidly at pools than at riffles until, at about bankfull discharge, the magnitude of these entrainment parameters becomes greater in pools than in riffles (Leopold and Wolman, 1960; Keller, 1971; Richards, 1976; Lisle, 1979). This reversal causes sediment to be transported through the pool and, provided sufficiently low transport energy, deposited on a downstream riffle. Thus, pools are believed to fill during falling hydrograph limbs and low flow and to scour during rising discharge and high flow, whereas riffles do the reverse, thereby maintaining pool-riffle morphology (Andrews, 1979; Parker and Peterson, 1980; Ashworth, 1987).

The studies cited above as well as several others tend to support aspects of the velocity reversal concept (e.g., Emmett et al., 1983; Dietrich and Whiting, 1989; Carling and Wood, 1994; Robert, 1997). Cao et al. (2003) find that flow reversal can occur in specific circumstances in association with a channel constriction. Similarly, Wilkinson et al. (2004) attribute maintenance of pool-riffle sequences to channel-width related phase shifts in location of shear stress maxima and minima as discharge changes.

Considerable uncertainties remain surrounding general acceptance of the velocity reversal hypothesis. In part, these originate with different approaches to the problem. Many studies rely on spatially averaged values of velocity and shear stress rather than local values, which directly affect bed load transport. An additional concern is use of the depth–slope product to calculate shear stress. This method assumes steady, uniform flow and may therefore provide misleading results for unsteady, nonuniform flows in pool-riffle sequences (Paola and Mohrig, 1996; Buffington and Montgomery, 1999a). Clifford and Richards (1992) point out that the velocity reversal hypothesis relies on three major assumptions: (1) that hydraulic power laws hold over the entire range of discharge; (2) these relations are similar for riffles and pools in straight and meandering channels; and (3) reversal occurs for all paired cross-sections designated as riffle and pool within a single pool-riffle unit.

Using computer simulations, Carling and Wood (1994) demonstrated that hydraulic behavior varies nonlinearly among pool-riffle pairs as hydraulic roughness, riffle spacing, and pool-to-riffle width ratio vary. This can produce important effects on spatial variation in competence and on the discharge at which a shear stress reversal may occur. In a review of recent studies, Sear (1996) concluded that not all pool-riffle sequences experience flow reversal, in part because of low competence at pool tails. Furthermore, patterns of scour and fill can be largely explained by differences between pools and riffles in bed surface particle structure. Booker et al. (2001) found that generalized descriptions of hydraulic patterns, such as velocity reversal, were overly simplistic, and routing of sediment by secondary circulation and near-bed flow can strongly influence pool-riffle maintenance. Clearly, additional detailed studies will be necessary to fully understand formation and maintenance of channel morphology in pool-riffle sequences, particularly in more complex channels (Clifford and Richards, 1992; Keller and Florsheim, 1993).

Most of our knowledge surrounding the velocity reversal concept as a general account of pool-riffle maintenance is based on studies conducted in free-formed pools in valley bottom streams, which are generally characterized by relatively low bed load transport rates and unobstructed flow. However, the velocity reversal concept may not be applicable to upland streams with relatively high bed load rates and large, in-channel obstructions such as large boulders or trees transported from upslope or the adjacent riparian zone. In such channels large increases in upstream sediment supply can alter pool-riffle morphology and may even bury these features (Lisle, 1982). Furthermore, processes responsible for maintenance of pools associated with in-channel flow obstructions may differ from processes controlling free-formed pools, in part because of additional large-scale turbulence associated with interaction of flow with the obstruction (Lisle, 1986).

In upland, alluvial streams, pool location is commonly forced by large, in-channel obstructions, and pool morphology can depend strongly on obstruction characteristics (Lisle, 1986; Smith et al., 1993; Hogan et al., 1998; Buffington et al., 2002). Association of pools with in-channel obstructions has been well described (Keller and Swanson, 1979; Lisle, 1986; Montgomery et al., 1995; Buffington et al., 2002). However, only a limited number of studies have directly investigated processes that maintain these pools.

In a flume study simulating bed scour around in-channel logs, Beschta (1983) reported that obstruction analogs created zones of exceptionally high turbulence capable of scouring and transporting gravel, even though mean, near-bed velocities indicated otherwise. This observation suggested that entrainment with rising discharge may be caused by a local increase in obstruction-related turbulence rather than increased average bed shear stress. Thompson et al. (1996, 1999) investigated the effects of a channel constriction and associated recirculating eddy on local competence reversal in a gravel-bed, alluvial stream and in a laboratory analog. At high flow in both settings, the constriction caused convergent, accelerating flow at the head of the pool and a recirculating eddy immediately downstream of the constriction. At the head of the pool and near the constriction, local water surface gradients were higher than at the riffle. In the flume, maximum velocities in the centre of the pool always exceeded maximum velocities at the riffle. However, mean pool velocities were lower because of larger cross-sectional flow area. Thus, pool scour was caused by a local competence reversal in only a portion of the pool.

1.1. Conceptual model of pool maintenance

The purpose of this chapter is to present data from field measurements of discharge, vertical velocity distribution, bed load transport, and streambed scour, fill, and grain-size distribution from a case study in support of a conceptual model explaining maintenance of obstruction-forced pools in gravel-bed streams with high sediment transport rates. Specifically, we are concerned with flow and sediment transport regimes that maintain channel morphology over a wide range of flow conditions, including very high discharge.

Results of a large number of studies analyzing scour at bridge abutments and piers are available in the literature; these are summarized in Breusers et al. (1977); Melville (1997). In the current chapter we draw upon findings and insights from this previous research to support a conceptual model of analogous scouring processes at naturally occurring, in-channel obstructions. We propose this obstruction-forced scour model as an alternative to the velocity reversal concept to explain maintenance of many pools in upland streams. In an earlier chapter we presented relations among channel dynamics, sediment entrainment, and bed load transport and grain size distribution at the same site (Hassan and Woodsmith, 2004).

According to this obstruction-forced scour model, pool scour adjacent to large, in-channel obstructions, such as large boulders or trees, closely resembles scour at bridge piers and abutments (Lisle, 1986; Smith, 1990; Buffington et al., 2002). Accordingly, pool morphology is maintained by scour and sediment transport driven by a combination of mean local bed shear stress and instantaneous lift and drag forces associated with large-scale turbulence generated by interactions of flow with the obstruction.

1.2. Scour processes at bridge abutments and piers

In laboratory studies, and by analogy in the case study presented herein, downward flow at the upstream face of an obstruction induces vortices at the obstruction-bed interface

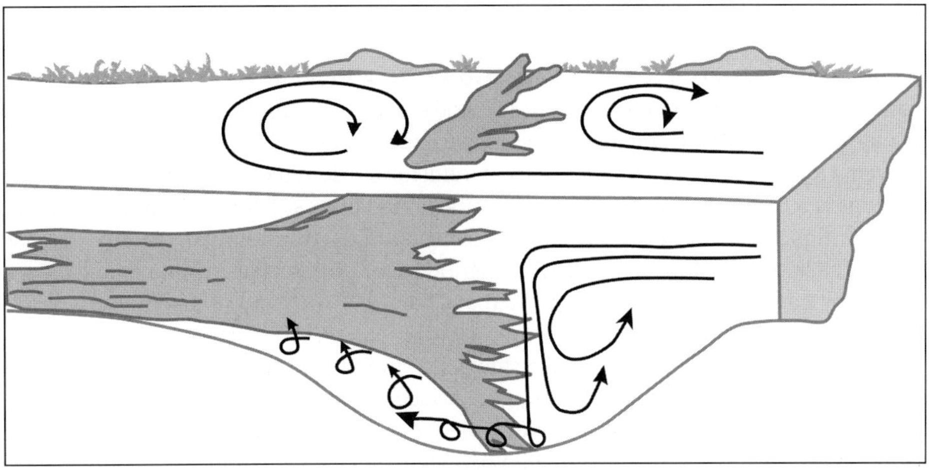

Figure 10.1. Generalized flow patterns and obstruction-related turbulence in a setting similar to the study site.

(Tison, 1961). Vortices with low-pressure centers are cast off near the base of the obstruction, lifting mobile sediment from the bed. Thus, large-scale horseshoe vortices and wake vortices are the primary mechanisms of local scour (Breusers et al., 1977) (Fig. 10.1). Bed material is mobilized by a combination of mean bed shear stress and turbulent agitation both upstream of the obstruction and in the lower portion of the scour hole (Melville, 1975). This combination of hydraulic forces scours pools at an average shear stress less than that required in the absence of obstruction-related scour. Scour may begin at velocities as low as 42% of the critical average velocity for material transport in the unobstructed part of the stream (Tison, 1961; Carstens, 1966; Breusers et al., 1977). Initially size and velocity of near-bed vortices increase rapidly as the new scour hole enlarges, then magnitude of the combined mean bed shear stress and turbulent force at the bed decreases with continued scour until an equilibrium form is reached (Melville, 1975).

Scour depth around engineered structures in gravel-bed channels is greatest at the unobstructed entrainment threshold, and then starts to decrease with stage as upstream sediment is mobilized and delivered to the scour hole. Above this transport threshold, depth undergoes cycles of scour and fill in response to sediment delivery from upstream (Chabert and Engeldinger, 1956; Breusers et al., 1977; Melville, 1984). As flow continues to increase, depth of scour may increase again if scouring in the pool exceeds the rate of sediment input (Melville, 1984; Melville and Sutherland, 1988). Following the discharge peak, pool depth stabilizes when upstream shear stress decreases to the point that bed load is no longer delivered to the pool. This model implies that deeper pools will occur in less mobile alluvial bed material that stabilizes at higher shear stress (Buffington et al., 2002).

2. Study area

Tom McDonald Creek, a tributary to Redwood Creek in north-coastal California, USA, drains an 18-km^2 catchment upslope of the study site, underlain by schist. The climate is Mediterranean with an estimated mean annual precipitation of 203 cm falling primarily between October and April. Floods from rainstorms dominate streamflows. Bankfull discharge is 3.6 m^3 s^{-1}. Gauging records for the period prior to our study are not available; therefore flood frequency was estimated using records from similar, nearby, gauged creeks. The largest peak flow for the study period based on post-flood trimlines and on-site and nearby gauging records was estimated at 25 m^3 s^{-1} with an estimated return period of 5 years (Smith, 1990).

Channel width averaged 10 m at the study site, and the gradient through the reach was 0.006 (Fig. 10.2). Median diameter (D_{50}) of surface material at the pool head, centre, and tail was 31, 3.8, and 17.5 mm, respectively. Median size of subsurface samples (10–40 cm below the bed surface), taken at a riffle immediately downstream of the study pool prior to the largest flood of the study period, was 16 mm. Influence of a large, in-channel obstruction adjacent to the deepest part of the pool (Fig. 10.2) on flow dynamics and channel morphology was obvious from observation. Effective width of this obstruction, perpendicular to flow was 3.6 m, roughly one-third of the active channel width.

The obstruction was a piece of large woody debris (LWD) from the forested catchment. Distribution of LWD in stream channels, its characteristics, and effects on channel morphology and sediment storage and transport have been studied in many places in

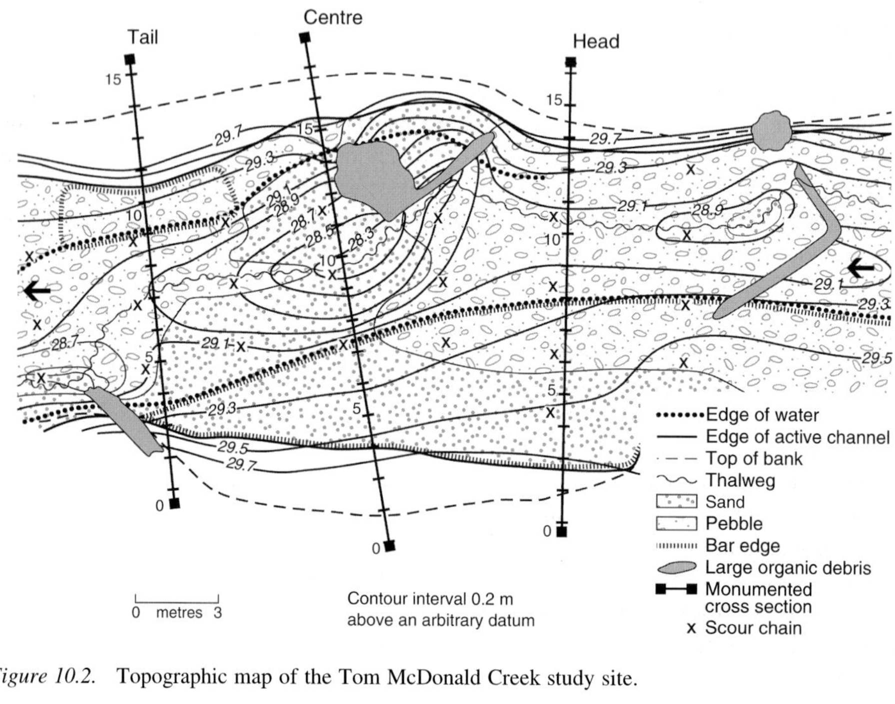

Figure 10.2. Topographic map of the Tom McDonald Creek study site.

the world, including the Pacific coastal region of North America, Mediterranean countries, the United Kingdom, Japan, and others (Keller and Swanson, 1979; Gurnell et al., 1995; Piegay and Gurnell, 1997; Woodsmith and Swanson, 1997). As noted by Piegay and Gurnell (1997), LWD-related effects on flow hydraulics and sediment dynamics commonly affect channel morphology, including pool size and distribution and channel stability.

3. Data collection and analysis

Measurements of flow velocity and direction, bed elevation, and bed load transport were made from three foot bridges following transects perpendicular to the thalweg. Transect 1 was located over the pool head (tail of the upstream riffle); transect 2 over the deepest part of the pool (pool centre), 7 m downstream of the pool head transect; and transect 3 was over the tail of the pool (head of the downstream riffle), 6 m downstream of the pool centre transect (Fig. 10.2). A water level recorder was installed 25 m downstream of the study pool and calibrated against discharge from cross-sectional measurements of velocity, width, and depth. One-minute average velocity and direction of streamflow were measured along vertical profiles spaced at 30 cm or larger intervals across the channel along each transect. Vertical measurement intervals were commonly 2.5 cm at the pool head and pool tail and 15 cm at the pool centre irrespective of flow depth, but larger spacing was sometimes required to complete profiles quickly in rapidly changing flow conditions. In addition to the vertical velocity profiles, several measurements of only the near-bed velocity were made at 30 cm or larger intervals along each transect. Velocity was measured using a standard Price current meter, thus measurements to within 5 cm of the bed were possible. Measurements closer to the bed using other instruments were attempted, but found to be unreliable, owing to interference from bed load and organic material in transport close to the bed. In order to investigate changes in discharge and shear stress relevant to bed load transport, measurements of near-bed velocity, from which shear stress was calculated, were averaged over 2.5 m of channel width, spanning the zone of shear stress maximum and bed load transport.

Local bed shear stress may be derived from a measured velocity gradient using a form of the "law of the wall" equation:

$$\tau_b = \rho\left[(u_2 - u_1)\frac{K}{\ln\left(\dfrac{z_2}{z_1}\right)}\right]^2 \tag{1}$$

where τ_b is the boundary shear stress at a point, ρ is the density of the fluid, u_2 and u_1 are the point mean velocities at distances z_2 and z_1 measured from and normal to the boundary, and K is the von Karman constant, usually taken to be 0.4 (Middleton and Southard, 1984). We found application of this equation to vertical velocity profiles to be problematic, owing to ambiguity in determining profile slope. Furthermore, shear stress calculated from velocity profiles incorporates shear stress attributable to form drag as well as that attributable to roughness of the bed surface (Dietrich and Whiting, 1989; Buffington and

Montgomery, 1999b). Similarly, shear stress estimates based on the depth–slope product were not used, owing to the difficulty of obtaining an accurate measurement of water surface slope given very complex flow patterns (Carling, 1983; Adams et al., 2000). Eq. 1 can be written as:

$$\tau_b = \rho(u_Z K)^2 \left[\ln\left(\frac{z}{z_0} \right) \right]^{-2} \tag{2}$$

where u_z is the velocity at height z above the bed, z_0, the roughness length, is a constant for fully rough flows equal to approximately $k_s/30$. k_s, the effective roughness height, can be approximated by 3.5 D_x, where D_x is a representative length scale of grains controlling resistance (Hey, 1979; Prestegaard, 1983). Estimating z_0 as 0.1 D_x, the equation further reduces to (Dietrich and Whiting, 1989; Whiting and Dietrich, 1990):

$$\tau_b = \rho(u_Z K)^2 \left[\ln\left(\frac{10z}{D_x} \right) \right]^{-2} \tag{3}$$

We calculated bed shear stress using Eq. 3. Near-bed velocity was estimated using the regression relationship of measured near-bed velocity against dimensionless discharge. z_0 was approximated as 0.1 $D_{50\text{-bedload}}$. We found that the common approach of using bed surface grain size as a measure of roughness length produced unrealistically large values of bed shear stress, comparable to total shear stress computed from the average depth–slope product (see Robert, 1997). Furthermore, our estimates of bed shear stress based on estimating z_0 as 0.1 $D_{50\text{-bedload}}$ agreed closely, over a wide range in discharge, with shear stress back calculated from our measured bed load transport rates, using the Meyer-Peter and Muller (1948) bed load equation. Regression coefficients for relationships of near-bed velocity and shear stress against dimensionless discharge were compared among the three transects employing an analysis of covariance approach (Snedecor and Cochran, 1980) and critical probability level of 0.05.

Direct, near-bed measurement of turbulence is problematic in a high-energy, gravel-bed stream with a mobile bed, owing to changing bed elevation and potential damage to instrumentation (Whiting and Dietrich, 1990; Bunte, 1996). Therefore, we did not attempt such measurements; however, we observed strong turbulent structures, which were obviously related to the obstruction at moderate and greater discharge. These included eddies upstream and downstream of the obstruction and vortices, boils, and upwelling near its edge.

Timing and magnitude of scour and fill were computed from 38 repeat sounding surveys along each of the three transects throughout the study period. These data were supplemented by an array of scour chains (Leopold et al., 1964) used to measure net change in bed elevation (Fig. 10.2). Scour chain data indicated maximum scour, net deposition, and, to some extent, the sequence of scour and fill events. However, a portion of the sediment deposited prior to scouring can be eroded, altering the record of its original thickness.

Bed load transport rates were measured at the three transects using a hand-held, 76 mm-orifice, Helley–Smith bed load sampler (Helley and Smith, 1971) fitted with an enlarged bag to better maintain sampling efficiency (Beschta, 1981). Measurements spanned a wide range in flow from the beginning of significant sediment transport to flows exceeding bankfull discharge. At each transect, eight samples were collected at 0.6-m intervals to represent transport across the channel. A minimum discharge of 1.26 $m^3 s^{-1}$ was required to mobilize measurable amounts of bed load. As is commonly noted in bed load studies, transport was localized over only a small portion of the channel width (e.g., Jackson and Beschta, 1982; Klingeman and Emmett, 1982; Dietrich and Whiting, 1989). Although bed load was concentrated in the zone of high shear stress within approximately 1–2 m of the thalweg, sampling spanned nearly the entire channel width. This avoided bed load bypassing the sampler owing to "topographic steering" (Nelson and Smith, 1989) or deflection by secondary flow (Booker et al., 2001). Therefore, even at high discharge, only trivial amounts of bed load were transported outside the sampled portion of channel. Bed load samples were dried and sieved in the laboratory for particle size analysis.

The Helley–Smith sampler is known to produce biased results in coarse materials (Sterling and Church, 2002). However, there is no consensus regarding the best sampler for obtaining true sediment transport rates in gravel-bed rivers (Ryan and Troendle, 1996; Hassan and Church, 2001). The size of most bed material at our pool centre and tail transects was within the range of high sampler efficiency reported by Emmett (1980); for these two transects we assumed that the Helley–Smith results were reasonable. Sampler efficiency was likely to be somewhat lower at the coarser pool head, imposing some limitation on the analysis of bed load data. To limit error in sampling coarse fractions, we truncated samples at 32 mm. Finally, we assumed that bias in our results was systematic, and therefore comparison among the three transects was reasonable. Similarity of regression relationships for bed load transport rate against discharge among the three transects was tested employing an analysis of covariance approach (Snedecor and Cochran, 1980) and critical probability level of 0.05.

4. Flow dynamics

From observation, flow dynamics and channel morphology were obviously influenced by the primary obstruction adjacent to the deepest part of the pool (Fig. 10.2). It follows that magnitude and distribution of bed shear stress, scour and fill, and sediment entrainment and transport must have also been affected. At all discharges the obstruction deflected flow and elevated shear stress along the base of the slip face of an upstream, left-bank lateral bar, scouring sediment and apparently halting downstream bar migration. In addition, location of a right-bank lateral bar was fixed by deposition in the low-shear stress environment in the lee of the obstruction (Fig. 10.2). Thus, as described by Lisle (1986) for similar obstruction-forced pools, the obstruction anchored the location of both lateral bars, thereby stabilizing channel morphology. These bars further modified streamflow patterns, particularly at moderate and high discharge.

Effects of the obstruction on streamflow patterns can be seen in flow velocity profiles (Fig. 10.3). Profiles selected for Figure 10.3 were representative of flow near the thalweg. These were neither measured at precisely fixed locations nor at the same discharge at all transects. Therefore, these profiles provide a general sense of flow patterns, rather than a rigorous comparison among locations and discharges. At the pool head, obstruction effects

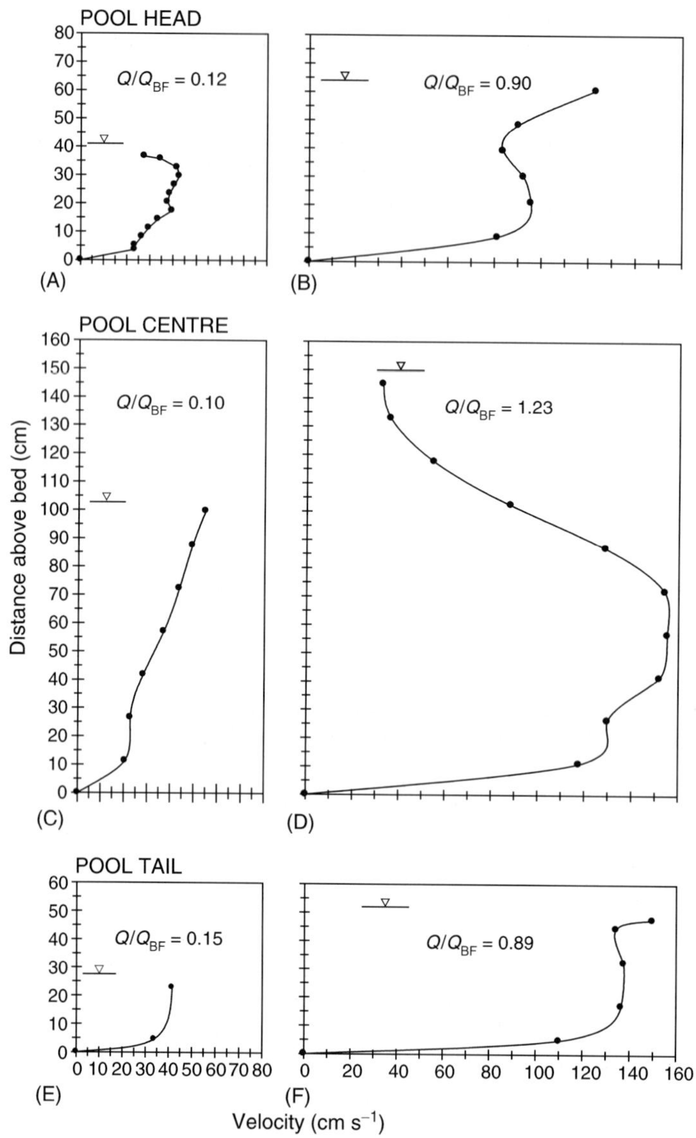

Figure 10.3. Velocity profiles. Q is discharge. Q_{BF} is bankfull discharge. Note scale differences. Triangles indicate water surface elevation.

were minimal at low discharge, therefore velocity profiles were similar to those typical for unobstructed flow, i.e. approximately logarithmic with maximum velocity depressed below the water surface (Fig. 10.3A). As discharge increased above one-half bankfull, backwater effects of the obstruction elevated the water surface on the right side of the channel. This and flow deflection to the left depressed streamwise velocities in the middle portion of the water column, distorting the logarithmic velocity profile (Fig. 10.3B).

Direct effects of the obstruction on streamflow patterns were most pronounced at the pool centre. These effects were evident even at moderate discharge, and included sharp leftward deflection of flow and creation of eddies, vortices, and boils. In addition, a large area of low-velocity, eddying current was created in the downstream lee of the obstruction. At low discharge, vertical velocity distribution was approximately logarithmic, although turbulent structures depressed velocity in the lower-centre of the water column, causing some distortion of the velocity profile (Fig. 10.3C). At high discharge, low shear stress in the lee of the obstruction promoted development of a right-bank lateral bar. This bar reinforced leftward flow deflection by the obstruction, thereby opposing flow deflection by the upstream left bank bar. Flow convergence at the base of the obstruction increased velocity in the lower portion of the water column, while opposing leftward- and rightward-deflected flow created backwater effects, which depressed velocity in the upper water column, thereby strongly distorting the velocity profile (Fig. 10.3D). At the pool tail flow resumed a more streamwise direction, and velocity profiles were generally logarithmic throughout a wide range in discharge (Fig. 10.3E,F).

5. Bed shear stress

Sediment transport and channel morphology were controlled by the magnitude and distribution of boundary shear stress across and along the channel, and the obstruction influenced shear stress distribution through modification of the flow pattern. Figure 10.4 presents variation in shear stress along the three transects at 0.16 Q_{BF}, 0.50 Q_{BF}, and 1.3 Q_{BF}. At low flow at the pool head, the maximum bed shear stress was located slightly to the left of the thalweg and near the toe of the left-bank lateral bar. Maximum shear stress was far less at the pool centre than at the pool head and tail and was located near the thalweg. At the pool tail the channel was wide with approximately uniform bed elevation, grain size distribution, and flow depth. Maximum bed shear stress occurred near the thalweg (Figs 10.2 and 10.4A).

At moderate discharge at the pool head, deflection by the left bank lateral bar forced shear stress maximum toward the right side near the thalweg. At the pool centre the zone of shear stress maximum widened toward the toe of the left-bank lateral bar. Magnitude of shear stress maximum remained markedly smaller than at the pool head or tail. At the pool tail the zone of shear stress maximum became somewhat more focused and shifted toward the toe of the right-bank lateral bar in the lee of the obstruction (Fig. 10.4B).

At flows near or greater than bankfull at the pool head, backwater effects of the obstruction caused location of the shear stress maximum to return toward the left, away from the thalweg and toward the toe of the left-bank lateral bar. Shear stress magnitude was greater than at the pool centre or tail. At the pool centre, shear stress maximum again became focused near the thalweg. At the pool tail flow deflection by the left-bank lateral bar further focused shear stress maxima toward the toe of the right-bank lateral bar (Fig. 10.4C). At all three transects at all measured flows, maximum shear stress location remained within about 1 m of the thalweg.

At all transects, near-bed velocity, averaged over 1 min and over 2.5 m of channel width (the zone of bed load transport), increased linearly with discharge up to flows of at least 1.3 Q_{BF} (Fig. 10.5). At higher discharge, data were more variable; therefore we used

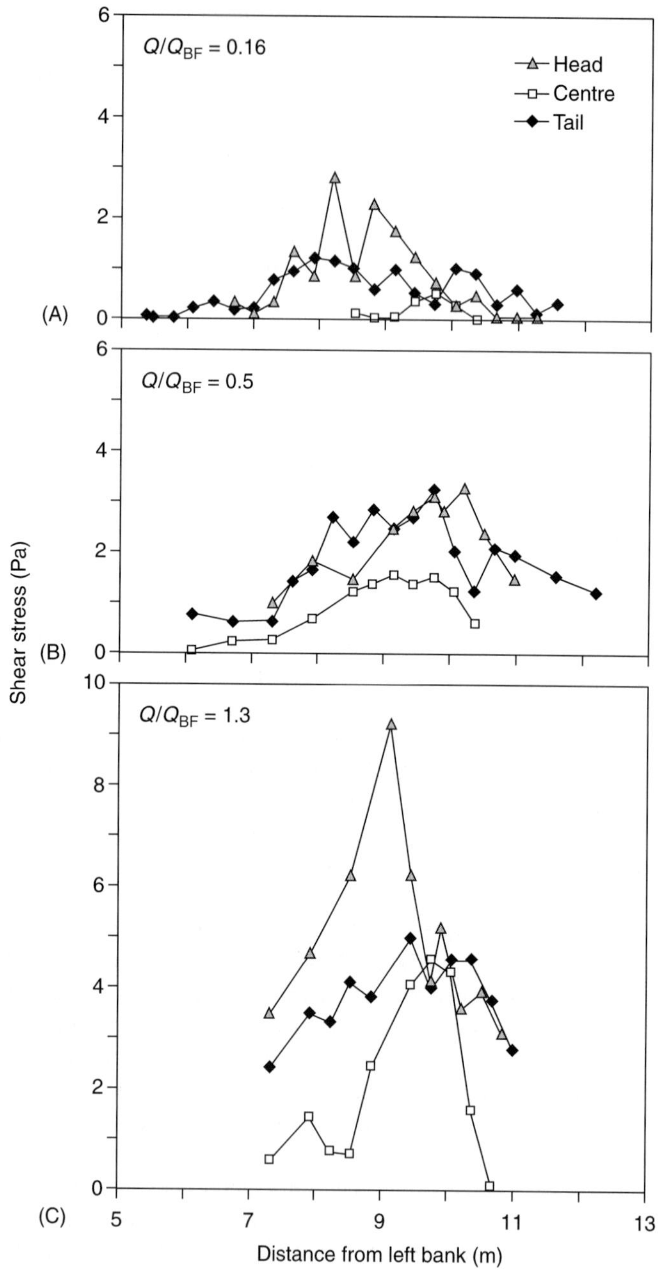

Figure 10.4. Cross-channel variation in bed shear stress at three values of discharge (Q) scaled by bankfull discharge (Q_{BF}).

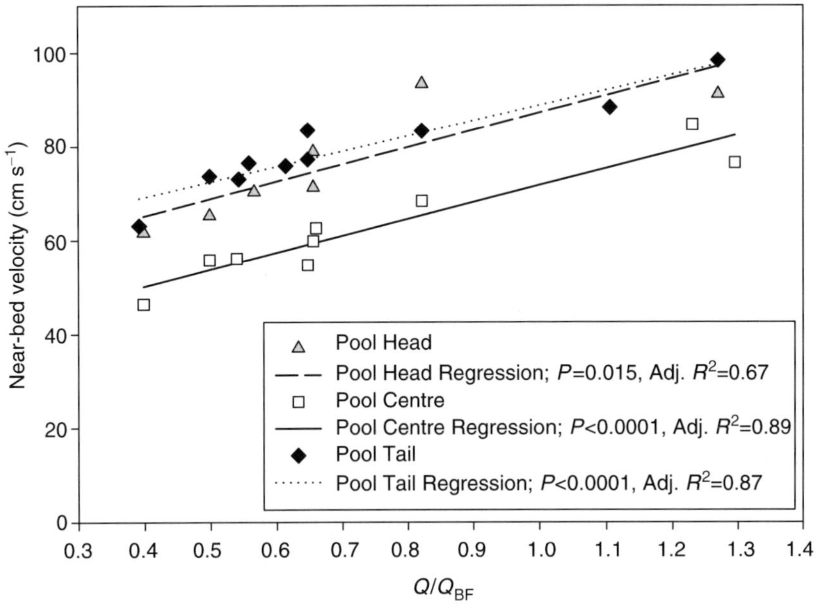

Figure 10.5. Increase in near-bed velocity with dimensionless discharge at the three transects. Straight lines are least squares regression fits to the data.

$Q/Q_{BF} < 1.3$ to define this relationship. Slopes of the regression relationships illustrated in Figure 10.5 were not significantly different among any of the transects, and the intercept for the pool centre relation was significantly less than those for the pool head and tail. Thus, near-bed velocity at the pool centre remained less than that at the pool head and tail, while increasing with discharge at the same rate up to flows at least as large as 1.3 times bankfull (Fig. 10.5).

Similarly the 0.5 power of bed shear stress ($\tau^{0.5}$) increased linearly with dimensionless discharge (Fig. 10.6). Shear stress was transformed to the 0.5 power to stabilize variance and to remove the effect of the increase in shear stress with the square of velocity. Intercepts for the least squares regression relationship illustrated in Fig. 10.6 were statistically different among all transects. Slopes were statistically different for the pool centre and pool tail only. The rate of increase in $\tau^{0.5}$ at the pool tail was the greater. Therefore, bed shear stress at the pool centre remained less than at the pool head and tail, while increasing with dimensionless discharge at the same or lesser rate (Fig. 10.6).

6. Scour and fill

Significant aggradation of both the upstream, left-bank (Fig. 10.7A) and downstream, right-bank (Fig. 10.7C) lateral bars occurred during 24 h surrounding the largest flood peak of the study period. Peak discharge of this event was approximately 7 times bankfull (Fig. 10.8). Aggradation of the upstream bar reduced cross-sectional area at the pool head transect (Fig. 10.7A), and to a much lesser degree and shorter duration, the pool centre transect (Fig. 10.7B). Cross-sectional area of the pool centre, disregarding bank erosion,

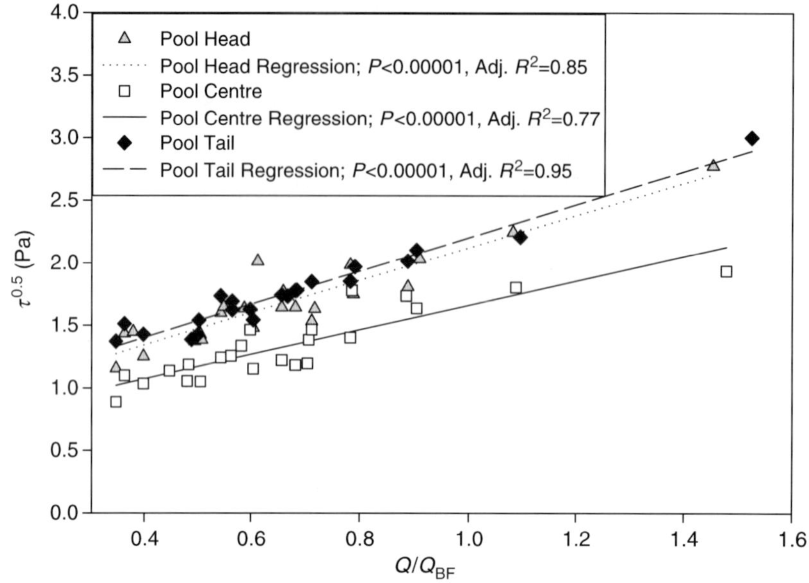

Figure 10.6. Increase in shear stress with dimensionless discharge at the three transects. Straight lines are least squares regression fits to the data.

was remarkably stable throughout a wide range of discharge, including this large flood, which peaked a few hours after the 860217 (17 February 1986) survey (Figs 10.7B, 10.8). Furthermore, subsequent flows eroded much of these bar deposits, returning the bed at the pool centre to nearly the pre-flood profile, disregarding bank erosion (Figs 10.7B, 10.8)

The pool did not store important volumes of sediment relative to the volume transported as bed load. Estimated pool volume was 14.5 m^3, providing a potential storage site for a maximum of 34,000 kg of bed load, assuming a maximum dry density for closely packed silty sand and gravel of 2360 kg m^{-3} (Holtz and Kovacs, 1981). The maximum fill in the pool, in excess of an arbitrary datum, was estimated to be 35% of its volume, deposited primarily on the pool margins during 24 h surrounding the peak discharge. Even this maximum volume of fill did not significantly change pool centre depth (Fig. 10.7B).

During this same period, 84,000 kg of bed load, the equivalent of 250% of the pool volume and 700% of the volume of fill, were transported through the pool. Maximum scour of 17% of the pool volume occurred during the rising limb of the large flood. During this same period, 38,000 kg of bed load were exported, equivalent to 110% of the pool volume and 660% of the volume of scour. These volumes do not include severe bank erosion between days 780 and 785 (Figs 10.7B, 10.8).

Figure 10.7. Cross-sectional soundings. Elevation is relative to an arbitrary datum. The 850514 (14 May 1985) survey provides a common reference in all plots. Post flood bank erosion on the right side of the pool centre is evident in the latest survey.

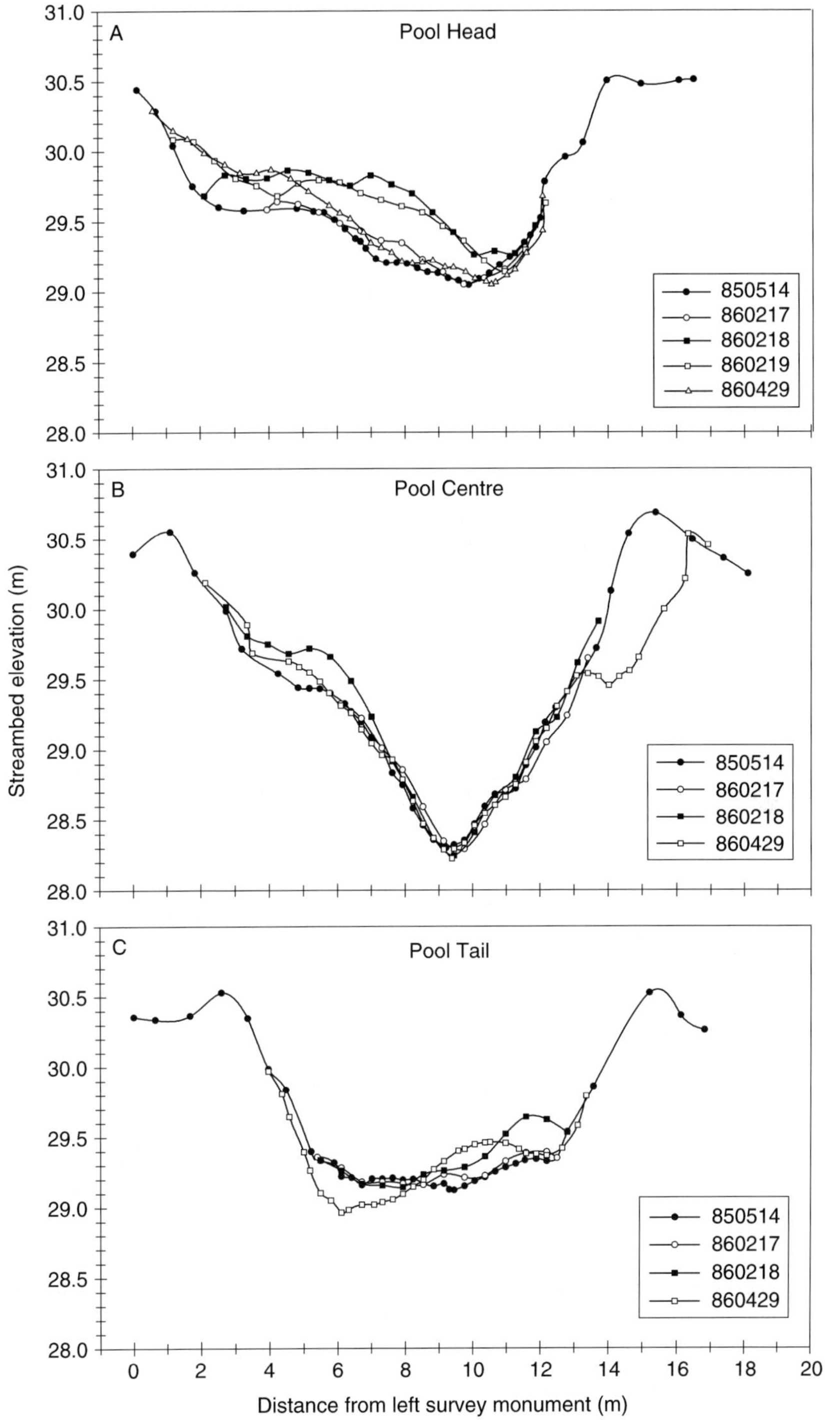
A
Pool Head
B
Pool Centre
C
Pool Tail
850514
860217
860218
860219
860429
Streambed elevation (m)
Distance from left survey monument (m)
31.0
30.5
30.0
29.5
29.0
28.5
28.0
0
2
4
6
8
10
12
14
16
18
20

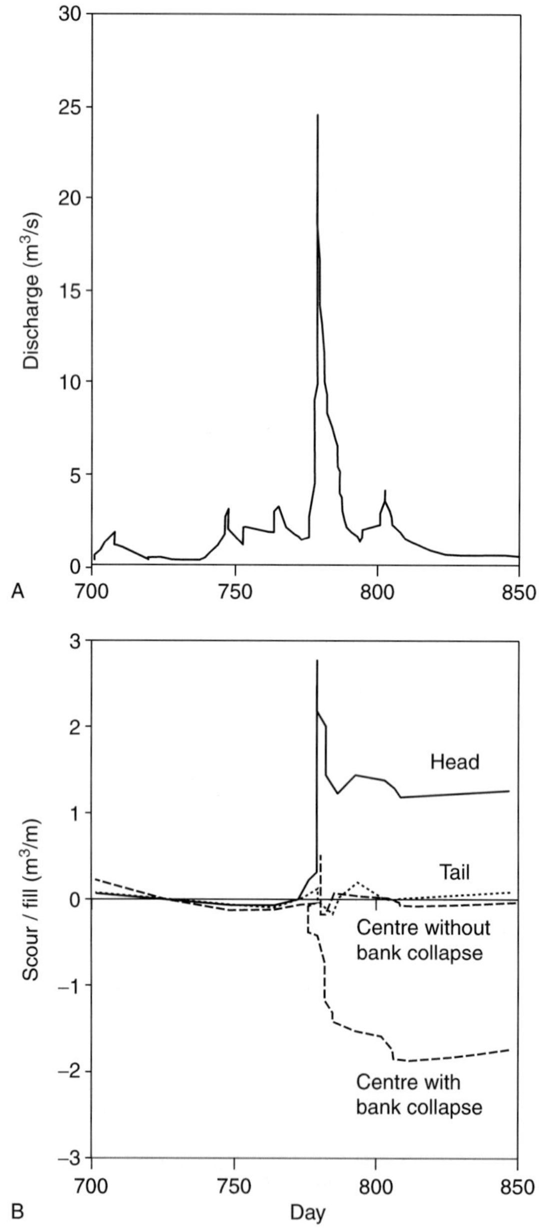

Figure 10.8. Flow hydrograph and scour (−) and fill (+) of the cross-sectional area of the streambed during the primary flood of the study period. Days are expressed numerically, the hydrograph peak occurred on February 17, 1986 (860217, day 778).

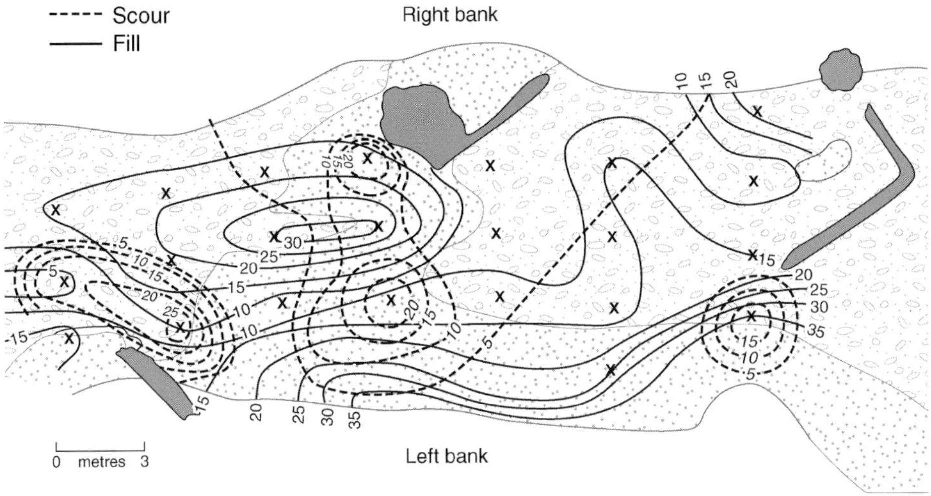

Figure 10.9. Scour and fill, as measured by scour chains, primarily during the major flood of the study period. Scour chain locations are marked with an X. Contour interval is 5 cm.

Scour chain data indicated zero net average change in the bed surface elevation at the pool centre (Fig. 10.9). However, soundings, unlike scour chain data, included the banks and indicated major lateral erosion of the right bank, resulting from flow deflection by the obstruction during the receding limb of the largest flood (Fig. 10.8). Disregarding bank erosion, both scour chains and soundings indicated very minor net change to the pool centre bed profile during the study period (Figs 10.7–10.9).

7. Sediment entrainment

To examine the relation between fractional sediment transport and shear stress for each bed load size fraction, the size-specific scaled fractional transport rate was plotted against particle size over a range of shear stress magnitude (Fig. 10.10A–C). The fractional transport rate was scaled using the subsurface material grain size distribution averaged over all three transects and truncated at 32 mm, because of limitations of the Helley–Smith sampler and uncertain sampling of material larger than 32 mm diameter. All curves show a break in slope where transport rate begins to decline with increasing grain size, i.e. a shift from full to partial mobility (Fig. 10.10A–C). Dark lines in Figure 10.10 connect break points in fractional mobility, which represent the largest fully mobile grain size for each value of shear stress. The increase in this grain size with shear stress implies selective transport.

Following the Wilcock and McArdell (1993) method, the critical shear stress needed to initiate transport for a given fraction was taken from the Parker et al. (1982) reference transport relations, covering most of the range of transported material (Fig. 10.10D). The reference shear stress is that needed to entrain an individual size fraction and produce

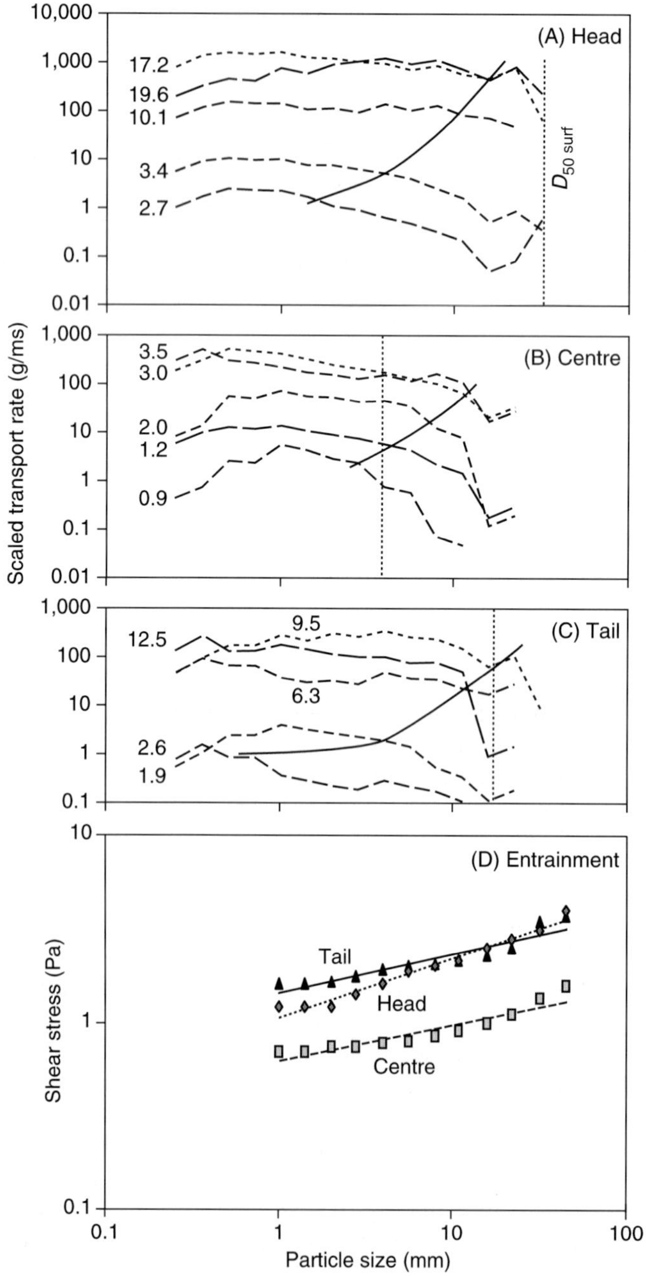

Figure 10.10. Scaled fractional sediment transport rate versus particle size for selected flows, shown as the associated bed shear stress (Pa): (A) pool head; (B) pool centre; (C) pool tail. The median size of the bed surface is plotted for comparison. (D): the relation between particle size and bed shear stress needed to initiate movement of individual particles.

a small reference transport rate. The fitted functional relations are as follows:

$$\tau_{ri} = 1.05 \pm 0.09 D_i^{0.32 \pm 0.016} \quad \text{for the pool head} \tag{4}$$

$$\tau_{ri} = 0.62 \pm 0.08 D_i^{0.19 \pm 0.022} \quad \text{for the pool centre} \tag{5}$$

$$\tau_{ri} = 1.43 \pm 0.17 D_i^{0.21 \pm 0.024} \quad \text{for the pool tail} \tag{6}$$

where D_i is the particle size in mm, and τ_{ri} is the reference shear stress in Pascal. These incipient motion analyses suggest that mean bed shear stress entrainment thresholds for the range of grain sizes measured were lower (entrainment at lower shear stress) at the pool centre than at the pool head or tail (Fig. 10.10D) (Hassan and Woodsmith, 2004).

We further examined bed load entrainment by analyzing the relationship between the intermediate diameter of the largest clast in each bed load sample and the dimensionless shear stress (Fig. 10.11). Due to sampling bias of the Helley–Smith sampler (discussed in

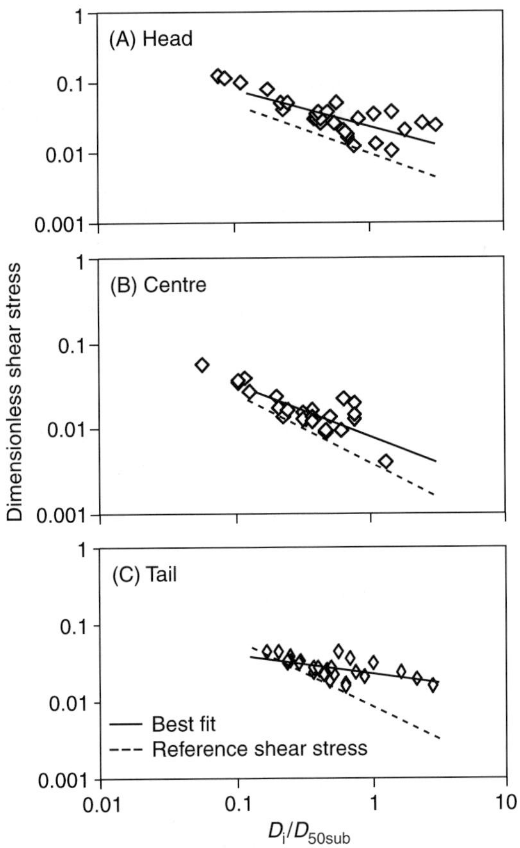

Figure 10.11. Dimensionless shear stress as a function of relative grain size (scaled to the median diameter of the subsurface material). In this plot we used the largest clast found in the sampler. The best-fit line and the reference shear stress (Eqs. 4–6) are shown.

Sterling and Church, 2002; Hassan and Woodsmith, 2004), these analyses should be viewed as an approximation of entrainment conditions. The relations for the data presented in Figure 10.11 are as follows,

$$\tau_{ri^*} = 0.023 \pm 0.004(D_{i\text{-max}}/D_{50sub})^{-0.51\pm0.07} \quad \text{for the pool head} \tag{7}$$

$$\tau_{ri^*} = 0.008 \pm 0.003(D_{i\text{-max}}/D_{50sub})^{-0.60\pm0.12} \quad \text{for the pool centre} \tag{8}$$

$$\tau_{ri^*} = 0.022 \pm 0.005(D_{i\text{-max}}/D_{50sub})^{-0.27\pm0.10} \quad \text{for the pool tail} \tag{9}$$

where the dimensionless shear stress $(\tau_{ri^*}) = \tau_b[(\rho_s\text{-}\rho)gD_i]^{-1}$, ρ_s is the sediment density, g is gravitational acceleration, $D_{i\text{-max}}$ is the largest particle in the bed load sample, and D_{50sub} is the median size of the subsurface bed material. A slope of -1.0 would imply equal mobility (Parker and Klingeman, 1982), therefore a strong relative size effect on entrainment of individual particles is suggested, implying that selective entrainment occurs along the three transects.

For comparison, we plotted Eqs. (4–6), the reference transport relations, using the same units as in the dimensionless Eqs. (7–9). In all cases the reference transport line plots near the lower limit of the data (Fig. 10.11). The reference transport line represents the lower limit for entrainment, below which no sediment transport is expected. Similar results were reported in Church and Hassan (2002) for data from Harris Creek, British Columbia, Canada.

8. Bed load transport rates

High bed load transport rates at the pool tail ($G_{\text{-tail}}$) were clearly associated with high discharge (Table 10.1). Nevertheless, highly variable $G_{\text{-tail}}/G_{\text{-head}}$ ratios for consecutive samples, collected within several minutes of one another together with minor change in pool cross-sectional area suggested that dynamic upstream sediment supply was a primary control on bed load transport. Regular patterns of pool filling ($G_{\text{-tail}}/G_{\text{-head}} < 1$) at low discharge and pool scour ($G_{\text{-tail}}/G_{\text{-head}} > 1$) at discharge above bankfull were not apparent in our data (Table 10.1). For example, several days after the 860217 flood at $Q/Q_{BF} = 0.35$, $G_{\text{-tail}}$ exceeded $G_{\text{-head}}$ by a factor of nearly 8; however, only 5 days later at $Q/Q_{BF} = 0.54$, $G_{\text{-head}}$ exceeded $G_{\text{-tail}}$ by a factor of 2.5 (Table 10.1). For $2 < Q/Q_{BF} < 3$, $G_{\text{-tail}}/G_{\text{-head}}$ varied from 0.4 to 17.4. These examples are extreme cases, nevertheless there was no consistent pattern of pool scour and fill, as measured by $G_{\text{-tail}}/G_{\text{-head}}$, in response to discharge magnitude. Similarly, there was no clear correlation between $G_{\text{-tail}}/G_{\text{-head}}$ and hydrograph limb. The four largest values of $G_{\text{-tail}}$ occurred within the first week following the 860217 flood peak (Table 10.1).

Bed load transport rate increased with discharge at all transects. Sensitivity (slope) of the rating relation changed at a critical discharge of about Q_{BF} (Fig. 10.12). Below Q_{BF}, sediment transport rate increased rapidly as discharge increased, but increased more slowly at flows above Q_{BF}. Piecewise regression, was employed to compare the rating equations among transects using Q_{BF} as a critical value (Fig. 10.12) (also see Ryan et al., 2002). Increase in bed load transport rate with discharge was not significantly different in the pool centre from that at the pool head or tail ($P = 0.05$). However, close proximity of

Table 10.1. Bed load Flux. Q is water discharge. Q_{BF} is bankfull discharge. G is bed load transport rate (kg h^{-1}). Rows are arranged in order of increasing discharge. Bed load export through the pool is indicated by $G_{\text{-tail}}/G_{\text{-head}} > 1.0$. Dates are expressed numerically as year, month, and day.

Q ($m^3\ s^{-1}$)	Q/Q_{BF}	$G_{\text{-tail}}/G_{\text{-head}}$	$G_{\text{-tail}}$ (kg h^{-1})	Hydrograph limb	Date
1.26	0.35	7.9	7.7	Falling	860304
1.32	0.36	0.9	4.8	Falling	860120
1.45	0.40	1.4	7.5	Falling	860320
1.78	0.49	0.9	5.9	Rising	860116
1.82	0.50	1.1	7.1	Rising	860122
1.83	0.50	1.4	21	Falling	860318
1.98	0.54	0.4	27	Rising	860309
2.06	0.57	0.5	9.1	Rising	860214
2.06	0.57	0.9	310	Falling	860308
2.18	0.60	0.9	9.2	Falling	860206
2.21	0.61	1.6	58	Falling	860311
2.39	0.66	1.4	46	Falling	860316
2.43	0.67	3.0	100	Rising	860202
2.48	0.68	2.1	69	Falling	860311
2.49	0.68	5.1	1300	Falling	860307
2.60	0.71	5.2	32	Falling	860205
2.85	0.78	3.1	30	Falling	860204
2.88	0.79	47.7	3200	Falling	860315
3.23	0.89	3.6	170	Peak	860203
3.29	0.90	0.5	1300	Falling	860226
3.99	1.10	0.6	1100	Falling	860313
5.56	1.53	4.5	3200	Rising	860216
5.86	1.61	1.0	4100	Falling	860224
7.41	2.03	0.4	890	Falling	860216
7.78	2.14	17.4	41,000	Rising	860221
10.6	2.93	0.6	1900	Rising	860217
12.1	3.32	0.5	8900	Rising	860219
12.9	3.55	2.2	56,000	Rising	860219

the three transects may have masked spatial variations, making this observation difficult to interpret.

9. Discussion

We observed that the primary obstruction in the study reach influenced flow dynamics and channel morphology through flow deflection and convergence and creation of turbulent structures including eddies, vortices, and boils forced by flow-obstruction interaction. To quantify obstruction effects we compared velocity distribution, bed shear stress, streambed scour and fill, and bed load transport rate and grain-size distribution among

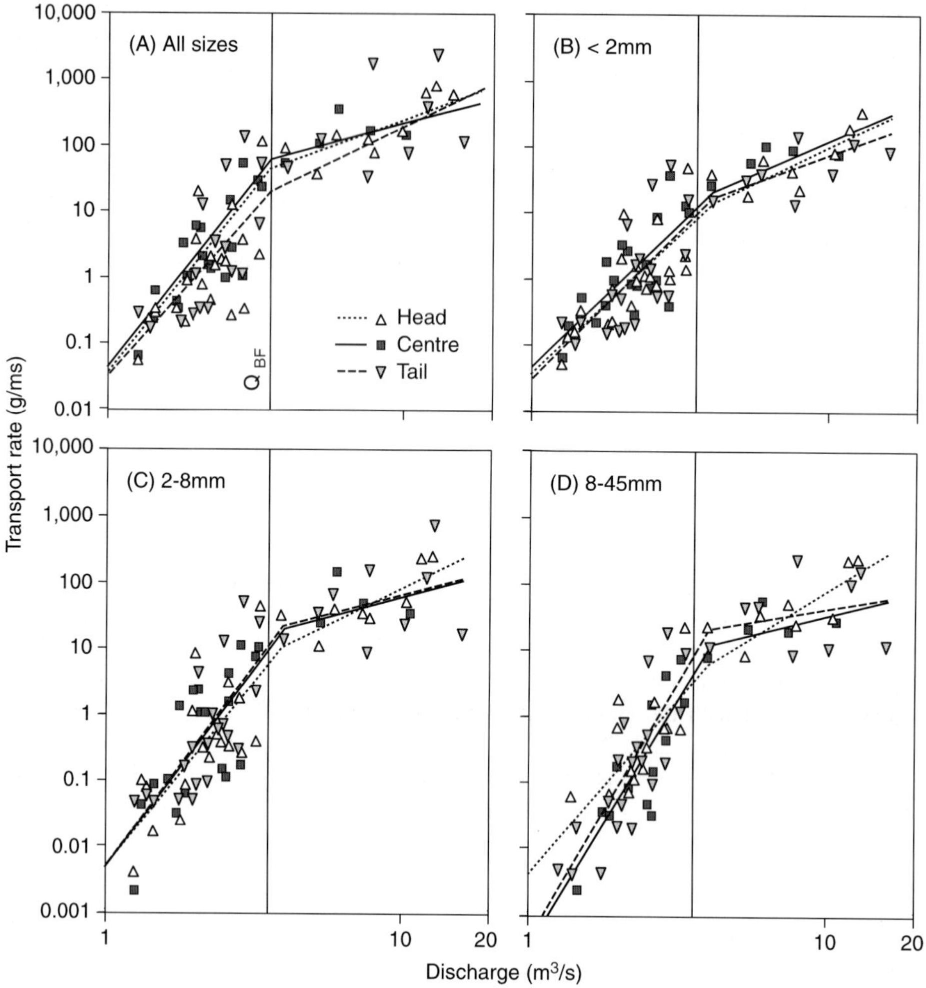

Figure 10.12. Total and size specific bed load rating curves for the pool head, pool centre, and pool tail of the form, (A): $\log G = a + b \log Q$ and (B): $\log G = a + b \log Q + m(\log Q - Q_{BF})|Q > Q_{BF}$, where G is the bed load transport rate in $g\,m^{-1}\,s^{-1}$ and Q_{BF} is the bankfull discharge, above which the rating sensitivity changed. The change in the sensitivity was selected using breakpoint analysis, and then the two-part regression line was fitted.

transects at the obstruction-forced pool head, centre, and tail. Of particular interest in this study were the inferred effects of large-scale flow turbulence, generated at the large, in-channel obstruction, on sediment entrainment, transport, and deposition.

Near-bed velocity was consistently lower at the pool centre than at the pool head and tail, and increased linearly with dimensionless discharge at all three transects at statistically similar rates. Likewise, bed shear stress, estimated from near-bed velocity and bed load grain size, was lower at the pool centre than at the pool head and tail, and increased linearly with dimensionless discharge at all three transects at rates that gave no indication of convergence.

Soundings indicated that scour and fill did not follow a systematic pattern of pool filling and riffle scour at discharge less than bankfull with the reverse occurring at higher flow. Cross-sectional area of the pool centre was remarkably constant throughout a wide range of discharge, despite the occurrence of very large flow events. During low-flow periods, sand-size and smaller sediment did accumulate in the shallows of the pool centre, away from the obstruction, but this commonly accounted for less than 5% of the pool cross-sectional area and was scoured at moderate discharge. Significant aggradation of lateral bars upstream and downstream of the obstruction occurred near the peak of the major flood, primarily affecting the pool head and tail transects. Subsequent flows removed much of these bar deposits. Both scour chain and sounding data indicated very minor net change to the average bed elevation across the pool centre transect, disregarding bank erosion.

Fractional transport rates indicated that only at the pool centre did bed load transport occur at mean bed shear stress values less than about 1.8 Pa. Furthermore, transport rates approached 100 g m^{-1} s^{-1} at the pool centre at notably lower average shear stress than at the pool head or tail. Incipient motion analyses suggested that mean bed shear stress entrainment thresholds were lower (entrainment at lower shear stress) at the pool centre than at the pool head or tail.

Bed load flux through the pool centre ($G_{\text{-tail}}/G_{\text{-head}}$) had little relationship to position on the storm hydrograph or discharge magnitude. These data, and the relatively stable pool centre morphology regardless of discharge, suggested that bed load transport was responding to variation in upstream sediment supply, rather than systematic, threshold discharge-driven scour and fill of the pool. This inference was made cautiously, owing to bias associated with the Helley–Smith bed load sampler (Wilcock, 1992; Ryan and Troendle, 1996; Hassan and Church, 2001; Sterling and Church, 2002) and the inherent temporal variability in bed load transport resulting from wave-like translation of bed material, passage of bedforms, armouring, particle clustering, and the stochastic nature of bed load entrainment and transport (Kuhnle and Southard, 1988; Pitlick, 1988; Gomez et al., 1989; Bunte, 1996). For example, Pitlick (1988) took consecutive 60-s samples of bed load in the Fall River, Colorado ($D_{50} = 1$ mm) during nearly steady flow. Composite samples, similar to those in the present study showed a two-fold variation in bed load rate. Similarly, based on data from magnetic tracer studies in Squaw Creek, Montana ($D_{50} = 22$ mm) during diurnal, snowmelt-driven discharge fluctuations, Bunte (1996) reported 5-min fluctuations in bed load rate that most commonly were less than two-fold. Although the assessment is obviously crude, even if one disregards $G_{\text{-tail}}/G_{\text{-head}}$ ratios between 0.4 and 3.0 in Table 10.1, there remains no systematic pattern of threshold discharge- or hydrograph structure-driven $G_{\text{-tail}}/G_{\text{-head}}$ ratio. However, these patterns of bed load flux were consistent with known mechanisms of scour at bridge abutments and piers (Breusers et al., 1977; Melville, 1997). The extent to which finer surface texture and lower roughness in the pool reduced entrainment thresholds and increased bed load transport rates (Dietrich et al., 1989; Buffington and Montgomery, 1999b) is unknown.

We inferred from these results that total entrainment and transport force in the pool centre was underestimated by average near-bed velocity or bed shear stress alone. The large obstruction anchored the location of scouring turbulent structures and constricted the channel, causing flow convergence, thereby creating conditions favorable to bed load entrainment and transport and more stable pool morphology than expected for nonobstruction-related pools. The role of smaller-scale turbulence in nonobstruction related sediment entrainment is well established (Kalinske, 1947; Sutherland, 1967; Baker and Ritter, 1975; Cheetham, 1979; Carling, 1983; Whiting and Dietrich, 1990; Nelson et al., 1995). We were not able to quantify obstruction related turbulence with direct measurements. Such measurements remain problematic in gravel-bed streams, owing to the mobile boundary and risk of damage to instruments (Whiting and Dietrich, 1990; Bunte, 1996).

10. Conclusions

Our results, observations of turbulent structures, and analogy to well known processes of scour at bridge abutments and piers support a conceptual model explaining maintenance of obstruction-forced pools at naturally occurring flow obstructions in gravel-bed streams. Analogous hydraulic conditions of fluvial scour around bridge abutments and piers are common, where bed material is mobilized by a combination of time-averaged bed shear stress and turbulent agitation in the scour hole (Melville, 1975; Breusers et al., 1977). In this case study, as in the case of bridge abutments and piers, turbulent effects generated by interaction of streamflow with the obstruction contribute an additional component of total force acting to entrain and transport sediment. Pool maintenance results from a combination of mean bed shear stress and these large-scale turbulent drags and lift forces. At discharge above the transport threshold for the upstream bed, bed load moves through the pool centre in response to sediment delivery from upstream, causing multiple pulses of bed load through the pool during a single storm hydrograph. Following the discharge peak, bed load flux drops when upstream shear stress decreases to the point that bed load is no longer delivered to the pool (Lisle, 1986; Smith, 1990; Buffington et al., 2002).

This model provides an alternative to the velocity reversal hypothesis, which describes pool maintenance in free-formed alluvial pools, rather than obstruction-forced pools. Our results do not refute the velocity reversal hypothesis, rather we argue simply that flow hydraulics in obstruction-forced pools can differ from those in free-formed pools. Clearly, competence in any pool must at times be sufficient to transport material delivered from upstream in order to maintain pool volume. However, our data indicate that maintenance of the study pool, formed by scour at a large in-channel obstruction, is attributable to processes other than shifting of the location of maximum near-bed velocity or bed shear stress to the pool centre at high discharge. We conclude that obstruction-forced pools do not necessarily undergo alternate scour and fill patterns during high and low flow sequences as suggested by the velocity reversal hypothesis. Rather, pool volume can remain remarkably stable throughout a wide range in discharge.

We emphasize the unique nature of obstruction-forced pools where large-scale turbulent structures generated by interactions of flow with the obstruction appear to generate entrainment and tractive forces that alter the shear stress-sediment transport relationship relative to upstream and downstream riffles. Although turbulence was not directly measured, it remains the likely factor to make up for relatively low values of mean shear stress in pools, thereby maintaining necessary competence for pool maintenance. This combination of mean bed shear stress and instantaneous turbulent force creates and maintains the pool at a site where scour may not have occurred in the absence of an obstruction.

Our results suggest that scour forced by large, in-channel obstructions can be a dominant mechanism, perhaps as important as mean hydraulic variables and caliber and volume of sediment supply, explaining local channel morphology and local bed load dynamics. However, pool scour varies with size and shape of obstructions and with drainage basin and channel characteristics (Buffington et al., 2002). Extrapolation of results of this study to other obstruction-pool geometries is untested.

Results of this study are of interest not only to fluvial geomorphologists, but also to engineers and land managers involved with aquatic habitat restoration through manipulation of channel morphology. These results provide insight into the mechanisms responsible for scour and maintenance of pools associated with either naturally occurring or artificially placed obstructions. Local hydraulics, channel morphology, and sediment transport can mutually adjust to large, in-channel obstructions in such a way that the associated pool morphology is maintained through a wide range in discharge. This dynamic stability may provide critical habitat with a wide diversity of water depth and velocity for aquatic organisms, including refuge habitat during high-flow conditions.

Acknowledgements

This research was supported by USDI, Redwood National Park; USDA, Forest Service, Pacific Northwest Research Station; and the Natural Sciences and Engineering Research Council of Canada through Research grant 249673 to M. Hassan. Jason Rempel performed part of the sediment transport analyses, and Eric Leinberger prepared the figures. Tim Max, USDA Forest Service, Pacific Northwest Research Station provided statistical advice. John Buffington, Tom Lisle, and Dave Montgomery provided insightful reviews. Their comments and suggestions greatly improved the manuscript.

References

Adams, S.B., Frissell, C.A., Rieman, B.E., 2000. Movements of nonnative brook trout in relation to stream channel slope. Trans. Am. Fish. Soc. 129, 623–638.

Andrews, E.D., 1979. Scour and fill in a stream channel, East Fork River, western Wyoming. U.S. Geol. Surv. Prof. Pap. 1117, 49.

Ashworth, P.J., 1987. Bedload transport and channel change in gravel-bed rivers. Ph.D. Thesis, University of Stirling, Scotland, 352 pp.

Baker, V.R., Ritter, D.F., 1975. Competence of rivers to transport coarse bedload material. Geological Survey, Professional Paper 86, 975–978.

Beschta, R.L., 1981. Increased Bag Size Improves Helley–Smith Bedload Sampler for Use in Streams with High Sand and Organic Matter Transport, Erosion and Sediment Transport Measurement, Vol. 133. IAHS Publisher, Vienna, pp. 17–25.

Beschta, R.L., 1983. Effects of Large Organic Debris on Channel Morphology, D.B. Simons Symposium on Erosion and Sedimentation. Simons, Li, & Association, Fort Collins, Colorado, pp. 8.63–8.78.

Booker, D.J., Sear, D.A., Payne, A.J., 2001. Modeling three-dimensional flow structures and patterns of boundary shear stress in a natural pool-riffle sequence. Earth Surf. Processes Land. 26, 553–576.

Breusers, H., Nicollet, G., Shen, H., 1977. Local scour around cylindrical piers. J. Hydraul. Res. 15, 211–252.
Buffington, J.M., Montgomery, D.R., 1999a. Effects of hydraulic roughness on surface textures of gravel-bed rivers. Water Resour. Res. 35, 3507–3522.
Buffington, J.M., Montgomery, D.R., 1999b. Effects of sediment supply on surface textures of gravel-bed rivers. Water Resour. Res. 35, 3523–3530.
Buffington, J.M., Lisle, T.E., Woodsmith, R.D., Hilton, S., 2002. Controls on the size and occurrence of pools in coarse-grained forest rivers. River Res. Appl. 18, 507–531.
Bunte, K., 1996. Analyses of the temporal variation of coarse bedload transport and its grain size distribution (Squaw Creek, Montana, USA). RM-GTR-288, USDA, Forest Service, Rocky Mountain Forest and Range Experiment Station, Fort Collins, CO.
Cao, Z., Carling, P., Oakey, R., 2003. Flow reversal over a natural pool-riffle sequence: a computational study. Earth Surf. Processes Land. 28, 689–705.
Carling, P.A., 1983. Threshold of coarse sediment transport in broad and narrow natural streams. Earth Surf. Processes Land. 8, 1–18.
Carling, P.A., Wood, N., 1994. Simulation of flow over pool-riffle topography: a consideration of the velocity reversal hypothesis. Earth Surf. Processes Land. 19, 319–332.
Carstens, M.R., 1966. Similarity laws for localized scour. J. Hydraul. Div., Am. Soc. Civ. Eng. 92, 13–16.
Chabert, J., Engeldinger, P., 1956. Etude des affouillements autour des piles des ponts. Laboratoire National d'Hydraulique, Chatou, France.
Cheetham, G.H., 1979. Flow competence in relation to stream channel form and braiding. Geol. Soc. Am. Bull. 90, 877–886.
Church, M., Jones, D., 1982. Channel Bars in Gravel-bed Rivers. In: Hey, R.D., Bathurst, J.C., Thorne, C.R. (Eds), Gravel-bed Rivers. Wiley, New York, pp. 291–338.
Church, M., Hassan, M.A., 2002. Mobility of bed material in Harris Creek. Water Resour. Res. 38, 101029/2001WR000753.
Clifford, N.J., Richards, K.S., 1992. The reversal hypothesis and the maintenance of riffle-pool sequences: a review and field appraisal. In: Carling, P.A., Petts, G.E. (Eds), Lowland Floodplain Rivers: Geomorphological Perspectives. Wiley, Chichester, UK, pp. 43–70.
Dietrich, W.E., Whiting, P., 1989. Boundary shear stress and sediment transport in river meanders of sand and gravel. In: Ikeda, S., Parker, G. (Eds), River Meandering. American Geophysical Union, Washington, pp. 1–50.
Dietrich, W.E., Smith, J.D., Dunne, T., 1979. Flow and sediment transport in a sand bedded meander. J. Geol. 87, 305–315.
Dietrich, W.E., Kirchner, J.W., Ikeda, H., Iseya, F., 1989. Sediment supply and the development of the coarse surface layer in gravel-bedded rivers. Nature 340, 215–217.
Emmett, W.W., 1980. A field calibration of the sediment trapping characteristics of the Helley–Smith bedload sampler, U.S. Geological Survey, Professional Paper 1139. Washington, DC, p. 44.
Emmett, W.W., Leopold, L.B., Myrick, R.M, 1983. Some characteristics of fluvial processes in rivers. Proc. Second International Symposium on River Sedimentation. Water Resources and Electric Power Press, Nanjing, China, pp. 730–754.
Freshwater Ecosystems 5, 143–166.
Gomez, B., Naff, R.L., Hubbell, D.W., 1989. Temporal variation in bedload transport rates associated with the migration of bedforms. Earth Surf. Processes Land. 14, 135–156.
Gurnell, A.M., Gregory, K.J., Petts, G.E., 1995. The role of coarse woody debris in forest aquatic habitats: implications for management. Aquat. Conserv. Mar. Fresh. Ecosyst. 5, 143–166.
Hassan, M.A., Church, M., 2001. Rating bed load transport in Harris Creek: seasonal and spatial variation over a cobble-gravel bed. Water Resour. Res. 37, 813–825.
Hassan, M.A., Woodsmith, R.D., 2004. Bed load transport in an obstruction-formed pool in a forest, gravel-bed stream. Geomorphology 58, 203–221.
Helley, E.J., Smith, W., 1971. Development and Calibration of a Pressure Difference Bedload Sampler. Open File Report 73-108, U.S. Geological Survey, Washington, D.C.
Hey, R.D., 1979. Flow resistance in gravel-bed rivers. J. Hydraul. Div., Am. Soc. Civ. Eng. 107, 365–379.
Hogan, D.L., Bird, S.A., Hassan, M.A., 1998. Spatial and temporal evolution of small coastal gravel-bed streams: the influence of forest management on channel morphology and fish habitat. In: Klingeman, P.C., Beschta, R.L., Komar, P.D., Bradley, J.B. (Eds), Gravel-Bed Rivers in the Environment. Water Resources Publications, LLC, Highland Ranch, Colorado, pp. 365–392.

Holtz, R.D., Kovacs, W.D., 1981. An Introduction to Geotechnical Engineering. Prentice-Hall, Englewood Cliffs, NJ, p. 733.
Jackson, W.L., Beschta, R.L., 1982. A model of two-phase bedload transport in an Oregon Coast Range stream. Earth Surf. Processes Land. 7, 517–527.
Kalinske, A.A., 1947. Movement of sediment as bed in rivers. Trans. Am. Geophys. Union 28, 615–620.
Keller, E.A., 1971. Areal sorting of bed load material: the hypothesis of velocity reversal. Geol. Soc. Amer. Bull. 82, 753–756.
Keller, E.A., Florsheim, J.L., 1993. Velocity-reversal hypothesis: a model approach. Earth Surf. Processes Land. 18, 733–740.
Keller, E.A., Swanson, F.J., 1979. Effects of large organic material on channel form and fluvial processes. Earth Surf. Processes Land. 4, 361–380.
Klingeman, P.C., Emmett, W.W., 1982. Gravel bedload transport processes. In: Hey, R.D., Bathurst, J.C., Thorne, C.R. (Eds), Gravel-bed Rivers. Wiley, New York, pp. 141–179.
Kuhnle, R.A., Southard, J.B., 1988. Bed load transport fluctuations in a gravel bed laboratory channel. Water Resour. Res. 24, 247–260.
Leopold, L.B., Wolman, M.G., 1960. River Meanders. Geol. Soc. Am. Bull. 71, 769–794.
Leopold, L.B., Wolman, M.G., Miller, J.P., 1964. Fluvial Processes in Geomorphology. W.H. Freeman, San Francisco, CA, p. 522.
Lisle, T.E., 1979. A sorting mechanism for a riffle-pool sequence. Geol. Soc. Am. Bull. 90, 1142–1157.
Lisle, T.E., 1982. Effects of aggradation and degradation on riffle-pool morphology in natural gravel channels, Northwestern California. Water Resour. Res. 18, 1643–1651.
Lisle, T.E., 1986. Stabilization of a gravel channel by large streamside obstructions and bedrock bends, Jacoby Creek, Northwestern California. Geol. Soc. Am. Bull. 97, 999–1011.
Melville, B.W., 1975. Local scour at bridge sites. Ph.D. Thesis, University of Auckland, N.Z.
Melville, B.W., 1984. Live-bed scour at bridge piers. J. Hydraul. Eng. 110, 1234–1247.
Melville, B., 1997. Pier and abutment scour: integrated approach. J. Hydraul. Eng. 123, 125–136.
Melville, B.W., Sutherland, A.J., 1988. Design method for local scour at bridge piers. J. Hydraul. Eng. 114, 1210–1226.
Meyer-Peter, E., Muller, R., 1948. Formulas for Bed-load Transport, 3rd Annual Conference. International Association of Hydraulic Research, Stockholm, pp. 39–64.
Middleton, G.V., Southard, J.B., 1984. Mechanics of sediment movement, short course NBno. 3. Soc. Econ. Paleontologists and Mineralogists, Tulsa, OK, p. 401.
Montgomery, D.R., Buffington, J.M., Smith, R.D., Schmidt, K.M., Pess, G., 1995. Pool spacing in forest channels. Water Resources Research, 31, 1097–1105.
Montgomery, D.R., Buffington, J.M., 1997. Channel-reach morphology in mountain drainage basins. Geol. Soc. Am. Bull. 109, 596–611.
Nelson, J.M., Smith, J.D., 1989. Evolution and stability of erodible channel beds. In: Ikeda, S., Parker, G. (Eds), River Meandering. American Geophysical Union, Washington, pp. 321–377.
Nelson, J.M., Shreve, R.L., McLean, S.R., Drake, T.G., 1995. Role of near-bed turbulence structure in bed load transport and bed load mechanics. Water Resour. Res. 31, 2071–2086.
Paola, C., Mohrig, D., 1996. Paleohydraulics revisited: paleoslope estimation in coarse-grained braided rivers. Basin Res. 8, 243–254.
Parker, G., Klingeman, P.C., 1982. On why gravel bed streams are paved. Water Resour. Res. 18, 1409–1423.
Parker, G., Peterson, A.W., 1980. Bar resistance of gravel-bed streams. J. Hydraul. Div., Am. Soc. Civ. Eng. 106, 1559–1575.
Parker, G., Klingeman, P.C., McLean, D.L., 1982. Bed load and size distribution in paved gravel-bed streams. J. Hydraul. Div. ASCE 108, 544–571.
Piegay, H., Gurnell, A.M., 1997. Large woody debris and river geomorphological pattern: examples from S.E. France and S. England. Geomorphology 19, 99–116.
Pitlick, J., 1988. Variability of bed load measurement. Water Resour. Res. 24, 173–177.
Prestegaard, K.L., 1983. Bar resistance in gravel-bed streams at bankfull stage. Water Resour. Res. 19, 472–476.
Ranch, Colorado, pp.365–392.
Richards, K.S., 1976. The morphology of riffle-pool sequences. Earth Surf. Processes Land. 1, 71–88.

Robert, A., 1997. Characteristics of velocity profiles along riffle-pool sequences and estimates of bed shear stress. Geomorphology 19, 89–98.
Ryan, S.E., Troendle, C.A., 1996. Bed Load Transport Patterns in Coarse-Grained Channels under Varying Conditions of Flows, Sedimentation Technologies for Management of Natural Resources in the 21st Century; Proceedings of Sixth Federal Interagency Sedimentation Conference, Las Vegas, NV, pp. VI-22–VI-27.
Ryan, S.E., Porth, L.S., Troendle, C.A., 2002. Defining phases of bedload transport using piecewise regression. Earth Surf. Processes Land. 27, 971–990.
Sear, D.A., 1996. Sediment transport processes in pool-riffle sequences. Earth Surf. Processes Land. 21, 241–262.
Smith, R.D., 1990. Streamflow and bedload transport in an obstruction-affected, gravel-bed stream. Ph.D. Thesis, Oregon State University, Corvallis, Oregon, p. 181.
Smith, R.D., Sidle, R.C., Porter, P.E. and Noel, J.R., 1993. Effects of experimental removal of woody debris on the channel morphology of a forest, gravel-bed stream. Journal of Hydrology 152, 153–178.
Snedecor, G.W., Cochran, W.G., 1980. Statistical Methods. Iowa St. Univ. Press, Ames, Iowa, p. 507.
Sterling, S., Church, M., 2002. Sediment trapping characteristics of a pit trap and a Helley–Smith sampler in a cobble-gravel bed river. Water Resour. Res. 38, 101029/2000WR000052.
Sutherland, A.J., 1967. Proposed mechanism for sediment entrainment by turbulent flows. Journal of Geophysical Research 72, 6183–6194.
Thompson, D.M., Wohl, E.E., Jarrett, R.D., 1996. A revised velocity-reversal and sediment-sorting model for a high-gradient, pool-riffle stream. Phys. Geogr. 17 (2), 142–156.
Thompson, D.M., Wohl, E.E., Jarrett, R.D., 1999. Velocity reversals and sediment sorting in pools and riffles controlled by channel constrictions. Geomorphology 27, 229–241.
Tison, L.J., 1961. Local scour in rivers. J. Geoph. Res. 66, 4227–4232.
Whiting, P.J., Dietrich, W.E., 1990. Boundary shear stress and roughness over mobile alluvial beds. J. Hydraul. Eng. 116, 1495–1511.
Wilcock, P.R., 1992. Flow competence: a criticism of a classic concept. Earth Surf. Processes Land. 17, 289–298.
Wilcock, P.R., McArdell, B.W., 1993. Surface-based fractional transport rates: mobilization thresholds and partial transport of a sand-gravel sediment. Water Resources Research 29, 1297–1312.
Wilkinson, S.N., Keller, R.J., Rutherfurd, I.D., 2004. Phase-shifts in shear stress as an explanation for the maintenance of pool-riffle sequences. Earth Surf. Processes Land. 29, 737–753.
Woodsmith, R.D., Buffington, J.M., 1996. Multivariate geomorphic analysis of forest streams: implications for assessment of land use impacts on channel condition. Earth Surf. Processes Land. 21, 377–393.
Woodsmith, R.D., Swanson, F.J., 1997. The influence of large woody debris on forest stream geomorphology. In: Wang, S.S.Y., Langedoen, E.J., Shields, F.D.J. (Eds), Management of Landscapes Disturbed by Channel Incision. University of Mississippi, Oxford, MS, pp. 133–138.

Catchment Dynamics and River Processes: Mediterranean and Other Climate Regions
Celso Garcia, Ramon J. Batalla, Editors

Chapter 11

Hydrological effects of dams and water diversions on rivers of Mediterranean-climate regions: examples from California

G. Mathias Kondolf[1,*] and Ramon J. Batalla[2]

[1]*Department of Landscape Architecture and Environmental Planning, University of California, Berkeley, CA 94720-2000, USA*

[2]*Department of Environment and Soil Sciences, University of Lleida, 25198 Lleida, Spain*

Abstract

Rivers in Mediterranean-climate and other semi-arid regions tend to be more heavily impounded and thus their hydrology more strongly affected than rivers in humid climates because demand for water is greater (to supply irrigated agriculture) and runoff is out-of-phase with demand. The impounded runoff index (ratio of reservoir capacity divided by mean annual runoff) is 0.8 on the Sacramento and 1.2 on the San Joaquin Rivers of California, much higher than rates: encountered in humid Atlantic climate regions. As a result of these high levels of impoundment, the overall magnitude and seasonal distribution of flows has changed substantially. Flood peaks tend to be reduced: the Q_2 declined on average 53 and 81% in the Sacramento and San Joaquin River basins, respectively. On many rivers, summer baseflows have increased to supply irrigation diversions downstream, creating a flatter hydrograph that no longer supports dynamic channel processes and the aquatic ecosystem that depends upon such channel dynamics. Vegetation has encroached in the formerly active channels of many rivers in response to reduced flood scour and sediment supply.

Keywords: impounded runoff, river regime, reservoirs, Mediterranean basins

1. Introduction

Flow variability is an important characteristic of river systems, with implications for river geomorphology, ecology, and human uses (Puckridge et al., 1998). Many aquatic and riparian-dwelling organisms are adapted to the seasonal and inter-annual variations in flow that characterize their native river habitats, including periodic high flows (Junk et al., 1989; Poff et al., 1997). Flow variability tends to be greater in arid- than humid-climate regions (McMahon, 1979; Finlayson and McMahon, 1988), with implications that

*Corresponding author. *E-mail address:* kondolf@berkeley.edu (G. Mathias Kondolf).

infrequent, episodic events exert a greater influence on river form and ecology (Wolman and Gerson, 1978; Hecht, 1994), and reservoir storage is proportionally greater (Thoms and Sheldon, 2000). Mediterranean-climate rivers are similar to arid-climate rivers in their high variability in flow and sediment load, but have higher rainfall and a pronounced seasonality that is predictable, at least in their overall characteristics, with cool, wet winters, and warm, dry summers (Conacher and Sala, 1998; Gasith and Resh, 1999). Seasonal precipitation and flow variability are high, with rainfall concentrated in winter months. Inter-annual flow variability is likewise high, driven by high variability in precipitation. The variability in annual precipitation can be illustrated by comparing the range in annual precipitation values for selected stations with similar mean annual rainfall in Mediterranean-climate vs. humid Atlantic climate regions of North America. Cavendish, Vermont, with a mean precipitation of 1120 mm y^{-1}, has a much smaller range of annual values than Healdsburg, California, with 1097 mm y^{-1}, and a coefficient of variation (CV) of only 0.14 contrasted with a CV of 0.33 for Healdsburg (Table 11.1).

The seasonal pattern of precipitation in Mediterranean-climate regions means that water availability and demand are out-of-phase. Precipitation and runoff occur (almost exclusively) in winter, while plants are dormant and demand for irrigation (and hydroelectric generation for air conditioning) is lowest. Thus, seasonal and inter-annual water storage is needed to meet the basic needs of human populations, to support industrial-scale agriculture, and in some cases for flood control. As a result, Mediterranean-climate rivers tend to be more highly regulated than humid climate rivers of comparable size. For example, Spain (whose climate is Mediterranean except in the northwestern provinces) has 1200 large dams (more than any other country in Europe and 2.5% of the population of dams of the world), which collectively impound 40% of the country's average annual runoff (http://www.dams.org). This is a much higher rate of impoundment that typically encountered in more humid regions. For example, if we consider similarly sized German rivers, reservoir capacity is from 5 to 18% of the annual

Table 11.1. Precipitation variation for illustrative Mediterranean and Atlantic climate stations.

Location	Elevation (meters a.s.l.)	Climate type	Annual mean (mm)	Annual minimum (mm)	Annual maximum (mm)	Standard deviation	Coefficient of Variation
Big Sur, CA	61	Mediterranean	1000	460	2260	1560	0.36
Healdsburg, CA	33	Mediterranean	1100	350	2440	1410	0.34
Springfield, MO	384	Humid	1070	640	1610	889	0.21
Cavendish, VT	242	Atlantic	1120	770	1430	625	0.14
Kingston, RI	33	Atlantic	1260	780	1780	830	0.17

Source: National Climatic Data Center database, http://www.ncdc.noaa.gov/oa/ncdc.html

runoff on the Elbe, Rhine, and Wesser Rivers (P. Ergenzinger and C. de Jong, Free University of Berlin, personal communication, December 2001), and the reservoir capacity in relation to annual runoff on the Potomac River (draining the Appalachian Mountains of eastern North America) is less than 20%.

As reviewed by Puckridge et al. (1998), flow variability in rivers has been characterized by a range of statistics, including slopes of flood frequency curves, seasonal distribution of mean monthly flows, coefficients of variation of annual maxima and minima, differences between mean and median flows, and other measures of skewness. Effects of reservoirs on flow regime have been documented in many studies, using a variety of statistical techniques (e.g. Petts, 1984; Williams and Wolman, 1984; Richter et al., 1996; Thoms and Sheldon, 2000). This chapter presents an analysis of hydrologic effects of dams and diversions on the Sacramento–San Joaquin River system of Mediterranean-climate. California, in which we test the degree to which the relative degree of impoundment can explain the degree of change in flow. These hydrologic changes have geomorphic and ecological consequences, which we discuss briefly, but whose development is beyond the scope of this chapter.

2. Area description

Coastal California experiences a Mediterranean-climate, and rivers draining the Sierra Nevada range have a combined Mediterranean-montane runoff regime, influenced by snowmelt runoff. Eastern portions of the state, in the rain shadow of large mountain ranges, experience a semi-arid climate, but precipitation does not follow the classic Mediterranean seasonal pattern of winter rains and dry summers. The state's largest river system, the Sacramento–San Joaquin, drains over 160,000 km^2 (more than 40% of the state's land area). The Sacramento River flows south–southeastward along the axis of the Great Central Valley, meeting the north–northwestward-flowing San Joaquin at their inland delta, and thence the combined rivers flow westward through Suisun, San Pablo, and San Francisco Bays to debouch into the Pacific Ocean at the Golden Gate. Most of the runoff is from the Sierra Nevada mountain range, a north–northwest trending tilted fault block with elevations exceeding 4 km, which receives substantial precipitation thanks to orographic lifting of moist Pacific air in winter storms. Much of this precipitation falls as snow in high elevations, and the snowmelt runoff from the Sierra constitutes the principal developed water source in California. The major tributaries to the Sacramento and San Joaquin Rivers have been impounded by since the late 19th and early 20th century, to divert water for gold mining in the foothills mining districts and for agriculture on the valley floor (California State Lands Commission, 1993). Through the 20th century new reservoirs have been constructed and small ones replaced with larger ones, such that by the end of the century there were over 1400 dams in the state large enough to fall under the regulatory purview of the Division of Safety of Dams (dams more than 4.6 m high and/or impounding more than 61,700 m^3) (California Department of Water Resources, 1988). The Sacramento–San Joaquin River system, source of much of the irrigation and municipal water for the state, is heavily plumbed.

3. Methods

We compiled hydrologic data on 14 major rivers in Sacramento–San Joaquin River system and computed changes in flow regime due to reservoir regulation and diversion. Our principal data source was the US Geological Survey published flow data, available online at http://water.usgs.gov. Data included mean daily flows, mean monthly flows, annual runoff, and annual peak flows. As a rough indicator of the degree to which reservoirs can alter flow regime downstream, we calculated the Impounded Runoff index, IR (expressed as a decimal or percentage) for major tributaries to the Sacramento River, as per Batalla et al. (2004),

$$\text{IR} = \text{reservoir capacity}/\text{mean annual inflow}$$

We obtained reservoir capacities from California Department of Water Resources (1988) and mean annual runoff from suitably located stream gauges, using pre-dam runoff values for the IR calculation. The ratio of storage capacity to mean annual flow has been used previously, for example, by Brune (1953) as the independent variable from which to predict reservoir trap efficiency. IR can be viewed in terms of average residence time, although the high inter-annual variability in flow means that it will never be a true residence time. For example, an IR of 1.0 implies an average residence time of 1 year, but because at least half of the years have flow less than this amount and some years greatly exceed the average, actual residence times should be less. In our calculations, we used simply the stated total reservoir capacity, as this was available for all reservoirs. This analysis could potentially be improved in the future by distinguishing between active storage (the volume of water that can actually be released from the reservoir) and dead storage (the volume of water that lies below the elevation of the reservoir outlet), as we would expect the former to be a better indicator of the potential for hydrologic modification.

For pre-dam mean annual runoff, we sought a flow value that would best represent flow prior to human alteration. In some cases, suitable gauging records existed pre- and post-reservoir construction. In these cases, we calculated our mean annual runoff value from the pre-dam data. In other cases, we relied on "unimpaired flow" data calculated by the dam operator. In rivers with significant diversions above the gauge (e.g. Tuolumne and San Joaquin below Friant Dam) we added in canal flows where necessary to obtain a mean annual runoff that would more closely reflect pre-dam conditions. The hydrologic effect of a reservoir can be expected to attenuate downstream, so we used post-dam records for gauges just downstream of the large storage reservoirs in the foothills, except for the Yuba River, where the downstream reservoir is exceeded in capacity by an upstream reservoir. To illustrate seasonal changes in flow on river reaches below reservoirs, we plotted mean monthly flows from pre- and post-dam periods. We calculated the pre-dam–post-dam correlation coefficient for each gauge series ($\Phi_{\text{pre,post}}$) by dividing the covariance of the two data sets by the product of their standard deviations:

$$\Phi_{\text{pre,post}} = \text{cov}_{\text{pre,post}}/(\sigma_x \cdots \sigma_y)$$

where $1 \geq \Phi \geq -1$, $\text{cov}_{\text{pre,post}} = 1/n[\sum(\text{pre}_i - \mu_x)(\text{post}_i - \mu_y)]$, σ is the standard deviation, n the number of data, and μ the mean value of the distribution. Values near 1 indicate that the post-dam flow regime closely matches the pre-dam regime in timing, and thus the seasonal pattern of flow has not changed. However, changes in the magnitude of flows would not be detected by this variable if the seasonal distribution was similar to pre-dam. Values of $\Phi_{\text{pre,post}}$ near 0 indicate that the post-dam monthly regime is independent of the natural pre-dam pattern, and values near -1 reflecting inversion of the flow regime.

We conducted flood frequency analyses (annual maxima series) (Dunne and Leopold, 1978) for the 14 rivers. We focused on relatively frequent floods (i.e. Q_2, Q_{10}). Flows with a return period of about 2 years are reported in the geomorphic literature to be the channel-forming discharges for many rivers, but in semi-arid climates channel form tends to be influenced by longer return period discharges. (Wolman and Gerson 1978). Interpretation of the Q_2 and Q_{10} flows is straightforward and empirical, since no extrapolation outside the actual range of data is necessary. To explore the relation between degree of impoundment (IR) and degree of hydrologic changes, we plotted the changes in flood magnitudes against IR.

4. Results and analysis

4.1. Degree of regulation

Rivers in the Sacramento–San Joaquin basin are highly impounded. The largest reservoirs are Shasta on the Sacramento (capacity 5.62 $m^3 \times 10^9$), Oroville on the Feather (capacity 4.36 $m^3 \times 10^9$), and New Melones on the Stanislaus (capacity 2.99 $m^3 \times 10^9$). Overall basin-wide IRs are 0.80 for the Sacramento and 1.2 for the San Joaquin, with individual river IRs ranging from 0.53 (Stony Creek) to 4.63 (Putah Creek) (Table 11.2, Fig. 11.1). Although reservoirs are found at nearly every elevation, storage is concentrated in large reservoirs located in the foothills. For example, there are 28 impoundments in the Stanislaus River basin, but the large foothills reservoir, New Melones, accounts for 85% of the storage. On the Stanislaus, Tuolumne, and Merced Rivers, larger reservoirs in the 1960s and 1970s (Table 11.2) replaced reservoirs built in the 1920s (with under 0.3 $m^3 \times 10^9$ capacity).

Roughly half of these dams were built and are operated by federal agencies such as the US Army Corps of Engineers (Black Butte, Camp Far West, New Hogan) or the US Bureau of Reclamation (Shasta, Whiskeytown, Folsom, New Melones, and Friant) or the California Department of Water Resources (Oroville), with the rest owned by local irrigation districts or power utilities (Monticello, New Bullards Bar, Pardee, Camanche, New Don Pedro, and New Exchequer). The many smaller upstream dams also tend to be owned by local irrigation districts or utilities.

The purposes of these reservoirs vary and are usually multiple, but most of the storage is devoted to irrigation. Many dams are jointly operated, with the US Army Corps of Engineers responsible for operating the flood pool (the storage devoted to flood control during the winter months) and another agency managing the rest of the storage capacity.

4.2. Changes in mean monthly flow

Mean monthly flows show a range of changes post-dam, from essentially no change on the Bear River below Camp Far West Reservoir, to the flattening of the annual hydrograph

Table 11.2. Reservoirs and impoundment ratios for major rivers, Sacramento–San Joaquin River system (California).

River	Dam	Year	Drainage Area (km^2)	USGS Gauge		Avg. Annual Runoff[a] ($m^3 \times 10^9$)	Large "foothills" Reservoir Capacity[b]		Total reservoir Storage Capacity[c]	
				Name	Avg. over:		$m^3 \times 10^9$	IR	$m^3 \times 10^9$	IR
Sacramento	Shasta	1945	17262	at Keswick	1946–1962	7.278	5.617	0.77	5.384	0.74
West-side tributaries										
Clear Creek	Whiskeytown	1963	521	Near Igo	1941–1962	0.369	0.297	0.81	0.327	0.89
Stony Creek	Black Butte	1963	1919	Below Black Butte Dam Near Orland	1956–1990	0.568[4]	0.197	0.35	0.303	0.53
Putah Creek	Monticello	1957	1492	Near Winters	1931–1956	0.427	1.976	4.63	1.976	4.63
East-side tributaries										
Feather	Oroville	1968	9342	at Oroville	1902–1967	5.215	4.364	0.84	6.714	1.29
Yuba[e]	Englebright	1940	2870	Below Englebright	1942–1995	2.178	0.087	0.04	1.739	0.80
	New Bullards Bar	1970					1.199	0.55		
							1.284	0.59		
Bear	Camp Far West	1963	738	Near Wheatland	1930–1963	0.380	0.126	0.33	0.289	0.76
American	Folsom	1956	4890	at Fair Oaks	1905–1954	0.354	1.246	0.37	2.261	0.67
Mokelumne	Pardee	1929	1489	Below Camanche Dam	1931–1963	0.744	0.259	0.35	1.032	1.39
	Camanche	1963	1603				0.531	0.71		
							0.790	1.06		

Calaveras	New Hogan	1963	940	Below New Hogan Dam, Near Valley Springs	1961–1990	0.205	0.400	1.96	0.396	1.93
Stanislaus	Melones	1926	2331	Below Goodwin Dam	1957–1978	0.202[f]	0.145	0.12		
	New Melones	1979	2331	Near Knight's Ferry			2.986	2.48	3.518	2.93
Tuolumne	Don Pedro	1923	3994	Below La Grange Dam,	1971–1995	1.772[g]	0.309	0.17		
	New Don Pedro	1971	3994	Near La Grange			2.505	1.41	3.444	1.94
Merced	Exchequer	1926	2694	Below Merced Falls,	1901–1925	1.290	0.347	0.27		
	New Exchequer	1967	2694	Near Snelling			1.273	0.99	1.305	1.01
San Joaquin	Friant[h]	1941	4242	Below Friant	1908–1940	2.095	0.643	0.31	1.140	0.54

[a] Pre-dam data unless otherwise indicated; averages as reported in USGS 1995.
[b] Reservoir capacities from DWR (1988).
[c] Total reservoir capacity includes only those reservoirs with storage capacity greater than 0.648 $m^3 \times 10^6$.
[d] Average. annual runoff as measured for full period as published in USGS 1990, adjusted for diversions.
[e] On the Yuba River, the largest reservoir is not the most downstream. Englebright Dam (1940) has a drainage area of 2870 km^2 but a reservoir capacity of only 0.057 $m^3 \times 10^9$, much smaller than New Bullards Bar, which is located on the North Fork and has a capacity of 0.785 $m^3 \times 10^9$. Average value reported for entire POR only (i.e. 1942–1995) in USGS (1995).
[f] Average annual runoff for pre-dam period of record at Knight's ferry of 0.426 $m^3 \times 10^9$ (1957–1978), plus average diversion to South San Joaquin Canal of 0.261 $m^3 \times 10^9$ (1914–1995) and average diversion to Oakdale Canal of 0.102 $m^3 \times 10^9$ (1914–1995), as published by USGS (1995).
[g] Average annual runoff for post dam period below La Grange Dam of 0.528 $m^3 \times 10^9$ (1971–1995) because comparable pre-dam values were not reported in USGS 1995; plus average diversion to Modesto Canal of 0.250 $m^3 \times 10^9$ (1909–1995), and average diversion to Turlock Canal of 0.385 $m^3 \times 10^9$.
[h] Millerton Reservoir drainage area and capacity from USGS (1995).

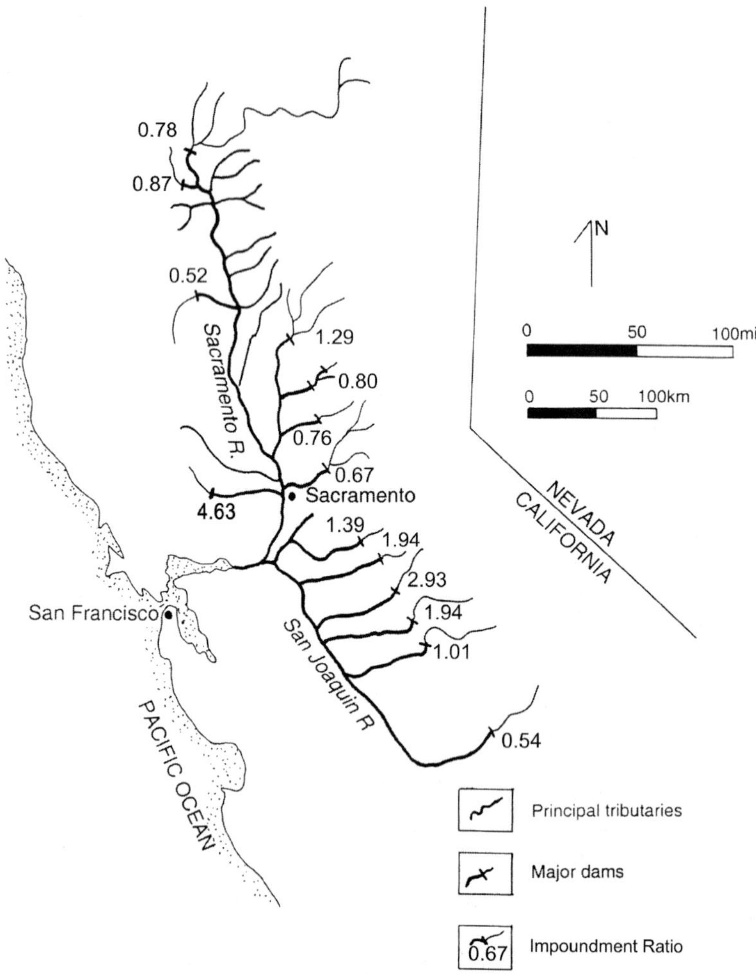

Figure 11.1. Map of principal tributaries of the Sacramento and San Joaquin Rivers, showing locations of major foothills dams and total impoundment ratio (IR). See Table 11.2 for list of rivers.

through eliminating winter peaks and increasing summer base flows by an order of magnitude on Putah Creek (Fig. 11.2). The value of IR is a good though imperfect predictor of change, yielding a regression relation ($r^2 = 0.53$, $N = 12$, $p < 0.05$) (Table 11.3, Fig. 11.3). Only 12 points are used in the regression because these are the only ones with adequate pre-dam mean monthly data (Table 11.3). Only 11 of the 12 points are visible in Figure 11.3 because the Sacramento and Yuba rivers plot directly on one

---

Figure 11.2. Pre- and post-dam mean monthly flows for 12 major rivers in the Sacramento–San Joaquin River system. The same rivers are listed in Tables 11.2 and 11.3, except for Stony Creek and Calaveras River, which are not shown because their pre-dam data series were very short.

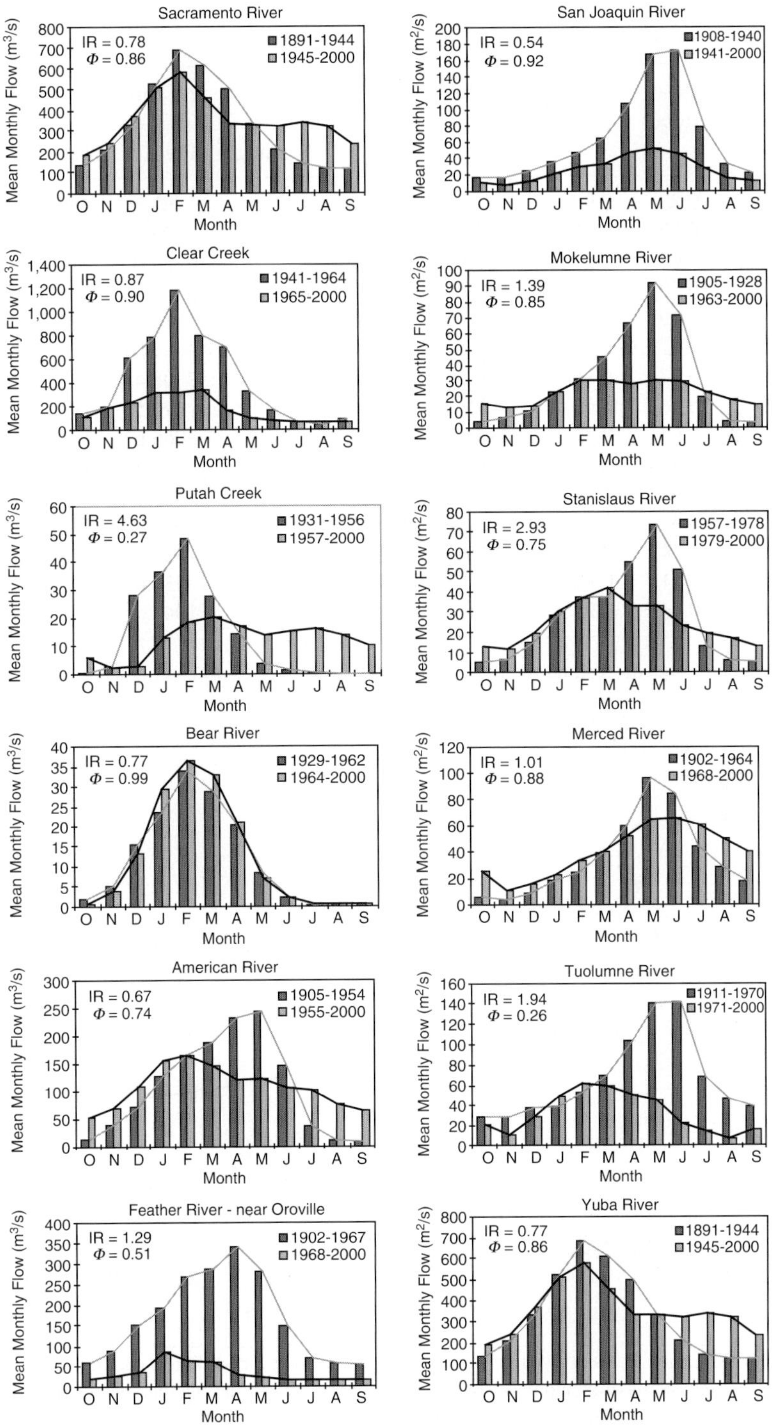
Sacramento River
IR = 0.78
Φ = 0.86
1891-1944
1945-2000
San Joaquin River
IR = 0.54
Φ = 0.92
1908-1940
1941-2000
Clear Creek
IR = 0.87
Φ = 0.90
1941-1964
1965-2000
Mokelumne River
IR = 1.39
Φ = 0.85
1905-1928
1963-2000
Putah Creek
IR = 4.63
Φ = 0.27
1931-1956
1957-2000
Stanislaus River
IR = 2.93
Φ = 0.75
1957-1978
1979-2000
Bear River
IR = 0.77
Φ = 0.99
1929-1962
1964-2000
Merced River
IR = 1.01
Φ = 0.88
1902-1964
1968-2000
American River
IR = 0.67
Φ = 0.74
1905-1954
1955-2000
Tuolumne River
IR = 1.94
Φ = 0.26
1911-1970
1971-2000
Feather River - near Oroville
IR = 1.29
Φ = 0.51
1902-1967
1968-2000
Yuba River
IR = 0.77
Φ = 0.86
1891-1944
1945-2000
Mean Monthly Flow (m3/s)
Mean Monthly Flow (m2/s)
Month
O N D J F M A M J J A S

Table 11.3. Changes in mean monthly flow pattern and peak flows for selected rivers in the Sacramento–San Joaquin River system (California).

River	IR (total)	$\Phi_{pre,post}$	Q_2 (m^3/s)			Q_{10} (m^3/s)		
			Pre	Post	% change	Pre	Post	% change
Clear Creek	0.87	0.90	203.9	107.6	−47	467.3	269.0	−42
Stony Creek[a]	0.52	[b]	566.4	192.6	−66	1925.8	495.6	−74
Putah Creek	4.63	0.27	719.9	44.2	−94	2224.0	216.7	−90
Sacramento	0.78	0.86	3109.5	2095.7	−33	5833.9	3568.3	−39
Feather (near Oroville)	1.29	0.51	1965.4	1243.8	−37	6010.4	4387.6	−27
Yuba	0.77	0.86	962.9	495.6	−49	3398.4	2265.6	−33
Bear	0.76	0.99	278.8	175.6	−37	675.5	793.0	17
American	0.67	0.74	1116.7	463.5	−58	3065.6	2226.8	−27
Mokelumne	1.39	0.85	263.4	53.8	−80	594.7	147.3	−75
Calaveras	1.94	[b]	354.0	55.2	−84	807.0	229.0	−72
Stanislaus	2.93	0.75	266.2	87.8	−67	991.2	186.9	−81
Tuolumne	1.94	0.26	589.1	99.1	−83	1324.0	239.3	−82
Merced	1.01	0.88	382.3	59.5	−84	934.6	201.1	−78
San Joaquin	0.54	0.92	396.1	49.0	−88	1041.0	386.4	−63

[a] Pre-dam flood frequency estimated from USACE (1987) (Stony Creek) and USACE, 1983 (Calaveras River).

[b] Not available due to limited pre-dam gauging record.

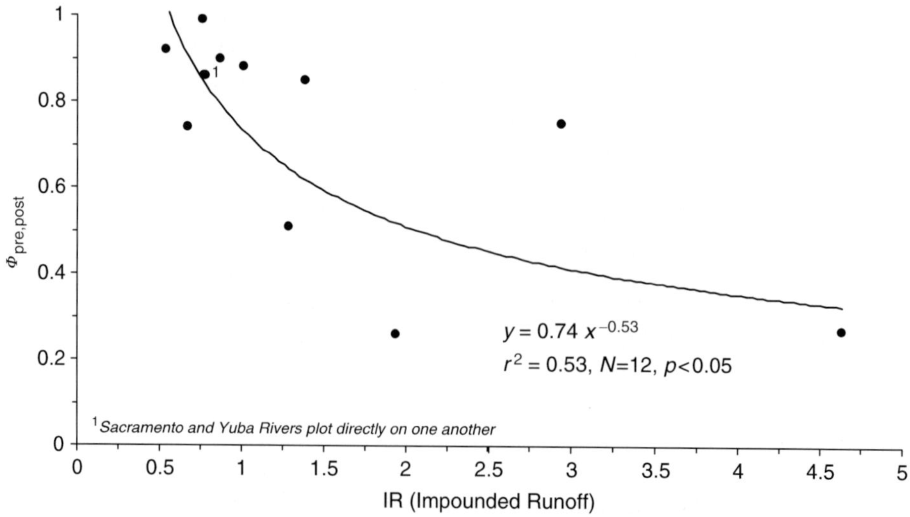

Figure 11.3. Changes in monthly flow regime (expressed as the correlation coefficient $\Phi_{pre,post}$) as a function of the degree of impoundment (IR) for 12 major rivers in the Sacramento–San Joaquin River system (the data set of 14 rivers less Stony Creek and Calaveras Rivers, for which inadequate pre-project data exist to characterize pre-dam mean monthly flows).

another. The tributaries with the greatest change in seasonality are the Tuolumne River and Putah Creek, which have Φ values of 0.26 and 0.27, respectively, and IRs of 1.94 and 4.63, respectively. On the Tuolumne, the seasonal change is the result of diversion of snowmelt runoff and elimination of the spring-early summer high flows, with a net reduction in total flow. On Putah Creek, diversions occur downstream of the gauge, so the seasonal change results from elimination of winter runoff peaks and sustained, augmented releases in the summer.

Inspection of the mean monthly flow plots shows reduction in seasonal high flows for both winter-rainfall-runoff-dominated rivers (e.g. Sacramento, Clear, Putah, Bear, and Yuba) and spring-snowmelt-runoff-dominated rivers (e.g. San Joaquin, Mokelumne, Stanislaus, Merced, Tuolumne, American, and Feather). In some rivers, the flood waters stored by the reservoir are released during the summer for diversion by irrigators downstream, so the gauge records may reflect simply a seasonal redistribution of flow, with no net reduction in annual flow (e.g. Putah). In other cases, irrigation diversions are made upstream of the gauge, so the annual flow is reduced as well. Friant Dam on the San Joaquin River is the most compelling example of this, with two canals (with capacity of about 150 $m^3 s^{-1}$) diverting directly from the dam. Values of Φ ranged from 0.26 to 0.99, with a median value of 0.85.

4.3. Changes in flood frequency

Flood frequency plots show progressive reduction in annual peak flows with expansion of reservoir capacity, as illustrated by the Mokelumne River below Camanche Dam, where downward shifts in the curve resulted from construction first of Pardee Reservoir in 1929, then the larger Camanche Reservoir downstream in 1963 (Fig. 11.4). Q_2 decreased from a pre-dam value of 260 $m^3 s^{-1}$ to just over 50 $m^3 s^{-1}$ after completion of Camanche.

Overall, Q_2 declined from 33 to 94% in these tributaries, with an average reduction of 65%. In general, the southern Sierran tributaries of the San Joaquin River experienced greater reduction in Q_2 than the northern Sacramento River tributaries, a pattern attributable in part to the greater reservoir storage in the San Joaquin. Q_{10} showed a similar pattern, with reductions ranging from 17 to 90%, and an average reduction of 57%. Although specific reservoir purpose and consequent operations rules varied among reservoirs, and thus the expected alteration in flood regimes, we hypothesized that the simple IR variable could provide some prediction of the degree of reduction in flood magnitudes. Plots of reduction in post-dam Q_2 and Q_{10} against IR show a trend toward greater reduction with higher IR, but the scatter is high (Fig. 11.5).

5. Discussion and conclusions

Flow variability has been a main motivation for, and casualty of, reservoir construction in Mediterranean-climate regions. The degree of hydrologic modification in these regions has not been widely documented or the implications fully appreciated. The highly seasonal pattern of discharge, so characteristic of Mediterranean-climate rivers and so inconvenient for human uses of rivers, has been reduced or even reversed in some cases,

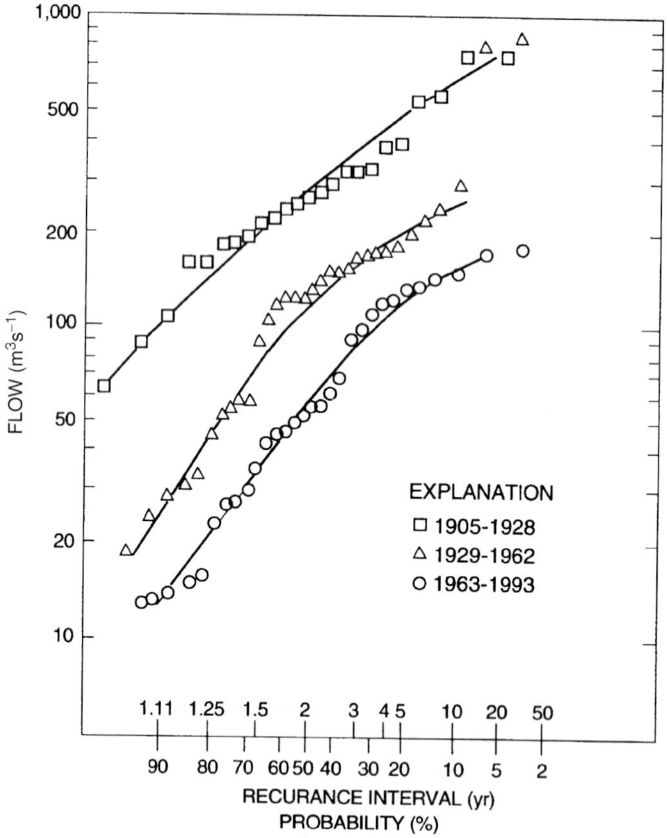

Figure 11.4. Flood frequency curves for the Mokelumne River below Camanche Dam for three time periods: before Pardee Dam (1905–1928), after Pardee Dam but before Camanche Dam (1929–1962), and after Camanche Dam (1963–1993) (Adapted from the Federal Energy Regulatory Commission, 1993).

resulting in a pronounced flattening of the annual hydrograph. The potential for such changes is greater on rivers with higher storage relative to annual runoff (IR), though the operating rules of specific reservoirs vary enough that IR is not sufficient to predict the degree of hydrologic modification.

For 22 regulated reaches in the Ebro River basin of Spain, Batalla et al. (2004) documented decreases in magnitude of Q_2 and Q_{10} floods averaging over 30%, compared with decreases of over 60% we document here for Q_2 and Q_{10} in the Sacramento–San Joaquin River system. In a general way, this can be attributed to greater reservoir storage in the Sacramento and San Joaquin River basins than in the Ebro River Basin (IR values of 0.8 and 1.2 vs. 0.6, respectively), but the relations between IR and flood reduction in individual rivers are variable. Post/pre-dam Q_2 and Q_{10} ratios plotted against IR yielded r^2 values of 0.52 and 0.60, respectively for the Ebro basin (Batalla et al., 2004), contrasted with r^2 values of 0.32 and 0.42 obtained here for the Sacramento–San Joaquin River system (Fig. 11.4). (The Ebro relation was linear, while a better fit for the Sacramento was obtained with a power function. Post-dam flood magnitude decreased rapidly for values of

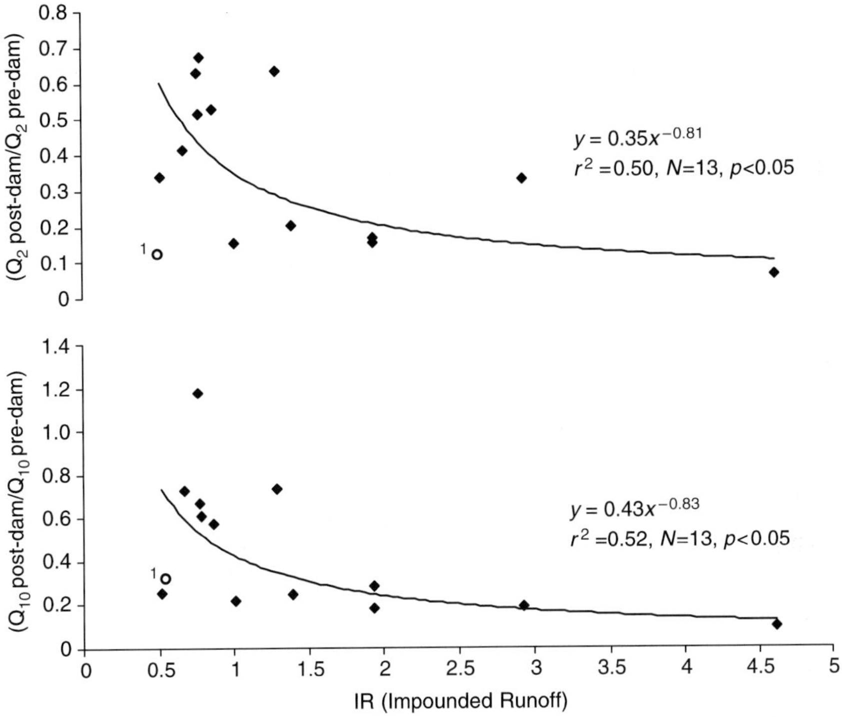

Figure 11.5. Reduction in flood magnitudes after dam construction in rivers of California, plotted against IR, for Q_2 and Q_{10}.

IR up to 1, more slowly for greater IR values, as the floods have already been substantially reduced.

The reduction in seasonality of runoff, accompanied by reduction in peak flows, has utterly changed the nature of many Mediterranean-climate rivers, for instance in Spain (Batalla et al., 2004) and California, as shown in this chapter. The reductions in flow magnitude result in reduced stream power available for sediment transport, which combined with the trapping of coarse sediment by the reservoirs, results in reduced sediment transport in reach downstream of these dams. As noted by Inbar (1992), Mediterranean-climate regions are mainly in active tectonic areas, characterized by steep relief, a factor that would favor high sediment yields (Milliman and Syvitski, 1992). Thin vegetative cover, typical of these regions, would also favor high sediment yields (Langbein and Schumm, 1958). Thus, Mediterranean-climate regions tend to have relatively high sediment yields, though by no means the highest in the world, as factors such as basin relief, uplift rates, and lithology tend to be dominant determinants of sediment yield (Milliman and Syvitski, 1992). The extreme reductions in sediment supply and transport capacity below reservoirs in Mediterranean climate regions thus means that the relative change wrought by dams should be greater in Mediterranean-climate rivers than most humid-climate rivers.

In Mediterranean-climate regions, the high variability in runoff means that large, infrequent floods tend to carry a greater percentage of runoff than would be the case in a comparably sized humid climate river, and since sediment transport is a power function of discharge, these floods carry proportionately higher sediment loads. As a result, channel processes in Mediterranean (and other semi-arid regions) tend to be more episodic (Wolman and Gerson, 1978; Hecht, 1994) than in comparable humid-climate and snowmelt rivers, where flows are more stable and small floods (e.g. $Q_{1.5}$) largely control channel processes and form (Leopold et al., 1964). Accordingly, reservoir-induced shifts from episodic to stable flow regime represent a profound transformation in riverine process for Mediterranean-climate rivers. In consequence, impounded rivers in California have experienced encroachment of vegetation into the formerly active channel and channel narrowing, as exemplified by the Trinity River (Kondolf and Wilcock, 1996).

Reservoir-induced reduction in flow variability and sediment load has ecological implications as well. Native species of Mediterranean-climate rivers and floodplains are adapted to the highly seasonal flow, sediment transport, and bed disturbance, and can survive prolonged dry periods, allowing them to colonize these habitats. However, once flow variability is reduced by regulation, exotic species that were formerly excluded by the highly variable flow regime may have a competitive advantage under the new, more stable flow regime. In California, the percentage of exotic species is greater in reaches below dams than elsewhere (Moyle, 2003).

Acknowledgements

The research on which this chapter was based was funded by the California Bay-Delta Ecosystem Restoration Program, and an NSF Biocomplexity Incubation grant to Adina Merelender, lead PI. Matt Deitch prepared Table 11.1, and Laura Pagano helped in manuscript preparation. The manuscript was substantially improved thanks to the review comments of Martin C. Thoms and Nicola Surian.

References

Batalla, R.J., Gomez, C.M., Kondolf, G.M., 2004. River impoundment and changes in flow regime, Ebro River basin, northeastern Spain. J. Hydrol. 290, 117–136.

Brune, G.M., 1953. The trap efficiency of reservoirs. Trans. Am. Geophys. Union 34 (3), 407–418.

California State Lands Commission, 1993. California's rivers: a public trust report. The Trust for Public Land, Sacramento.

California Department of Water Resources, 1988. Dams within the jurisdiction of the State of California, Bulletin 17–88. California Department of Water Resources, Sacramento.

Conacher, A.J., Sala, M., 1998. Land degradation in Mediterranean environments of the world, Nature, Extent, Causes and Solutions. Wiley, Chichester.

Dunne, T., Leopold, L.B., 1978. Water in Environmental Planning. Freeman and Company, New York.

Federal Energy Regulatory Commission, 1993. Final environmental impact statement, proposed modifications to the lower Mokelumne river project, California, FERC Project No. 2916-004. Federal Energy Regulatory Commission, Office of Hydropower Licensing, Washington, DC.

Finlayson, B.L., McMahon, T.A., 1988. Australia *v* the world: a comparative analysis of streamflow characteristics. In: Warner, R.F. (Ed.), Fluvial Geomorphology of Australia. Academic Press, Sydney, pp. 17–40.

Gasith, A., Resh, V.H., 1999. Streams in Mediterranean climate regions: abiotic influences and biotic responses to predictable seasonal events. Annu. Rev. Ecol. Syst. 30, 51–81.

Hecht, B., 1994. South of the spotted owl: restoration strategies for episodic channels and riparian corridors in Central California, Western Wetlands. In: Kent, D.M., Zentner, J.J. (Eds), Selected Proceedings of the 1993 Conference of the Society of Wetland Scientists. University of California, Davis, CA, pp. 104–117.

Inbar, M., 1992. Rates of fluvial erosion in basins with a Mediterranean type climate. Catena 19, 393–409.

Junk, W.J., Bayley, P.B., Sparks, R.E., 1989. The flood pulse concept in river–floodplain systems. Can. Spec. Publ. Fish. Aquat. Sci. 106, 110–127.

Kondolf, G.M., Wilcock, P.R., 1996. The flushing flow problem: defining and evaluating objectives. Water Resour. Res. 32 (8), 2589–2599.

Langbein, W.B., Schumm, S.A., 1958. Yield of sediment in relation to mean annual precipitation. Am. Geophys. Union Trans. 39, 1076–1084.

Leopold, L.B., Wolman, M.G., Miller, J.P., 1964. Fluvial Processes in Geomorphology. W.H. Freeman and Sons, San Francisco, CA.

McMahon, T.A., 1979. Hydrological characteristics of arid zones. In: International Association of Hydrological Sciences (Ed.), The Hydrology of Areas of Low Precipitation, Publication No. 128. IAHS-AISH, Canberra, pp. 105–120.

Milliman, J.D., Syvitski, J.P.M., 1992. Geomorphic/tectonic control of sediment delivery to the ocean: the importance of small mountainous rivers. J. Geol. 100, 525–544.

Moyle, P.B., 2003. Inland Fishes of California. University of California Press, Berkeley, CA.

Petts, G.E., 1984. Impounded Rivers: Perspectives for Ecological Management. Wiley, Chichester.

Poff, N.L., Allan, J.D., Bain, M.B., Karr, J.R., Prestegaard, K.L., Richter, B.D., Sparks, R.E., Stromberg, J.C., 1997. The natural flow regime. Bioscience 47, 769–784.

Puckridge, J.T., Sheldon, F., Walker, K.F., Boulton, A.J., 1998. Flow variability and the ecology of large rivers. Mar. Freshwater Res. 49, 55–72.

Richter, B.D., Baumgartner, J.V., Powell, J., Braun, D.P., 1996. A method for assessing hydrologic alteration within ecosystems. Conserv. Biol. 10, 1163–1174.

Thoms, M.C., Sheldon, F., 2000. Water resource development and hydrological change in a large dryland river. The Barwon–Darling River, Australia. J. Hydrol. 228, 10–21.

USACE (US Army Corps of Engineers), 1983. New Hogan Dam and Lake Calaveras River, California: Master Water Control Manual, San Joaquin River basin, California. USACE, Sacramento.

USACE (US Army Corps of Engineers), 1987. Black Butte Dam and Lake, Stony Creek, California: Appendix III to Master Water Control Manual, Sacramento River basin, California. USACE, Sacramento.

Williams, G.P., Wolman, M.G., 1984. Downstream effects of dams on alluvial rivers. U.S. Geol. Surv. Prof. Paper 1286.

Wolman, M.G., Gerson, R., 1978. Relative scales of time and effectiveness of climate in watershed geomorphology. Earth Surf. Proc. Land. 3, 189–208.

Catchment Dynamics and River Processes: Mediterranean and Other Climate Regions
Celso Garcia, Ramon J. Batalla, Editors

Chapter 12

The geomorphology and management of a dynamic, unstable gravel-bed river: the Feshie–Spey confluence, Scotland

Alan Werritty[1,*], Trevor B. Hoey[2] and Andrew R. Black[1]

[1]Department of Geography, University of Dundee, Dundee DD1 4HN, Scotland, UK
[2]Department of Geography and Geomatics, Centre for Geosciences, University of Glasgow, Glasgow G12 8QQ, Scotland, UK

Abstract

The alluvial fan that has developed at the confluence of the Rivers Feshie and Spey over the past 13,000 years provides an exceptional example of an unstable, gravel-bed river in the Scottish Highlands protected under UK and EU environmental law. River engineering extending back to the early 19th century has only registered a modest impact on this dynamic system up to the 1980s. Since then a flood rich period (l989–1994) generated a series of avulsions caused by local aggradation of the main channel. Initially the channel switched to the western side of the fan (in 1989–90) and was restored to its former position following the re-instatement of flood banks and channel regrading in 1991. By 1997 the main channel had shifted to the eastern side of the fan following a further avulsion and the re-occupation of palaeo-channels, triggered in part by further repairs of the flood banks on the west side of the fan but mainly in response to natural processes. In doing so the channel now occupies a position akin to that proposed in a river engineering scheme proposed in 1991 but not implemented. Re-sectioning of the Spey to provide a flood-relief channel immediately downstream of the confluence was completed in 1992, but this has not adversely impacted on water levels in the Loch Insh marshes (a internationally protected wetland upstream of the Feshie–Spey confluence). In seeking to reconcile the conflicting demands of nature conservation and generating an economic return from the land, a map of the geomorphological sensitivity of the site (based on Brunsden's "landscape change safety factor" concept) has been developed. This reports three zones with contrasting sensitivities and recommended management strategies: (1) highly sensitive and dynamic areas (HS-5) where no engineering works should be attempted; (2) areas of medium sensitivity (M-20) where permitted engineering works need careful management and monitoring; and (3) areas of low sensitivity (LS-100) where appropriate river engineering should be permitted. The assessment of environmental risk is based on the probability of each of the zones being de-stabilised by floods with return periods of 5, 20 and 100 years, respectively. Paradoxically, this imprecise guidance to river engineers provides the best framework for combining conservation sensitive management with cost-effective engineering.

Keywords: alluvial fan, instability, floods, avulsion, conservation, river engineering

*Corresponding author. *E-mail address:* awerritty@dundee.ac.uk (A. Werritty).

1. Background

The majority of rivers in upland Britain have gravel beds and, depending on the combination of local controls, display varying degrees of lateral and vertical stability. Whether or not a particular channel reach is stable is controlled by the frequency and severity of recent floods, the local channel gradient and the availability of sediment from adjacent hill slopes and floodplains (Werritty and Ferguson, 1980; Werritty and Leys, 2001). High-energy rivers capable of mobilising and reworking gravel-sized bed material exhibit braiding or wandering behaviour and can generate bed level changes over decadal timescales. However, human activity (such as river engineering, regulation and impoundment) combined with land-use controlled reductions in sediment supply over many decades has meant that relatively few upland rivers in the UK now register such channel instability. This means that few sites now exist in Britain where unstable, dynamic rivers provide truly natural landforms and habitats to provide a scientific and cultural record of past fluvial environments (Werritty and Brazier 1991a). Where such sites do exist it is important to conserve them in as natural a state as is possible.

Given their natural tendency to wander and braid, the management issues associated with unstable, dynamic rivers are inherently different from those of stable, lower energy rivers. Firstly, dynamic rivers comprise rapidly shifting channels with flows often across exposed gravel sheets rather than in well-defined channels. Where engineering schemes are designed to stabilise these channels and release higher value riparian land, these rivers are often confined into single channels with strengthened banks. The success of such schemes is typically short-term and limited with the natural geomorphology and ecology of the river often damaged (Werritty and Brazier, 1991a). Secondly, dynamic rivers often display a cyclic behaviour over decades and centuries with plan forms becoming more complex after major floods and recovering after minor ones (Ferguson and Werritty, 1983) and bed elevation alternating between aggradation and degradation in response to the natural rhythms of sediment transfer (Paige and Hickin, 2000). Management of such rivers is difficult as, unless the full range of historical behaviour is known, there is always the risk of interpreting local transient behaviour as being representative of the norm.

Large alluvial fans occur at sites where a tributary delivers a large sediment load to the main stream leading to localised aggradation and the development of an unstable confluence that can extend several kilometres upstream. The confluence of the Rivers Feshie and Spey in the Western Cairngorm Mountains in Scotland (National Grid Reference NN 843064) provides an excellent example of a dynamic confluence alluvial fan of national geomorphological and ecological significance that also controls the water levels in Loch Insh 1 km upstream (Werritty and Brazier, 1991b). The site supports pastoral agriculture and woodland and has been actively managed for over a century. The Feshie–Spey fan has been designated as a "Site of Special Scientific Interest" under the UK *Wildlife and Countryside Act*, 1981 and is further protected by being a "Special Conservation Area" under the EU *Habitats Directive*. Recent floods have destabilised parts of the fan prompting one landowner into localised engineering at the expense of the natural heritage interest. This is illustrative of the wider issues that exist in many parts of the world where active alluvial fans require cost-effective engineering to protect vital transports links (e.g. the main road along the west coast of South Island, New Zealand, Davies and McSaveney, 2001) or permit appropriate urban expansion (e.g. the town of Eilat in Southern Israel, Schick et al. 1999).

In this study, we examine the natural processes operating at the Feshie–Spey fan, responses in terms of recent management efforts and use the concept of geomorphic sensitivity to reconcile the conflicting demands of scientific conservation and an economic return from the land.

2. Channel change at the River Feshie–River Spey confluence

The Feshie confluence comprises a currently active fan inset within a much larger relict fan formed in response to deglaciation *c.* 13,000 years ago (Werritty and Brazier, 1991b). Adjacent to the currently active main channel are relict palaeochannels and terraces across which the Feshie occasionally flows during floods. Downstream of the fan apex (which has migrated northwards as a result of river engineering in the 19th century) the channel divides into a series of highly active distributaries. Changes in channel plan form have been examined over 250 years from maps (since 1750), documentary sources (since 1814) and aerial photographs (since 1948). A complex history emerges in which the main distributary shifted eastwards during the 1860s to a well-defined and apparently stable corridor which it occupied until 1989. For a brief period (less than two years and in response to major floods in 1989 and 1990) the main channel reverted to its pre-1862 position on the western side of the fan occupying a shorter (and thus steeper) route to the Spey before returning to its pre-1989 position in the centre of the fan following the re-instatement of breached flood banks in 1991 (Fig. 12.1). Since 1992 the main distributary has shifted eastwards by a process of avulsion to a relict part of the fan unoccupied since 1750. The above history is typical of alluvial fans with long periods of apparent stability punctuated by avulsion and sudden channel shifts (Zarn and Davies, 1994).

This pattern of stability and sudden channel change occurs in response to localised aggradation during which rising river bed elevation increases the probability of avulsion and channel shifting. As the active channel becomes increasingly perched above the adjacent fan surface it becomes more vulnerable to avulsion during a major flood. Aerial photographs from 1990 and 1995 clearly show this process taking place on the eastern side of the fan during the particularly flood rich period in the late 1980s and early 1990s. This avulsion was completed by October 1998 when long profile surveys reported that the former main channel (progressively abandoned since 1992) was locally >1.0 m higher than the new distributary system formed following the avulsion (Fig. 12.2). It is likely that this new distributary system will continue to operate as the main channel for the foreseeable future.

Localised instability will also occur if the main channel is unable to remove all of the sediment supplied to the confluence by the tributary during floods. At this site the gradient of the Spey is 0.002 and flood flows on the Spey are naturally attenuated by Loch Insh. This has enabled a series of alternate gravel bars to develop downstream of the confluence from 1899 stabilising into vegetated islands by 1946 (Fig. 12.3). By the early 1990s these islands provided the control over water levels in the Spey (Johnson et al., 1991).

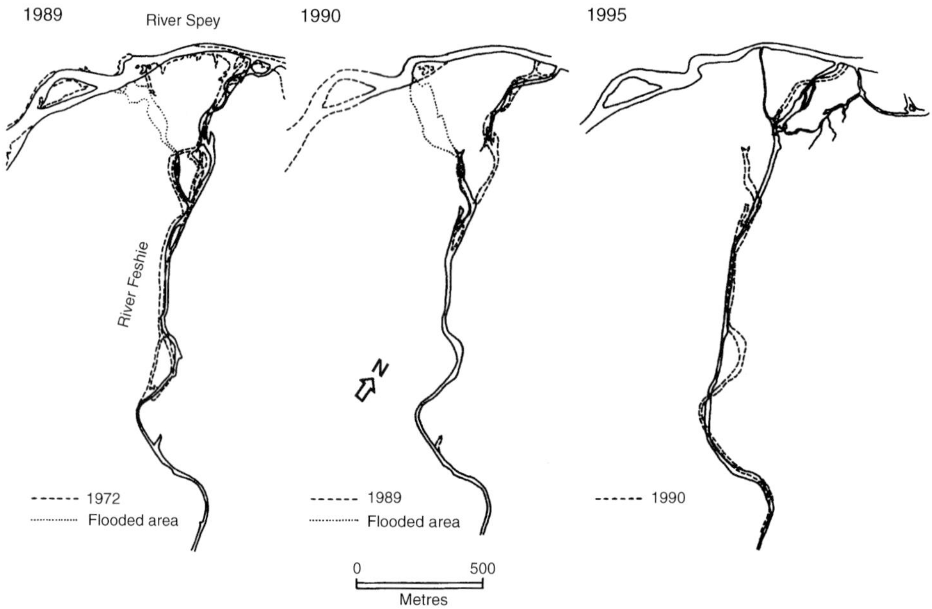

Figure 12.1. Changes in the position of the Feshie–Spey confluence 1989–1995. Based on successive pairs of aerial photographs.

3. History of river engineering

Following the successful construction of flood embankments on both the Feshie and the Spey in the early 19th century, there have been repeated proposals to either regrade the Spey or channelise the lowest-most reaches of the Feshie in order to alleviate the impact of floods (Werritty and Brazier, 1991b; see Cuthbertson, 1990 for details). Although the documentary evidence is incomplete and thus difficult to assess, subsequent large-scale maps and field surveys suggest that none of the major schemes from 1862 onwards has been fully implemented. However, minor realignment of the flood embankments in the vicinity of the islands immediately downstream of the Feshie confluence with the Spey was undertaken sometime between 1899 and 1946.

Following the floods in 1989 and 1990, riparian landowners called for engineering works to provide protection from flooding and bank erosion. A report commissioned by Highland Regional Council (Cuthbertson, 1990) identified sediment accumulation in the Spey, flood bank deterioration and channel erosion as the main problems and proposed the following non-exclusive solutions: a resectioning of the Spey, realigning the position of the Feshie confluence, and reinstating the flood banks. Two engineering schemes were then proposed by the Scottish Office and a riparian landowner (referred to below as Scheme A and Scheme B – Fig. 12.4) and assessed in terms of their hydraulic and hydrological impacts by numerical modelling of water levels (Johnson et al., 1991). This assessment was undertaken on behalf of the Nature Conservancy Council Scotland (now Scottish Natural Heritage) concerned to protect the seasonal flooding of the

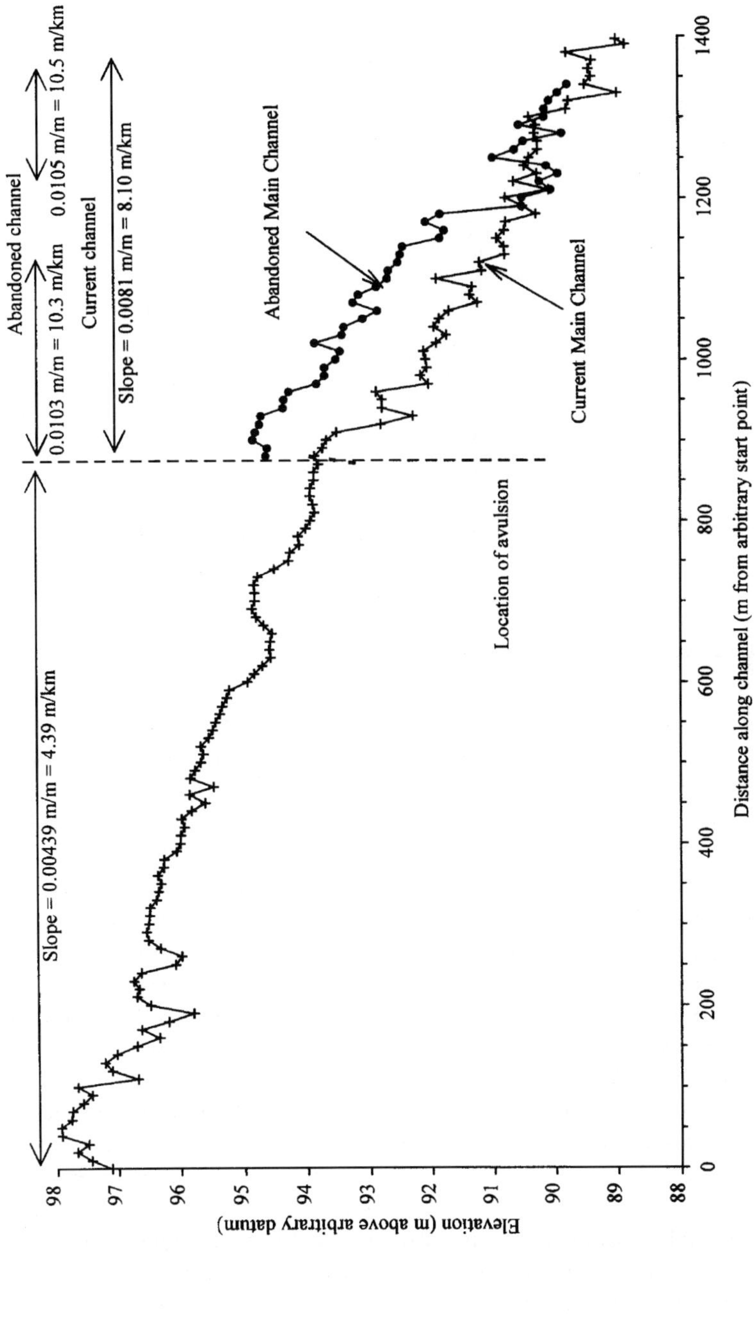

Figure 12.2. Long profiles of the Feshie fan channels in 1998.

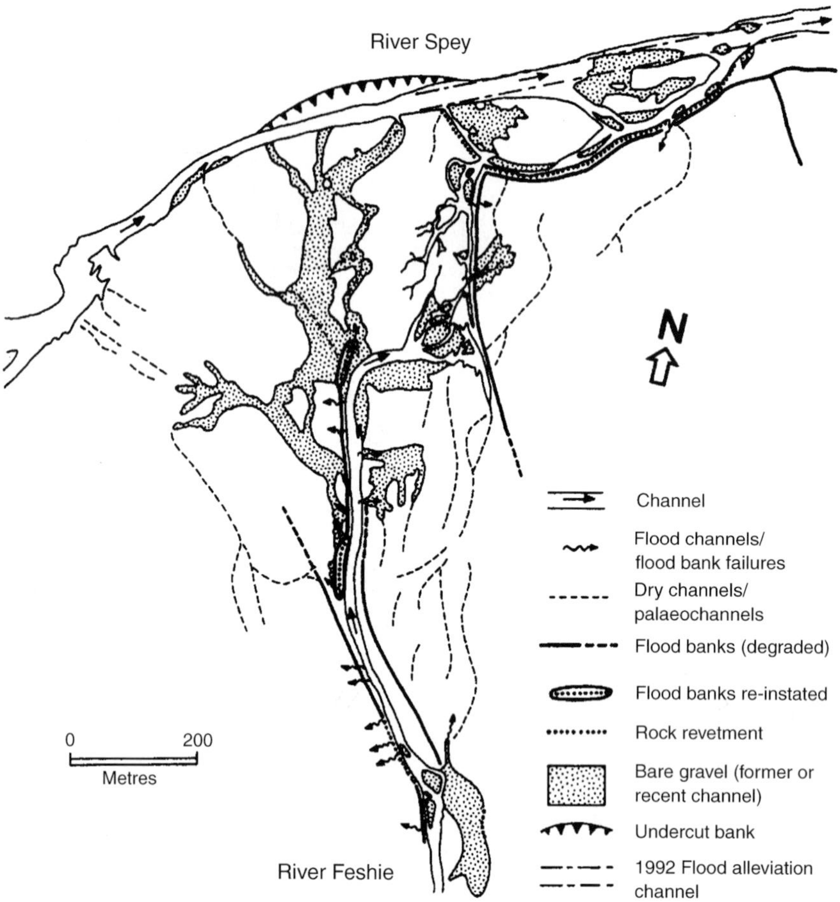

Figure 12.3. Geomorphological map of the Spey–Feshie confluence – June 1999.

Insh Marshes, an internationally protected wetland immediately upstream of Loch Insh. The scheme finally implemented in 1992 was a modified version of B adjusted to reduce the adverse impact of water levels on the Insh Marshes. This involved re-sectioning the channel downstream of the confluence through the stabilised islands which had formed since 1899 and provided the control for outflows from Loch Insh. Scheme A was not implemented and on the River Feshie there has only been re-instatement of breached and damaged flood banks since 1990. These repairs were undertaken on the west side of the channel in 1991 to ensure that the Feshie returned to its pre-1989 course and again in 1998 to maintain the channel in its post-1995 course.

4. Geomorphological and hydrological impacts of the river engineering

The geomorphic impacts of pre-1990 river engineering on the Feshie appear to have been relatively modest. Flood banks dating back to the early 19th century along both sides of

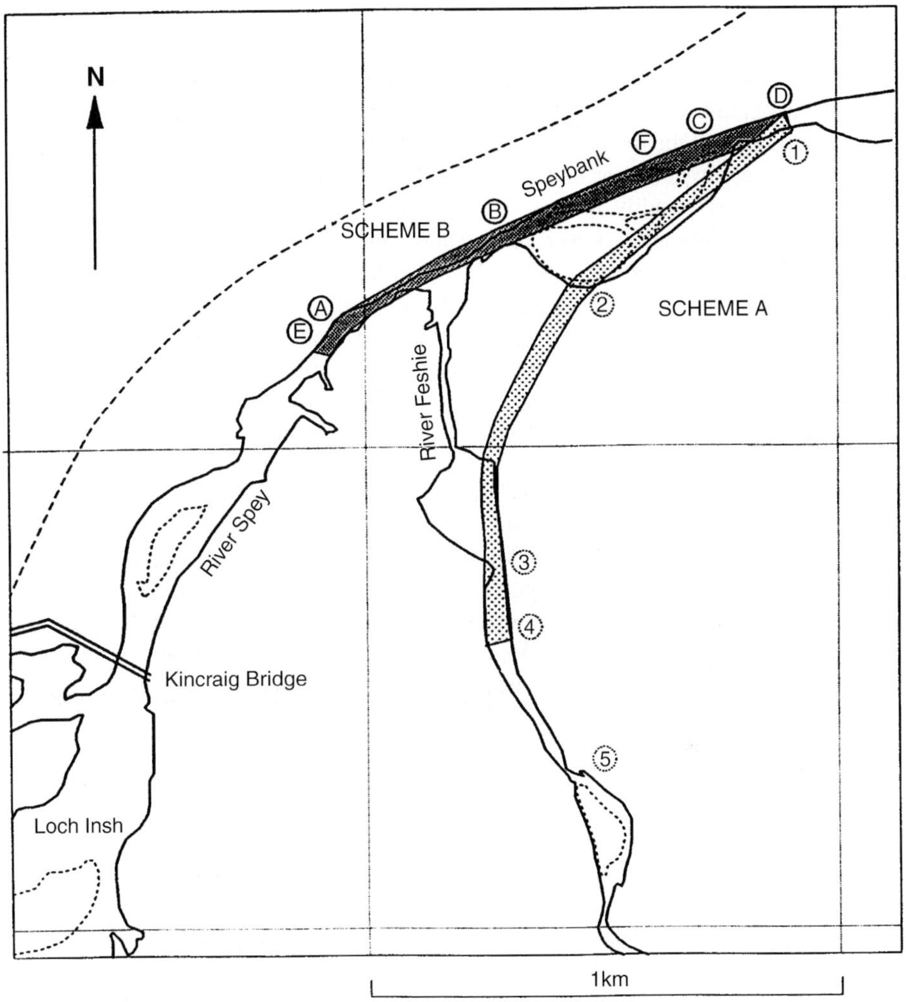

Figure 12.4. Proposed flood alleviation schemes at the Feshie–Spey confluence (from Johnson et al. 1991).

the river diminish consistently in height downstream and become buried by river sediment toward their downstream limits. Assuming that these flood banks were of roughly constant height when constructed this implies that the fan has naturally deposited an aggradational wedge within the confined area over nearly 200 years. During this period the confinement of the channel will have accelerated the natural process of avulsion by concentrating deposition on a small area of river causing local aggradation (Laronne and Duncan, 1990). This confinement of the main channel within a relatively narrow corridor exaggerated the natural tendency of increased bed elevation causing the river to become perched >1 m above the surrounding fan surface by 1990 (Fig. 12.2). As the protection provided by the flood banks decreased over time, their multiple failure in the floods of 1989 and 1990 was to be expected and avulsion (on this occasion to the west of the main channel) an inevitable and entirely natural process.

The flood banks on the western side of the Feshie were reinstated following the 1990 flood, accompanied by the removal of debris (mainly trees) and localised regrading of the river bed. This caused the main channel to revert to its pre-flood location by July 1991. As no comparable repair work was undertaken on the eastern flood bank, the likelihood of an avulsion to the east steadily increased during the 1990s enhanced by the presence of palaeochannels significantly lower in elevation than the bed of the reinstated main channel. Accordingly, between 1992 and 1997 a series of floods breached the already weakened eastern flood bank leading to the deposition of splays of sand and gravel through the conifer woodland and the re-occupation of palaeochannels. In turn these new eastern channels became increasingly important capturing the whole of the low flow discharge of the Feshie and by 1998 the avulsion was completed. Local constrictions caused by tree debris and remnants of flood banks meant that this new main channel initially had a very irregular plan form (Fig. 12.3). However, by the summer of 2000 many of these irregularities had been erased and the main channel increasingly conformed to the wandering river plan form characteristic of other reaches on the Feshie (Werritty and Brazier, 1991a). The historical reconstruction of localised aggradation causing avulsion and channel switching on other parts of the Feshie fan (outlined in Section 2) implies that this most recent avulsion would have occurred naturally. Nevertheless, the repair of the flood banks at the head of the avulsed channel and the minor in-channel works noted above probably accelerated the switch. The near 90° bend in the main channel as it enters the avulsed channel is very unusual (Fig. 12.3) and, taken in isolation, would normally imply some form of river management.

This most recent avulsion on the Feshie fan has yet to register a significant impact on the Spey which is still able to remove the coarse bedload delivered by the Feshie. The change in the position of the confluence has also yet to impact on the mid-channel islands. The 1992 construction of a flood relief channel on the Spey by regrading the mid-channel islands immediately downstream of the confluence (as proposed in engineering scheme B, Fig. 12.4) has proved only a partial success. By 2000 this relief channel was already partly infilled with river sediment and colonised by vegetation. If, as seems likely, this channel continues to infill and stabilise, the local flood alleviation benefits will have disappeared within another 10–20 years.

Engineering scheme B was adjusted during its implementation to minimise adverse impacts on the seasonal flooding of an internationally protected wetland. Nearly 10 years after the river works were completed, the impact of the scheme on the hydrology of the Insh Marshes has been assessed by analysing flows on the Spey upstream (at Invertruim) and downstream (at Kinrara) and water levels on the outlet of Loch Insh at Kincraig Bridge and at the Insh drain within the wetland (Werritty et al., 2000). Peak over threshold (partial duration) analysis failed to reveal any consistent step change in water levels at Kincraig Bridge following the cutting of the flood relief channel, although post-1992 levels are at the upper end of the overall range. There is evidence of a progressive change in peak levels from the mid-1980s but this is attributed to the "flood rich period" recorded across much of Scotland in the late 1980s and early 1990s (Black and Burns, 2002) rather than to engineering works at the confluence. Coincident with this increase in peak flows was a shift in low flows such that water levels in the Insh Marshes have become less sensitive to the overall flow in the Spey. These changes are viewed as entirely natural reflecting adjustments in the shape of the channel in the confluence area and complex linkages between flood banks, soil moisture, drain levels, and water levels within the Insh Marshes. Thus there is no evidence that the engineering works at the confluence have adversely impacted on the hydrology of the Insh Marshes.

5. Discussion and conclusions

Assessing the range of a river's natural dynamism at a site is crucial if a successful river management strategy is to be developed. Highly dynamic rivers are difficult to characterise using traditional concepts derived from regime theory and hydraulic geometry relations designed to explain the behaviour of rivers which are broadly stable and in long-term equilibrium. By contrast, the behaviour of the Feshie–Spey confluence over the last 250 years is one in which long periods of relative stability are punctuated by short bursts of instability when the crossing of a local threshold (aggradation leading to avulsion and channel switching) results in the main channel occupying a new location on the alluvial fan (Werritty and Brazier, 1991a). Following this abrupt change the new channel quickly establishes a plan form, cross-sectional geometry and local slope capable of conveying the flow and sediment delivered to it. The river's behaviour over the longer term is best characterised as metastable dynamic equilibrium (Schumm 1977; Renwick 1992): the metastable phases being each avulsion episode and the dynamic equilibrium phases the intervening periods of stability (cf. comparable behaviour by the nearby Allt Mor mountain torrent over similar timescales, McEwen and Werritty, 1988).

The development of a successful management strategy for the Feshie–Spey confluence depends on correctly assessing the stability of the site, or parts thereof. This can be done using the concept of geomorphological sensitivity based on the landscape change safety factor defined by Brunsden (1990) as:

Landscape change safety factor
= magnitude of barriers to change/magnitude of the disturbing forces

When this ratio falls below 1, landscape change is initiated and continues until the landform reaches a new stable configuration. At any one time, spatial variation in the landscape safety factor is found in different parts of any geomorphological system. This implies that when an event of a given magnitude occurs (for example, a flood with a 1 in 10 years recurrence interval) there will be some parts of the system, which change whilst other parts remain unaltered. In this study the above definition and usage is broadened so that the term "geomorphological sensitivity" includes both the propensity of the natural system to change and the likely extent of the impacts of management on the system.

Managing natural systems requires identification of the areas most "at risk" followed by appropriate intervention to minimise that risk. Because the natural processes determining risk levels are dynamic, the boundaries between areas displaying different risk levels are gradational rather than precise allowing the definition of the managed areas to change over time (Gemmell et al., 2001). Three such sensitivity zones have been identified and mapped for the Spey–Feshie confluence (Fig. 12.5) in a way which is indicative rather than

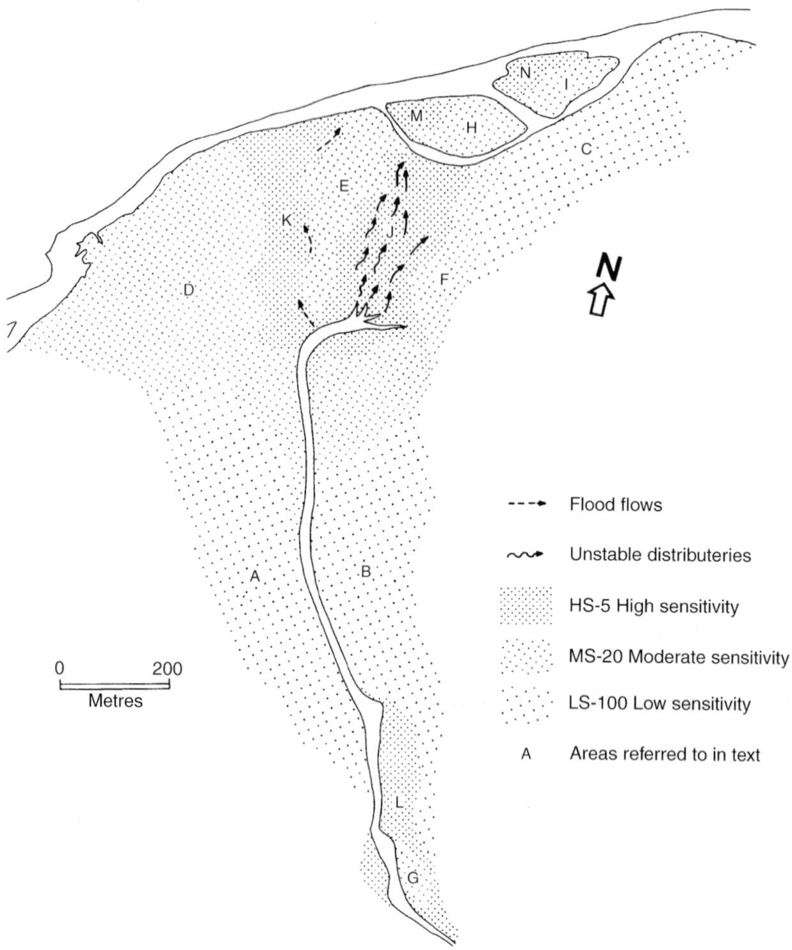

Figure 12.5. Geomorphological sensitivity for the Spey–Feshie confluence (sensitivity defined as in Table 12.1).

absolute and which shows boundaries as undefined, dynamic and evolving. Proposed management strategies are identified in Table 12.1. The first zone (HS-5: highly sensitive to change: areas J, K, L, M and N in Fig. 12.5) identifies areas in which the channel position is likely to change in response to events with a five-year recurrence interval. Any riparian development within HS-5 is strongly discouraged and the development of a management response involving landowners/occupiers and Scottish Natural Heritage is a high priority. The second zone (MS-20: moderate sensitivity to change: areas D, E, F, G, H and I in Fig. 12.5) identifies relatively stable areas at present, but subject to major change by events with a recurrence interval of 20 years. Potential sites for further avulsions can be identified and monitored, and management plans in response to these predictions should be developed with the appropriate landowners. The third zone (LS-100; low sensitivity to change: areas A, B and C in Fig. 12.5) identifies areas, which

Table 12.1. Definition of sensitivity classes and proposed management strategies.

HS-5	High sensitivity to change – channel position likely to change within five years. Plan management response/action now. Strongly discourage any riparian development especially bank/flood protection works. Work with riparian owners/occupiers to increase knowledge and understanding.
MS-20	Moderate sensitivity to change – relatively stable zone at present but may be subject to major change in response to events* with a recurrence interval of ≥20 years. Possible channel change/avulsion locations can be predicted and these sites should be monitored. Formulate management response based on predictions and monitoring. Discourage any development of riparian zone and advise potential developers of possible future problems.
LS-100	Low sensitivity to change – relatively stable zone not likely to change significantly over the 100 year timescale. However, some change may occur in response to events* with a recurrence interval of ≥100 years.

* Note that these "events" are not only hydrological (floods), but may also be sediment supply events. The passage of large "slugs" of sediment through upland river systems often occurs independent of large scale hydrological events, and these can cause channel instability in the absence of any hydrological change.

have proved stable in the recent past and are unlikely to undergo significant change except in response to rare events with a 100-year recurrence interval. Conservation sensitive flood alleviation schemes should be permitted in these areas.

Classifying and mapping geomorphological sensitivity in this way balances uncertainties in the future behaviour of dynamic sites with practical decisions on how such sites should be managed. Crucially important is the distinction between being able to predict what will happen with a given probability rather than specifying precisely when and where it will happen. For this reason, probability statements identifying dynamic areas at risk is favoured over more prescriptive and potentially misleading statements such as where the main channel might be located in 10 years time. It is also important to realise that management interventions in dynamic sites can have unintended consequences resulting in costly, but largely ineffective, engineering works. Thus less prescriptive and apparently imprecise geomorphological guidance for river managers is preferred as the way of providing long-term benefits to land owners (Newson et al., 2001). Accepting such a change in emphasis is a challenge for all involved including river engineers, land owners and regulatory bodies, not least because current legislation on river conservation and engineering may need to be revised (Day and Hudson, 2001).

Acknowledgements

This work has been carried out as research commissioned by the Nature Conservancy Council for Scotland and Scottish Natural Heritage. Access to the site was provided by Forest Enterprise plc and Badenoch Land Management. Data and technical assistance were provided by Bill Eakin, Derek Fraser, Dick Johnson and Kath Leys. Derek McGlashan and Jonathan Werritty assisted with fieldwork, and Jim Ford (Dundee University) drafted the maps.

References

Black, A.R., Burns, J.C., 2002. Re-assessing the flood risk in Scotland. Sci. Tot. Environ. 294, 169–184.

Brunsden, D., 1990. Tablets of stone: towards the Ten Commandments of geomorphology. Z. Geomorphol. 70, 1–37.

Cuthbertson, 1990. Flooding in Badenoch and Strathspey. Unpublished report to Highland Regional Council.

Davies, T.R., McSaveney, M.J., 2001. Anthropogenic fanhead aggradation, Waiho River, Westland, New Zealand. In: Mosley, M.P. (Ed.), Gravel-Bed Rivers V. NZ Hydrological Society, Wellington, pp. 531–549.

Day, T.J., Hudson, H.R., 2001. River management: the recent New Zealand experience. In: Mosley, M.P. (Ed.), Gravel-Bed Rivers V. NZ Hydrological Society, Wellington, pp. 555–579.

Ferguson, R.I., Werritty, A., 1983. Bar development and channel change in the gravely River Feshie, Scotland. In: Collinson, J.D., Lewin, J. (Eds), Modern and Ancient Fluvial Systems, Vol. 6. Special Publication International Association of Sedimentologists, Blackwells, Oxford, pp. 181–193.

Gemmell, S.L.G., Hansom, J.D., Hoey, T.B., 2001 The geomorphology, conservation and management of the Lower River Spey and Spey Bay SSSIs. SNH Research, Survey and Monitoring Report 57.

Johnson, R.C., Piper, B.S., Acreman, M.C., Gilman, K., 1991. Flood alleviation in Upper Strathspey: modelling and environment study. Report by the Institute of Hydrology to Nature Conservancy Council for Scotland, Vol. I.

Laronne, J.B., Duncan, M.J., 1990. Bed movement in the North Branch, Ashburton River. Internal Report WS 1185, Hydrology Centre, DSIR, Christchurch, New Zealand.

McEwen, L.J., Werritty, A., 1988. The hydrology and long-term geomorphic significance of a flash flood in the Cairngorm Mountains, Scotland. Catena 15, 361–377.

Newson, M., Thorne, C., Brookes, A., 2001. The management of gravel-bed rivers in England and Wales: from geomorphological research to strategy and operations. In: Mosley, M.P. (Ed.), Gravel-Bed Rivers V. NZ Hydrological Society, Wellington, pp. 581–606.

Paige, A.D., Hickin, E.J., 2000. Annual bed-elevation regime in the alluvial channel of Squamish River, southwestern British Columbia, Canada. Earth Surface Processes and Landforms 25, 991–1010.

Renwick, W.H., 1992. Equilibrium, disequilibrium and nonequilibrium landforms in the landscape. Geomorphology 5, 433–455.

Schick, A.P., Grodek, T., Wolman, M.G., 1999. Hydrological processes and geomorphic constraints on urbanization of alluvial fan slopes. Geomorphology 31, 325–335.

Schumm, S.A., 1977. The Fluvial System. Wiley, New York.

Werritty, A., Brazier, V., 1991a. The geomorphology, conservation and management of the River Feshie SSSI. Unpublished Report to the Nature Conservancy Council.

Werritty, A., Brazier, V., 1991b. Geomorphological aspects of the proposed Strathspey flood alleviation scheme. In: Johnson, R.C., Piper, B.S., Acreman, M.C., Gilman. K. (Eds), Flood Alleviation in Upper Strathspey: Modelling and Environment Study. Supplementary Reports by the Institute of Hydrology to Nature Conservancy Council for Scotland, Vol. II.

Werritty, A., Ferguson, R.I., 1980. Pattern changes on a Scottish braided river over 1, 30 and 200 years. In: Cullingford, R.A., Davidson, D.A., Lewin, J. (Eds), Timescales in Geomorphology. Wiley, Chichester, pp. 53–68.

Werritty, A., Leys, K.F., 2001. The sensitivity of Scottish rivers and upland valley floors to recent environmental change. Catena 42, 251–273.

Werritty, A., Hoey, T.B., Black, A.R., 2000. Geomorphological and hydrological changes at the River Feshie/Spey confluence and Insh Marshes SSSIs. Scottish Natural Heritage Commissioned Report F98AC101 (Unpublished).

Zarn, B., Davies, T.R., 1994. The significance of processes on alluvial fans to hazard assessment. Z. Geomorphol. 38, 487–500.

Subject index